T0233295

A FIRST
COURSE IN
LINEAR ALGEBRA

A FIRST COURSE IN LINEAR ALGEBRA

Minking Eie • **Shou-Te Chang**

National Chung Cheng University, Taiwan

 **World Scientific**

NEW JERSEY · LONDON · SINGAPORE · BEIJING · SHANGHAI · HONG KONG · TAIPEI · CHENNAI · TOKYO

Published by

World Scientific Publishing Co. Pte. Ltd.

5 Toh Tuck Link, Singapore 596224

USA office: 27 Warren Street, Suite 401-402, Hackensack, NJ 07601

UK office: 57 Shelton Street, Covent Garden, London WC2H 9HE

Library of Congress Cataloging-in-Publication Data

Names: Eie, Minking, 1952– | Chang, Shou-Te.

Title: A first course in linear algebra / by Minking Eie (National Chung Cheng University, Taiwan),
   Shou-Te Chang (National Chung Cheng University, Taiwan).

Other titles: Linear algebra

Description: New Jersey : World Scientific, 2016. | Includes index.

Identifiers: LCCN 2016013684| ISBN 9789813143104 (hardcover : alk. paper) |
   ISBN 9789813143111 (softcover : alk. paper)

Subjects: LCSH: Algebras, Linear--Textbooks. | Algebra--Textbooks.

Classification: LCC QA184.2 .E44 2016 | DDC 512/.5--dc23

LC record available at https://lccn.loc.gov/2016013684

British Library Cataloguing-in-Publication Data

A catalogue record for this book is available from the British Library.

Printed in Singapore

Preface

Linear algebra is the study of *vector spaces* and the *linear transformations* between them. Since high school we have encountered phrases like *vectors* and *scalars*. We can view an arrow from a point A to a point B on the real plane $\mathbb{R}^2$ (or the real space $\mathbb{R}^3$) as a vector. A vector can be multiplied by a real number (curiously called a *scalar*) t by extending the vector t times. We can add two vectors together using the *Parallelogram Law*.

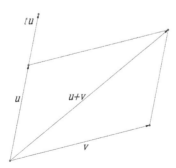

For a real number a and vectors v and w we denote the *scalar multiplication* of v by a by av and the *vector addition* of v and w by $v+w$. Suppose that $v = (a, b)$ and $w = (c, d)$ are two vectors in $\mathbb{R}^2$. We know that that the vector addition is

$$(a, b) + (c, d) = (a + c, b + d),$$

and the scalar multiplication is

$$t(a, b) = (ta, tb).$$

Let's point out some readily seen properties.

- For all vectors v and w, $v + w = w + v$.

- For all vectors u, v and w, $(u + v) + w = u + (v + w)$.

- The zero vector $O = (0, 0)$ is such that $v + O = v$ for all vector v.

- For each vector $v = (a, b)$, we have the vector $w = (-a, -b)$ such that $v + w = O$.

- For each vector v, $1v = v$.

- For real numbers a and b and each vector v, $(ab)v = a(bv)$.

- For each real number a and vectors v and w, $a(v + w) = av + aw$.

- For real numbers a and b and each vector v, $(a + b)v = av + bv$.

These seemingly trivial properties (and they alone) define the concept of *vector spaces* which can be applied in a much broader sense. Contrary to what most beginning students of linear algebra believe, this is not a course on $\mathbb{R}^2$, $\mathbb{R}^3$ or even $\mathbb{R}^n$. The techniques we develop will be applied to a much wider range of objects.

To view the spaces $\mathbb{R}^2$ or $\mathbb{R}^3$ as vector spaces is to miss many of their interesting facets. None of the properties mentioned are the analytic properties one studies in the courses of calculus or advanced calculus, even though $\mathbb{R}^2$ and $\mathbb{R}^3$ are objects of interest in those courses. One does not need to know the completeness of real numbers or to view the real spaces as metric spaces. One does not need to understand differentiation and integration. Instead, we only need to know how to add up or multiply two real numbers in order to work on $\mathbb{R}^2$ or on $\mathbb{R}^3$. If you will open up your mind, you will observe that the vector addition and (real) scalar multiplication can also be defined on the set $\mathbb{C}^2$, in a similar fashion to what is done on $\mathbb{R}^2$. Furthermore, observe what is required for the scalars. Again, we only need to know how to add and multiply scalars together for the eight properties above to make sense. If we allow the scalars to be complex numbers, it still makes sense to define a vector addition and a (complex) scalar multiplication on $\mathbb{C}^2$ in a similar fashion. In general, we will treat any object (with suitable scalars) satisfying the eight properties described as a vector space. In this course, we will discuss more examples of scalars and vector spaces.

The process of defining vector spaces is called the *axiomatic* approach (very popular for algebraists). This means that we will give the *axioms* for the objects we study. In other words, we lay down the basic and unbending rules or assumptions for whatever we study, and then we use them to explore the consequences (the *theorems*). It is amazing that what a rich and wonderful world of results will come out of just a handful of axioms.

This textbook is designed for a one-year (6-credit-hour) undergraduate course. We will be mainly dealing with finite dimensional vector spaces. We will start with a formal presentation of definitions, structures and properties regarding vector spaces. The material will seem dry to some extent until we study the appropriate mappings between vector spaces called the linear transformations. Matrices will be associated with linear transformations between finite dimensional vector spaces. Students will then understand the true significance of matrices and determinants. To answer a question on a matrix is often equivalent to answering a question on a particular linear transformation. To answer a question on a linear transformation may often depend on how much one understands the underlying vector spaces. Hence, one will be presented several faces of the same question, which enables one to find a solution sometimes from a most surprising angle.

However, to master the contents of this textbook, the important thing really is to try out the exercise. And this does not include reading other people's solutions of the exercises. The purposes of exercises are

- to familiarize oneself with the new concepts just learned;

- to establish examples one might use in future studies;

- to discover the intricacies one might miss at first reading;

- to discover new problems that one might look for answers in coming chapters.

Indeed, we cannot stress enough how important doing exercises is! We are sure there are more benefits from doing exercises that are not mentioned here. In this book, most of the exercises are designed to help students develop a deeper understanding of the content of the book. There is also a small amount of challenging exercises! We are sure students will feel greatly rewarded after they have solved those more difficult problems.

We hope the amount of exercises in this book will not be intimidating to our readers. Well-designed exercises are extremely helpful in helping students become familiar with the material. They also stimulate students' interests in wanting to learn more. We hope that our problem sets are well chosen and not excessive so that they fill the purpose of well-designed exercises.

A guide to the instructors: Each section of this book is designed for approximately three hours of lecture. Although this textbook is theoretically oriented, it contains a few optional sections dealing with applications, including §4.4, §5.4, §5.5, §7.4, §7.5 and §7.6. They can be skipped under time constraint. The rest of the book is self-contained.

We are sure this textbook contains many typos and small errors. Please send us emails if you have any questions or comments. Your input will be greatly appreciated. We can be reached at `minking@math.ccu.edu.tw` and `stchang@math.ccu.edu.tw`. Finally, we would like to thank Dr. Tung-Yang Lee for kindly taking up part of the typing, which helped us meet the deadline. We would also like to thank Dr. Suqi Pan, our editor at World Scientific, for carefully reading our manuscript and for all the helpful suggestions.

Minking Eie

Shou-Te Chang

May 22, 2016

Notation

$\mathbb{Z} = \{0, \pm 1, \pm 2, \pm 3, ...\}$ = the set of *integers*

$\mathbb{Z}_+ = \{1, 2, 3, 4, ...\}$ = the set of *positive integers*

$\mathbb{Z}_- = \{-1, -2, -3, -4, ...\}$ = the set of *negative integers*

$\mathbb{Q}$ = the set of *rational numbers*

$\mathbb{R}$ = the set of *real numbers*

$\mathbb{C}$ = the set of *complex numbers*

About the Authors

Mingking Eie is currently a Professor of Department of Mathematics, National Chung Cheng University in Chiayi, Taiwan. He received his Ph. D. in Mathematics in 1982 from University of Chicago and began his career as an associate research fellow at Institute of Mathematics, Academia Sinica of Taiwan. He was promoted to a research fellow in 1986 and became an Outstanding Researcher of National Science Council of Taiwan at the same time. In 1991, he left Academia Sinica and took up the current position. More than 50 research papers have been published about theory of modular forms of several variables, theory of Jacobi forms over Cayley numbers, Bernoulli identities and multiple zeta values. He has also published several sets of mathematical textbooks for senior high school students in Taiwan and a Calculus textbook in English for Business undergraduates.

Shou-Te Chang is currently an Associate Professor at the Department of Mathematics, National Chung Cheng University in Chiayi, Taiwan. She received her Ph. D. in Mathematics in 1993 from University of Michigan, Ann Arbor. Her research interest is in commutative algebra and homological algebra. She has published papers on Horrocks' question, generalized Hilbert-Kunz functions and local cohomology.

Contents

CHAPTER 1

Vector Spaces

It is probably the single most elementary and useful mathematical tool. We feel that Linear Algebra and Topology are the two prerequisites for any discipline of mathematics, while Linear Algebra is a prerequisite for any discipline that uses any mathematics at all.

In an algebraic course, it always involves the study of "algebraic structures" and the "morphisms" between the structures. For Linear Algebra, the "algebraic structures" are vector spaces.

1.1   A few words on sets and logics

Before we start, we need to review the basic language of mathematics: sets. We will also dwell slightly on how to write a proof. Remember that the business of mathematicians is to prove theorems.

Sets.   We will say a **set** is a collection of **elements**. If an element x is contained in a set A, we will write $x \in A$. If not, we write $x \notin A$. If we say two sets A and B are the same (equal), we mean that they contain exactly the same elements and we write $A = B$.

A set can be described by an account of all its elements or it can be described by a collection of sentences (conditions). Note that the conditions have to be specific enough to verify whether an element is inside this set or not.

Logical statements.   A logical statement such as the sentences describing the conditions of an element in a set or making up a definition, cannot tolerate ambiguity. It is essential to familiarize oneself with the logical operators and rules of syntax.

A sentence (condition) must be of the form "(S)ubject + (V)erb + (O)bject". In mathematics, $=, >, <, \subset, \supset, \in$ and $\notin$ are verbs, while $+, -, \cdot, \div, \cup$ and $\cap$ are conjunctives. For example, "$a > b$" is a simple sentence (condition), while "$a + b$" is not!

A logical statement consists of conditions and logical operators, put together under the rules of syntax. Let (A) and (B) be sentences (conditions). The most commonly seen logical statements are two following types:

- 'If (A) then (B)', '(B) if (A)', '(A) only if (B)', 'Suppose (A). Then (B).', 'Assume (A). Then (B).', '(B) as long as (A)', '(B) whenever (A)', '(A) implies (B)', '(A) $\Rightarrow$ (B)', '(B) $\Leftarrow$ (A)'.
  [(A) is called a *sufficient* condition for (B) and (B) is called a *necessary* condition for (A).]

- '(A) if and only if (B)', '(A) and (B) are equivalent', '(A) $\Leftrightarrow$ (B)'.
  [(A) is called a *necessary and sufficient* condition for (B).]

Obviously, there are other logical operators. We will gradually introduce them in this book. You can learn how to write proofs by imitation.

Relations and operations of sets. The following concepts on sets are extremely elementary, but since they are so essential, we'll still present them here in the form of definitions.

Definition 1.1.1. Let A and B be two sets.

- We say A is a **subset** of B or A **is contained in** B, denoted $A \subseteq B$, if $x \in A$ implies $x \in B$.

- We say A is a **superset** of B or A **contains** B, denoted $A \supseteq B$, if $B \subseteq A$.

- Moreover, $A = B$ if and only if $A \subseteq B$ and $B \subseteq A$.

Remark. The main "if" (not ones appearing in the subclauses) in the statement of any definition is an "if and only if". Basically, a definition is a way to say the same thing in two different ways.

Remark. In many textbooks, the notation $\subset$ is used instead of $\subseteq$. However, some authors prefer to use $\subset$ for strict containment, that is, to exclude the possibility that $A = B$. So one should take the precaution of finding out what the author meant by $\subset$. In this book, we'll use $\subsetneq$ for strict containment.

Fact. *The empty set $\varnothing$ is a subset of A for any set A.*

Definition 1.1.2. Let A and B be two sets.

(a) The **union** of A and B: $A \cup B = \{x : x \in A \text{ or } x \in B\}$.

(b) The **intersection** of A and B: $A \cap B = \{x : x \in A \text{ and } B\}$.

(c) The (relative) **complement** of A in B or the **difference** of A and B: $B \setminus A = \{x : x \in B \text{ and } x \notin A\} = \{x \in B : x \notin A\}$.

(d) The (absolute) **complement** of A: $\sim A = A^{\sim} = \{x : x \notin A\}$ (when the universal set is understood).

Next we make a list of basic properties regarding unions, intersections and complements.

Proposition 1.1.3. *Let A, B, C be sets, and we will call the universal set at this moment E. The following statements are all true:*

(a) $A \subseteq A \cup B$; $A \supseteq A \cap B$.

(b) $A \cup \varnothing = A$; $A \cap \varnothing = \varnothing$.

(c) *Commutativity:* $A \cup B = B \cup A$; $A \cap B = B \cap A$.

(d) *Associativity:* $A \cup (B \cup C) = (A \cup B) \cup C$; $A \cap (B \cap C) = (A \cap B) \cap C$.

(e) *Idempotents:* $A \cup A = A = A \cap A$.

(f) *If $A \subseteq B$, then $A \cup C \subseteq B \cup C$ and $A \cap C \subseteq B \cap C$.*

(g) *$A \cup B = B$ if and only if $A \subseteq B$; $A \cap B = B$ if and only if $A \supseteq B$.*

(h) *Distributivity:* $A \cap (B \cup C) = (A \cap B) \cup (A \cap C)$; $A \cup (B \cap C) = (A \cup B) \cap (A \cup C)$.

(i) $(A^{\sim})^{\sim} = A$.

(j) $\varnothing^{\sim} = E$; $E^{\sim} = \varnothing$.

(k) $A \cup A^{\sim} = E$; $A \cap A^{\sim} = \varnothing$.

(l) $B \setminus A = B \cap A^{\sim}$.

(m) $A \subseteq B$ *if and only if* $B^{\sim} \subseteq A^{\sim}$.

(n) *De Morgan laws:* $(A \cup B)^{\sim} = A^{\sim} \cap B^{\sim}$; $(A \cap B)^{\sim} = A^{\sim} \cup B^{\sim}$.

Proof. These are quite basic results. We will prove (g) as a demonstration on how to write proofs. We will leave the rest as a somewhat tedious exercise (see Problem (1)).

(g) There are two statements here: one on the union, and the other on the intersection. And for each statements there are two directions to prove.

We first show that $A \cup B = B$ if and only if $A \subseteq B$.

The "only if" part: Let $x \in A$. Then $x \in A \cup B = B$. Thus $A \subseteq B$.

The "if" part: To show that $A \cup B = B$, one needs to prove $A \cup B \subseteq B$ and $A \cup B \supseteq B$. Note that $A \cup B \supseteq B$ by (a). Now suppose $x \in A \cup B$. It

follows that $x \in A$ or $x \in B$. In the first case, $x \in A \subseteq B$ by assumption.
We conclude that $x \in B$. Thus $A \cup B \subseteq B$. We have shown that $A \cup B = B$.

Next we show that $A \cap B = B$ if and only if $A \supseteq B$.

The "only if" part: Let $x \in B$. Then $x \in B = A \cap B$. Hence $A \supseteq B$.

The "if" part: To show that $A \cap B = B$, note that $A \cap B \subseteq B$ by (a).
Now suppose $x \in B$. Then $x \in B \subseteq A$ by assumption. It follows that
$x \in A$ and $x \in B$. In other words, $x \in A \cap B$. Thus $B \subseteq A \cap B$. We have
shown that $A \cap B = B$. $\qquad\square$

We often use an **index set** (for example, the Λ in the next definition)
to index a set of elements. The index set is a way to tell us the possible
size of a set. The index set is usually an infinite set, since we probably
don't need to use one otherwise. We often call a set of sets as a **family** to
emphasize the elements are sets themselves.

Definition 1.1.4. Let $\mathscr{C} = \{S_i\}_{i \in \Lambda}$ be a family of sets.

(a) The union of $\{S_i\}_{i \in \Lambda}$: $\bigcup \mathscr{C} = \bigcup_{i \in \Lambda} S_i = \{x : x \in S_i \text{ for some } i \in \Lambda\}$.

(b) The intersection of $\{S_i\}_{i \in \Lambda}$: $\bigcap \mathscr{C} = \bigcap_{i \in \Lambda} S_i = \{x : x \in S_i \text{ for all } i \in \Lambda\}$.

Even for an infinite family of sets we have the following facts on union,
intersection and complements.

Proposition 1.1.5. *Let A be a set and let $\{S_i\}_{i \in \Lambda}$ be a family of sets. We
have the following results.*

(a) **Distributivity:**

- $A \cup \left(\bigcap_{i \in \Lambda} S_i\right) = \bigcap_{i \in \Lambda}(A \cup S_i)$;
- $A \cap \left(\bigcup_{i \in \Lambda} S_i\right) = \bigcup_{i \in \Lambda}(A \cap S_i)$.

(b) **De Morgan laws:**

- $\left(\bigcup_{i \in \Lambda} S_i\right)^{\sim} = \bigcap_{i \in \Lambda} S_i^{\sim}$;
- $\left(\bigcap_{i \in \Lambda} S_i\right)^{\sim} = \bigcup_{i \in \Lambda} S_i^{\sim}$.

Negation.   In mathematics we are often asked to check if a statement is true, or more precisely, we are asked to determine the set of x for which a statement $A(x)$ holds. To simplify the notation, we will write $A = \{x : A(x)$ is true$\}$ and $B = \{x : B(x)$ is true$\}$. Thus if we want to check if x satisfies $A(x)$ or $B(x)$, it is the same as to check whether $x \in A \cup B$. Or if we want to check if x satisfies both $A(x)$ and $B(x)$, it is the same as to check whether $x \in A \cap B$. Similarly, The set of those x for which $A(x)$ is false (or not $A(x)$ is true) is $A^\sim$.

Even a statement like "if $A(x)$ then $B(x)$" has an equivalent statement on sets: "$A \subseteq B$". By Proposition 1.1.3(1.1.3), this is also equivalent to $B^\sim \subseteq A^\sim$. Hence we have that

$$(*) \qquad \Big(\text{if } A(x) \text{ then } B(x)\Big) \iff \Big(\text{if} \sim B(x) \text{ then} \sim A(x)\Big).$$

Another equivalent statement to "if $A(x)$ then $B(x)$" is "$A^\sim \cup B = E$" where E is the underlying universal set. Since

$$A^\sim \cup B = B \cup A^\sim = (B^\sim)^\sim \cup (A^\sim),$$

we have $(*)$ again.

How to prove theorems.   To prove a statement such as "if $A(x)$ then $B(x)$", there are three approaches.

- We simply assume that $A(x)$ is true, and we use the help of axioms, definitions, various lemmas, propositions, theorems and corollaries in order to reach the conclusion that $B(x)$ is true.

- We assume that $B(x)$ is false, and then we apply axioms, definitions, various lemmas, propositions, theorems and corollaries in order to reach the conclusion that $A(x)$ is false.

- We may use *Reductio ad absurdum* (proof by contradiction). By this we mean that we assume that the whole premise is false, then we try to reach a contradiction.

How do you regard a sentence such as "if $A(x)$ then $B(x)$" is false? This means that $A^\sim \cup B \neq E$ or equivalently, that $\varnothing \neq (A^\sim \cup B)^\sim =$

$(A^\sim)^\sim \cap B^\sim = A \cap B^\sim$. Hence we may start by assuming that there is an x such that $A(x)$ is true but $B(x)$ is false. Such an x is also called a **counterexample** of the statement "if $A(x)$ then $B(x)$".

Example 1.1.6. The statement "if $A(x)$ then $B(x)$" is equivalent to "$A \subseteq B$". The negation of $A \subseteq B$ is "$A \not\subseteq B$". What does $A \not\subseteq B$ mean?

If $A \subseteq B$, then $x \in A$ implies that $x \in B$. A counterexample to the condition $A \subseteq B$ is an element $x \in A$ but $x \in B$.

Exercises 1.1

1. Prove Proposition 1.1.3.

2. (a) Give an example to show that $(A \cup B) \cap C \neq A \cup (B \cap C)$.

   (b) Show that $(A \cup B) \cap C = A \cup (B \cap C)$ if and only if $A \subseteq C$. Observe that this condition has nothing to do with B.

3. Here are some exercises on complements of sets. Prove the following assertions are true.

   (a) $A \subseteq B$ if and only if $A \setminus B = \emptyset$.

   (b) $A \setminus (A \setminus B) = A \cap B$.

   (c) $A \cap (B \setminus C) = (A \cap B) \setminus (A \cap C)$.

   (d) $A \cap B \subseteq (A \cap C) \cup (B \cap C^\sim)$.

   (e) $(A \cup C) \cap (B \cup C^\sim) \subseteq A \cup B$.

4. Let A and B be sets. The **symmetric difference** or **Boolean sum** of A and B is given by

$$A + B = (A \setminus B) \cup (B \setminus A).$$

   Show that

   (a) $A + B = B + A$,

   (b) $(A + B) + C = A + (B + C)$,

    (c) $A + \varnothing = A$, and

    (d) $A + A = \varnothing$

for any sets A, B and C.

5. Prove Proposition 1.1.5.

1.2   What constitutes a vector space?

The concepts of vector spaces actually consists of two parts: the **scalars** and the **vectors**.

Scalars. The scalars of a vector space must be elements from an algebraic structure called a **field**. Simply speaking, a field is a set on which we can perform $+$, $-$, $\times$ and $\div$.

Definition 1.2.1 (Fields). We say a set F is a **field** if the following properties are satisfied:

(A) (Addition) To every pair of elements a and b in F there corresponds a unique element $a + b$ in F, called the **sum** of a and b, such that

    (i) addition is **commutative**: $a + b = b + a$;

    (ii) addition is **associative**: $(a + b) + c = a + (b + c)$;

    (iii) there is an element 0 (the **additive identity**) such that $a + 0 = a$ for all a in F;

    (iv) to every a there is an element $-a$ (the **additive inverse**) such that $a + (-a) = 0$.

(B) (Multiplication) To every pair of elements a and b in F there corresponds a unique element ab in F, called the **product** of a and b, such that

    (i) multiplication is **commutative**: $ab = ba$;

    (ii) multiplication is **associative**: $(ab)c = a(bc)$;

    (iii) there is an element 1 (the **multiplicative identity**) such that $a1 = a$ for all a in F;

    (iv) to every **nonzero** a in F there is an element a^{-1} (or denoted $1/a$) (the **multiplicative inverse**) such that $aa^{-1} = 1$.

(C) Multiplication is **distributive** with respect to addition: $a(b + c) = ab + ac$ for all a, b, c in F.

Remarks. (1) By convention, $0 \neq 1$. Hence a field contains at least two elements.

    (2) It is customary to write $a - b$ for $a + (-b)$ and to write a/b for ab^{-1} where $b \neq 0$.

Of course, the properties listed in Definition 1.2.1 are nothing new to students since we can observe them in sets such as $\mathbb{Q}$, $\mathbb{R}$ and $\mathbb{C}$ with their natural addition and multiplication. In fact, these properties are so commonly seen that students often forget about their existence. But they are worth being pointed out and given names. Notice that if we slightly change the setting, many structures cease to remain fields. For example neither $\mathbb{Z}$ nor $\mathbb{Z}_+$ are fields. In the first structure many elements lack multiplicative inverses. In the second structure most elements lack both multiplicative and additive elements. In fact, in the second structure, there is no additive identity.

Let us look at a less familiar example so that we may understand the concept of fields more.

We define addition and multiplication on the set $\mathbb{Z}_2 = \{0, 1\}$ given by

+	0	1
0	0	1
1	1	0

and

·	0	1
0	0	0
1	0	1

.

Please verify that that $\mathbb{Z}_2$ is a field by going through all the properties given in the definition of fields.

Next let's give a list of some basic properties of fields. These properties can be derived from definition of fields. They may look trivial but they require some arguments.

Proposition 1.2.2. *The following properties are true for elements in a field F:*

(a) *(Cancellation Law) if $a + c = b + c$ then $a = b$; if $c \neq 0$ and $ac = bc$ then $a = b$;*

(b) *$0a = 0$ for all $a \in F$;*

(c) *$-(-a) = a$;*

(d) *$(-a)b = -(ab) = a(-b)$; $(-a)(-b) = ab$.*

Note that (d) is what every grade school pupil knows: Minus times minus results in a plus. Now you know why!

Proof. (a) Suppose $a + c = b + c$. Then

$$
\begin{aligned}
a &= a + 0, && \text{by the property of additive identity,} \\
  &= a + [c + (-c)], && \text{by the existence of additive inverse,} \\
  &= (a + c) + (-c), && \text{by associativity,} \\
  &= (b + c) + (-c), && \text{by assumption,} \\
  &= b + [c + (-c)], && \text{by associativity,} \\
  &= b + 0, && \text{by the property of additive inverse,} \\
  &= b, && \text{by the property of additive identity.}
\end{aligned}
$$

Suppose $ac = bc$ where $c \neq 0$. Then

$$
\begin{aligned}
a &= a1, && \text{by the property of multiplicative identity,} \\
  &= a[cc^{-1}), && \text{by the existence of multiplicative inverse,} \\
  &= (ac)c^{-1}, && \text{by associativity,} \\
  &= (bc)c^{-1}, && \text{by assumption,} \\
  &= b(cc^{-1}), && \text{by associativity,} \\
  &= b1, && \text{by the property of multiplicative inverse,} \\
  &= b, && \text{by the property of multiplicative identity.}
\end{aligned}
$$

(b) Again, using the axioms in Definition 1.2.1, we can see that

$$
\begin{aligned}
0a + 0 = 0a &= (0 + 0)a, && \text{by the property of additive identity,} \\
            &= 0a + 0a, && \text{by distributivity.}
\end{aligned}
$$

Now by Cancelation Law in (a), $0a = 0$.

(c) Since $a + (-a) = 0$, $a = -(-a)$, the additive inverse of $-a$.

(d) To show that $(-a)b = -(ab)$ is to show that $(-a)b$ is the additive inverse of ab. This is clear since

$$ab + (-a)b = [a + (-a)]b, \qquad \text{by distributivity,}$$
$$= 0b, \qquad \text{by the property of additive inverse,}$$
$$= 0, \qquad \text{by (b).}$$

One can show that $a(-b) = -(ab)$ by symmetry. Similarly,

$$(-a)(-b) = -[a(-b)], \qquad \text{by the first part of (d),}$$
$$= -(-ab), \qquad \text{by the first part of (d),}$$
$$= ab, \qquad \text{by (c).}$$

This completes the proof. □

Remark. The additive identity, the additive inverse, the multiplicative identity and the multiplicative inverse are unique in a field. These simple facts follow from the cancellation law in Proposition 1.2.2. See Problem (1).

Vector spaces. We now come to the main object of this book.

Definition 1.2.3 (Vector Spaces). A **vector space** V **over** the field F (elements in F will be called **scalars**) or an F**-vector space** is a set of elements called **vectors** which satisfies the following axioms:

(A) (Vector Addition) To every pair of vectors v and w in V there corresponds a unique element $v + w$ in V, called the **sum** of v and w, such that in V

(i) vector addition is **commutative**: $v + w = w + v$;

(ii) vector addition is **associative**: $u + (v + w) = (u + v) + w$;

(iii) there is a vector O (called the **origin**, the **zero** vector or the **trivial** vector) such that $v + O = v$ for all v in V;

(iv) to every v in V there is an element $-v$ (the **additive inverse** of v) such that $v + (-v) = O$.

(B) (Scalar Multiplication) To every scalar a in F and every vector v in V there corresponds a unique vector av, called the **scalar product** of a and v, such that

    (i) scalar multiplication is **associative**,: $(ab)v = a(bv)$;

    (ii) $1v = v$ for all v in V.

(C) For a, $b \in F$ and v, $w \in V$ we have

    (i) Scalar multiplication is **distributive** with respect to vector addition: $a(v + w) = av + aw$.

    (ii) Scalar multiplication is **distributive** with respect to scalar addition: $(a + b)v = av + bv$.

Remarks. (1) A vector space can never be an empty set. It contains at least one element, the trivial vector.

(2) We often write $v - w$ for $v + (-w)$. It does not make sense to write v/w since w^{-1} does not necessarily exist.

The vector space V is called a **real vector space** if $F = \mathbb{R}$, a **rational vector space** if $F = \mathbb{Q}$ and a **complex vector space** if $F = \mathbb{C}$.

In this book we will be mainly using real vector spaces as examples simply because it is "more" familiar to college freshmen, not because they are more important. Occasionally, complex or rational vector spaces will be discussed. However, bear in mind that practically all proofs and all theorems will work for vector spaces over any field unless otherwise noted. Eventually when you need to apply Linear Algebra, it is likely you are not going to apply it to real vector spaces alone.

Proposition 1.2.4. *Let V be an F-vector space. Let a be a scalar in F and let u, v and w be vectors in V. Then the following properties hold:*

    (a) (Cancellation Law) *If $u + w = v + w$ then $u = v$;*

    (b) $0v = O$;

    (c) $(-a)v = -(av) = a(-v)$;

    (d) $aO = O$.

Proof. The proof is very similar to that of Proposition 1.2.2.

(a) Assume $u + w = v + w$. Then

$$
\begin{aligned}
u &= u + O, && \text{by (A)(iii),} \\
&= u + [w + (-w)], && \text{by (A)(iv),} \\
&= (u + w) + (-w), && \text{by (A)(ii),} \\
&= (v + w) + (-w), && \text{by assumption,} \\
&= v + [w + (-w)], && \text{by (A)(ii),} \\
&= v + O, && \text{by (A)(iv),} \\
&= v, && \text{by (A)(iii).}
\end{aligned}
$$

(b) Again, we can see that

$$
\begin{aligned}
0v + O &= 0v, && \text{by (A)(iii)} \\
&= (0 + 0)v, && 0 \text{ being the additive identity of } F, \\
&= 0v + 0v, && \text{by (C)(ii).}
\end{aligned}
$$

Now by Cancelation Law in (a), $0v = O$.

(c) To show that $(-a)v = -(av)$ is to show that $(-a)v$ is the additive inverse of av. This is clear since

$$
\begin{aligned}
av + (-a)v &= [a + (-a)]v, && \text{by (C)(ii),} \\
&= 0v, && \text{by the property of additive inverse of scalars,} \\
&= O, && \text{by (b).}
\end{aligned}
$$

One can show that $a(-v) = -(av)$ by symmetry.

(d) Note that

$$
\begin{aligned}
aO + O = aO = a(O + O), && \text{by (A)(iii)} \\
= aO + aO, && \text{by (C)(i).}
\end{aligned}
$$

Now by Cancelation Law in (a), $aO = O$. $\qquad \square$

Remark. The zero vector and the additive inverse of a vector are both unique. See Problem 4.

Basic examples of vector spaces. Next we look at some of the most basic examples. These examples will be used to demonstrate many interesting properties of the vector spaces in future sections.

Example 1.2.5. Let F be any field. There is exactly one way to define vector addition and scalar multiplication on a singleton set $\{O\}$. The vector space $\{O\}$ is called the **trivial** vector space (over F).

Let's give a short review of the set $\mathbb{R}^n$. As a set

$$\mathbb{R}^n = \{(a_1, a_2, \ldots, a_n) : a_i \in \mathbb{R}\}.$$

The element $\mathbf{a} = (a_1, a_2, \ldots, a_n)$ is called an n-**tuple**. The number a_i is called the i-th **entry**, the i-th **component** or the i-th **coordinate** of the n-tuple $\mathbf{a}$. Naturally we may define the vector addition and scalar multiplication in $\mathbb{R}^n$ by letting

$$(a_1, a_2, \ldots, a_n) + (b_1, b_2, \ldots, b_n) = (a_1 + b_1, a_2 + b_2, \ldots, a_n + b_n), \quad \text{and}$$

$$t(a_1, a_2, \ldots, a_n) = (ta_1, ta_2, \ldots, ta_n), \qquad \text{where } t \in \mathbb{R}.$$

This makes $\mathbb{R}^n$ a real vector space. For this reason, the n-tuple is also called an n-**vector** or a **row** n-**vector**.

The sets $\mathbb{Q}^n$, $\mathbb{C}^n$ or even F^n may be defined similarly, and the vector addition and scalar multiplication may also be defined accordingly. Vectors in F^n may also be written as **column vectors**

$$\begin{pmatrix} a_1 \\ a_2 \\ \vdots \\ a_n \end{pmatrix}.$$

Example 1.2.6. (1) $\mathbb{Q}$, $\mathbb{Q}^2$ or even $\mathbb{Q}^n$, $n \in \mathbb{Z}_+$, are all rational vector spaces.

(2) $\mathbb{R}$, $\mathbb{R}^2$ or even $\mathbb{R}^n$, $n \in \mathbb{Z}_+$, are all real vector spaces as well as rational vector spaces.

(3) $\mathbb{C}$, $\mathbb{C}^2$ or even $\mathbb{C}^n$, $n \in \mathbb{Z}_+$, are all complex, rational or real vector spaces.

(4) F^n, $n \in \mathbb{Z}_+$, is an F-vector space.

(5) $\mathbb{C}$ is an $\mathbb{R}$-vector space but $\mathbb{R}$ is not a $\mathbb{C}$-vector space.

Next, we are going to give some more examples of vector spaces.

Example 1.2.7. An $m \times n$ **matrix** with **entries** in a field F is a rectangular array of the form

$$(1.1) \qquad \begin{pmatrix} a_{11} & a_{12} & \cdots & a_{1n} \\ a_{21} & a_{22} & \cdots & a_{2n} \\ \vdots & \vdots & \ddots & \vdots \\ a_{m1} & a_{m2} & \cdots & a_{mn} \end{pmatrix}$$

where each entry a_{ij} is an element in F. Note that this matrix has m rows and n columns. If $m = n$ we say the matrix is a **square** matrix. We denote by $M_{m \times n}(F)$ the set of all $m \times n$ matrices with entries in F. We can define vector addition and scalar multiplication naturally to make $M_{m \times n}(F)$ into an F-vector space. Let

$$A = \begin{pmatrix} a_{11} & a_{12} & \cdots & a_{1n} \\ a_{21} & a_{22} & \cdots & a_{2n} \\ \vdots & \vdots & \ddots & \vdots \\ a_{m1} & a_{m2} & \cdots & a_{mn} \end{pmatrix} \quad \text{and} \quad B = \begin{pmatrix} b_{11} & b_{12} & \cdots & b_{1n} \\ b_{21} & b_{22} & \cdots & b_{2n} \\ \vdots & \vdots & \ddots & \vdots \\ b_{m1} & b_{m2} & \cdots & b_{mn} \end{pmatrix}$$

be two matrices in $M_{m \times n}(F)$ and let $c \in F$. We define

$$(1.2) \quad A + B = \begin{pmatrix} a_{11} + b_{11} & a_{12} + b_{12} & \cdots & a_{1n} + b_{1n} \\ a_{21} + b_{21} & a_{22} + b_{22} & \cdots & a_{2n} + b_{2n} \\ \vdots & \vdots & \ddots & \vdots \\ a_{m1} + b_{m1} & a_{m2} + b_{m2} & \cdots & a_{mn} + b_{mn} \end{pmatrix} \quad \text{and}$$

$$(1.3) \qquad cA = \begin{pmatrix} ca_{11} & ca_{12} & \cdots & ca_{1n} \\ ca_{21} & ca_{22} & \cdots & ca_{2n} \\ \vdots & \vdots & \ddots & \vdots \\ ca_{m1} & ca_{m2} & \cdots & ca_{mn} \end{pmatrix}.$$

You can check easily that $M_{m \times n}(F)$ is an F-vector space. When $m = n$ we also denote $M_{m \times n}(F)$ by $M_n(F)$.

To simplify notation, the matrix in (1.1) is often written as

$$\left(a_{ij} \right)_{m \times n} \qquad \text{or simply} \qquad \left(a_{ij} \right).$$

The $m \times n$ at the lower right-hand corner indicates the size of the matrix: it has m rows and n columns. By convention, inside the parenthesis is the general form of the (i, j)-**entry**, that is, the entry at the i-th row and the j-th column. If a second matrix is also mentioned, it is often written as

$$\left(b_{kl}\right)_{m \times n} \quad \text{or simply} \quad \left(b_{kl}\right),$$

whose entry, by convention, indicates the (k, l)-entry. The identity in (1.2) can be rewritten as

$$\left(a_{ij}\right) + \left(b_{ij}\right) = \left(a_{ij} + b_{ij}\right) \quad \text{or} \quad \left(a_{ij}\right) + \left(b_{kl}\right) = \left(a_{ij} + b_{ij}\right),$$

while the identity (1.3) can be rewritten as

$$c\left(a_{ij}\right) = \left(ca_{ij}\right).$$

Example 1.2.8. Let $\mathscr{P}$ be the set of all the polynomials with real coefficients. It is also clearly a real vector space under the usual polynomial addition and multiplication by real constants.

Example 1.2.9. Let $\mathscr{F}(\mathbb{R}, \mathbb{R})$ be set of the functions from $\mathbb{R}$ to $\mathbb{R}$. Let f and g be functions in V. Define $f + g$ to be the function defined by

$$(f + g)(x) = f(x) + g(x).$$

If $c \in \mathbb{R}$, define cf to be the function defined by

$$(cf)(x) = cf(x).$$

It is easy to see that V is a real vector space.

Example 1.2.10. Consider the set $\mathscr{F}([0, 1], \mathbb{R})$ of all functions from $[0, 1]$ to $\mathbb{R}$ and the set $\mathscr{F}([0, 1], [0, 1])$ of all functions from $[0, 1]$ to $[0, 1]$. With function addition and the obvious scalar multiplication, $\mathscr{F}([0, 1], \mathbb{R})$ is a real vector space while $\mathscr{F}([0, 1], [0, 1])$ is not!

Exercises 1.2

1. Show that the additive identity, the additive inverses, the multiplicative identity and the multiplicative inverses are unique in a field.

2. Let $F = \{a + bi \in \mathbb{C} : a, b \in \mathbb{Q}\}$ with inherited addition and multiplication from $\mathbb{C}$. Show that F is a field.

3. Let $\mathbb{Z}_3 = \{0, 1, 2\}$ be equipped with addition and multiplication given by

+	0	1	2
0	0	1	2
1	1	2	0
2	2	0	1

and

·	0	1	2
0	0	0	0
1	0	1	2
2	0	2	1

.

(a) Show that $\mathbb{Z}_3$ is a field.

(b) Find -0, -1, -2, 1^{-1} and 2^{-1}.

4. Show that the zero vector and the additive inverse of a vector space are both unique in a vector space.

5. Give the matrices $(i + j)_{2\times 4}$, $(2^i)_{3\times 3}$ and $(a^{ij})_{3\times 2}$ explicitly.

6. Not all multiplications are commutative as we will see in the following problem.

Let $A = (a_{ij})_{m\times n}$ and $B = (b_{k\ell})_{n\times r}$ be two matrices over a field K. We may define the product $AB = (c_{ij})_{m\times r}$ by letting

$$c_{ij} = \sum_{k=1}^{n} a_{ik}b_{kj}, \qquad \text{for } i = 1, \ldots, m \text{ and } j = 1, \ldots, r.$$

If $n \neq r$, BA is not defined. In case $n = r$, BA is an $m \times m$ matrix. It is obvious $AB \neq BA$ when $m \neq n$.

Now suppose A and B are both square matrices of size n. Is it true that $AB = BA$ in general? Prove your assertion or give a counter-example.

1.3   Subspaces

Subspaces.   There are vector spaces within vector spaces. To completely understand the structure of a vector space is to understand the hierarchy among its substructure.

Definition 1.3.1. A **subspace**, or more specifically a **linear subspace**, of an F-vector space V over F is a subset of V which is an F-vector space itself under the inherited vector addition and scalar multiplication.

Remarks. (1) Any subspace of V contains the zero vector O.

(2) In any vector space V there are two obvious subspaces: the space V itself and the trivial subspace $\{O\}$.

Fortunately to verify a subset of a vector space is a subspace, one needs not to verify every axiom in Definition 1.2.3.

Theorem 1.3.2 (Tests for subspaces). *Let V be a vector space over F and W be a subset of V. Then the following three sets of conditions* (i), (ii) *and* (iii) *are equivalent.*

(i) *The subset W with inherited vector addition and scalar multiplication from V is a subspace of V .*

(ii) (a) $O \in W$.

(b) $v + w \in W$ *for all* $v, w \in W$.

(c) $aw \in W$ *for all* $a \in F$ *and* $w \in W$.

(iii) (a) $O \in W$.

(b) $av + bw \in W$ *for all* $a, b \in F$ *and* $v, w \in W$.

Proof. The implication "(i) $\Rightarrow$ (iii)" is clear.

"(iii) $\Rightarrow$ (ii)": The condition (a) is assumed. For all $a \in F$ and v, $w \in W$, $v + w = 1v + 1w \in W$ and $aw = 0v + aw \in W$ by (iii)(a) and (b).

"(ii) $\Rightarrow$ (i)": The condition (ii)(a) implies that W is nonempty. Furthermore, (ii)(a) and (b) imply the vector addition and scalar multiplication are both well-defined for F and W. All the required properties in Definition 1.2.3 are satisfied for all elements in V. Thus they are all satisfied for a subset of V.   □

Remark. In fact, in (ii) and in (iii) of Theorem 1.3.2, the condition that $O \in W$ can be replaced by the condition that W is nonempty. But in practice, usually the easiest way to show that W is nonempty is to show that $O \in W$.

Examples of subspaces. We now use Theorem 1.3.2 to test whether a subset of a vector space is a subspace or not.

Example 1.3.3. Remember that $\mathbb{C}$ can be regarded as a real vector space. As a subset $\mathbb{R}$ is also a real vector space and the vector addition and scalar multiplication are both inherited from $\mathbb{C}$. Hence $\mathbb{R}$ can be thought of as a subspace of $\mathbb{C}$ over $\mathbb{R}$.

The set of rational numbers is not a subspace of $\mathbb{C}$ or $\mathbb{R}$ over $\mathbb{R}$. The product of $\sqrt{2}$ (as a scalar) and 1 (as a vector) is no longer in $\mathbb{Q}$. So $\mathbb{Q}$ failed the test for subspaces. Similarly, $\mathbb{Z}$ is neither a subspace of $\mathbb{C}$ or $\mathbb{R}$.

Example 1.3.4. The subset $V = \{(a, 0) : a \in \mathbb{R}\}$ is a subspace of $\mathbb{R}^2$ over $\mathbb{R}$. Clearly it is nonempty since $(0, 0) \in V$. For all $a, b \in \mathbb{R}$ and $r \in \mathbb{R}$, we have $(a, 0) + (b, 0) = (a + b, 0) \in V$ and $r(a, 0) = (ra, 0) \in V$.

Definition 1.3.5. Let $A = \left(a_{ij}\right)_{m \times n} \in M_{m \times n}(F)$. We use A^t denote the **transpose** of A, where

$$A^t = \left(a_{ji}\right)_{m \times n}.$$

Example 1.3.6. If $A = \begin{pmatrix} 1 & 2 & 3 \\ 4 & 5 & 6 \end{pmatrix}$ then $A^t = \begin{pmatrix} 1 & 4 \\ 2 & 5 \\ 3 & 6 \end{pmatrix}$.

Definition 1.3.7. Let $A = \left(a_{ij}\right)_{n \times n} \in M_n(F)$. We say A is a **symmetric** matrix if $A^t = A$. That is, A is symmetric if and only if $a_{ij} = a_{ji}$ for all $i, j = 1, 2, \ldots, n$. We say A is a **skew-symmetric** matrix if $A^t = -A$. That is, A is skew-symmetric if and only if $a_{ij} = -a_{ji}$ for all $i, j = 1, 2, \ldots, n$. We say A is a **diagonal** matrix if $a_{ij} = 0$ for all $i \neq j$.

Example 1.3.8. Consider the real vector space $M_n(\mathbb{R})$.

(1) Is the subset of all symmetric matrices in $M_n(\mathbb{R})$ a subspace?

(2) Is the subset of all diagonal matrices in $M_n(\mathbb{R})$ a subspace?

(3) Is the subset of all skew-symmetric matrices in $M_n(\mathbb{R})$ a subspace?

(4) Is the set of all matrices in $M_n(\mathbb{R})$ with nonnegative entries a subspace?

Solution. Let $A = (a_{ij})$, $B = (b_{ij}) \in M_n(\mathbb{R})$ and $r \in \mathbb{R}$.

(1) The zero matrix is a symmetric matrix. If A and B are symmetric, then $A + B = (a_{ij} + b_{ij})$ is also symmetric since $a_{ij} + b_{ij} = a_{ji} + b_{ji}$. The matrix $rA = (ra_{ij})$ is also symmetric since $ra_{ij} = ra_{ji}$. The subset of symmetric matrices in $M_n(\mathbb{R})$ form a subspace.

(2) The zero matrix is a diagonal matrix. If A and B are diagonal, then $A + B = (a_{ij} + b_{ij})$ is also diagonal since $a_{ij} + b_{ij} = 0$ for all $i \neq j$. The matrix $rA = (ra_{ij})$ is also symmetric since $ra_{ij} = ra_{ji}$. The subset of diagonal matrices in $M_n(\mathbb{R})$ form a subspace.

(3) The zero matrix is a skew-symmetric matrix. If A and B are symmetric, then $A + B = (a_{ij} + b_{ij})$ is also skew-symmetric since $a_{ij} + b_{ij} = -a_{ji} - b_{ji} = -(a_{ji} + b_{ji})$. The matrix $rA = (ra_{ij})$ is also symmetric since $ra_{ij} = r(-a_{ji}) = -ra_{ji}$. The subset of skew-symmetric matrices in $M_n(\mathbb{R})$ form a subspace.

(4) The identity matrix I_n is a matrix with nonnegative entries. However, $(-1)I_n$ is not. Thus the set of all matrices with nonnegative entries is not a subspace.   ◇

Definition 1.3.9. The **trace** of an $n \times n$ matrix $A = \left(a_{ij}\right)$, denoted $\operatorname{tr} A$, is defined to be

$$a_{11} + a_{22} + \cdots + a_{nn}.$$

Example 1.3.10. The trace of the zero matrix is clearly 0. If A and $B \in M_n(\mathbb{R})$ and $r \in \mathbb{R}$, note that

$$\operatorname{tr}(A + B) = \operatorname{tr} A + \operatorname{tr} B \quad \text{and} \quad \operatorname{tr}(rA) = r \operatorname{tr} A.$$

The set of matrices of trace 0 passes the test for subspaces. Thus the subset of all matrices whose traces are 0 in $M_n(\mathbb{R})$ forms a subspace.

Example 1.3.11. Let $n \in \mathbb{Z}$. Then $\mathscr{P}_n = \{f(x) \in \mathscr{P} : \deg f \leq n\}$ is a subspace of $\mathscr{P}$. Note that by convention we define the degree of the zero polynomial to be $-\infty$.

Example 1.3.12. Consider the real vector space $\mathscr{F}(\mathbb{R}, \mathbb{R})$. Form a course of Calculus, we know that the sum of two continuous functions is still continuous, and so is a scalar multiple of a continuous function. The same is true for differentiable functions as well. Thus, the subset $\mathscr{C}(\mathbb{R}, \mathbb{R})$ of all continuous functions and the subset $\mathscr{D}(\mathbb{R}, \mathbb{R})$ of all differentiable functions in $\mathscr{F}(\mathbb{R}, \mathbb{R})$ are both subspaces. It is also clear that $\mathscr{D}(\mathbb{R}, \mathbb{R})$ is a subspace of the real vector space in $\mathscr{C}(\mathbb{R}, \mathbb{R})$.

Theorem 1.3.13. *The intersection of a nonempty collection of subspaces of a F-vector space V is a subspace of V.*

Proof. Let $\{W_i : i \in \Lambda\}$ be a collection of subspaces of V. Clearly the zero vector $O \in W_i$ for all $i \in \Lambda$. Hence $O \in \bigcap_{i \in \Lambda} W_i$.

Let $v, w \in \bigcap_{i \in \Lambda} W_i$ and $a \in F$. Then $v, w \in W_i$ for all $i \in \Lambda$. By the properties of subspaces $v+w$ and av are both in W_i for all $i \in \Lambda$. Thus $v+w$ and $av \in \bigcap_{i \in \Lambda} W_i$. The subset $\bigcap_{i \in \Lambda} W_i$ passes the test for subspaces. $\square$

A first peek at systems of linear equations. first we use an example to observe that the solutions of a system of homogeneous linear equations in n variables with real coefficients form a subspace of $\mathbb{R}^n$.

The subset of $(x_1, x_2, x_3, x_4) \in \mathbb{R}^4$ satisfying

$$(1.4) \qquad \begin{cases} 2x_1 - 2x_2 + 3x_3 + 5x_4 = 0 \\ -x_1 + 3x_2 + 2x_3 - 4x_4 = 0 \end{cases}$$

is a subspace of $\mathbb{R}^4$. Clearly, the zero vector $(0,0,0,0)$ is a solution of (1.4). If both (a_1, a_2, a_3, a_4) and (b_1, b_2, b_3, b_4) are solutions, then $(a_1, a_2, a_3, a_4) + (b_1, b_2, b_3, b_4)$ and $r(a_1, a_2, a_3, a_4)$ are also solutions for any $r \in \mathbb{R}$, since

$$\begin{cases} 2(a_1 + b_1) - 2(a_2 + b_2) + 3(a_3 + b_3) + 5(a_4 + b_4) \\ =(2a_1 - 2a_2 + 3a_3 + 5a_4) + (2b_1 - 2b_2 + 3b_3 + 5b_4) = 0 \\ - (a_1 + b_1) - 3(a_2 + b_2) + 2(a_3 + b_3) - 4(a_4 + b_4) \\ =(-a_1 - 3a_2 + 2_3 - 4a_4) + (-b_1 + 3b_2 + 2b_3 - 4b_4) = 0 \end{cases}$$

and

$$\begin{cases} 2ra_1 - 2ra_2 + 3ra_3 + 5ra_4 = r(2a_1 - 2a_2 + 3a_3 + 5a_4) = 0 \\ - ra_1 + 3ra_2 + 2ra_3 - 4ra_4 = r(-a_1 + 3a_2 + 2a_3 - 4a_4) = 0 \end{cases}$$

Hence the solution set forms a subspace by the subspace test.

On the other hand, the subset of $(x_1, x_2, x_3, x_4) \in \mathbb{R}^4$ satisfying

$$(1.5) \qquad \begin{cases} 2x_1 - 2x_2 + 3x_3 + 5x_4 = 4 \\ -x_1 + 3x_2 + 2x_3 - 4x_4 = 0 \end{cases}$$

is not a subspace of $\mathbb{R}^4$. One can easily verify that $(3, 1, 0, 0)$ is a solution while $2(3, 1, 0, 0)$ is not!

To solve the common solutions of a bunch of equations such as in (1.4) and (1.5) is the first experience of linear algebra to many people. This type of grouped equations is known as **systems of linear equations**. To solve a system of linear equations over F in general is to solve simultaneously m equations with coefficients in F in n variables, as in

$$\begin{array}{ccccccccc} a_{11}x_1 & + & a_{12}x_2 & + & \cdots & + & a_{1n}x_n & = & b_1 \\ a_{21}x_1 & + & a_{22}x_2 & + & \cdots & + & a_{2n}x_n & = & b_2 \\ & & & & \cdots\cdots\cdots & & & & \\ a_{m1}x_1 & + & a_{m2}x_2 & + & \cdots & + & a_{mn}x_n & = & b_m \end{array}$$

where $a_{ij} \in F$ for $i = 1, 2, \ldots, m$ and $j = 1, 2, \ldots, n$. In particular, if $b_1 = b_2 = \cdots = b_m = 0$, the system of linear equations

$$(1.6) \qquad \begin{array}{ccccccccc} a_{11}x_1 & + & a_{12}x_2 & + & \cdots & + & a_{1n}x_n & = & 0 \\ a_{21}x_1 & + & a_{22}x_2 & + & \cdots & + & a_{2n}x_n & = & 0 \\ & & & & \cdots\cdots\cdots & & & & \\ a_{m1}x_1 & + & a_{m2}x_2 & + & \cdots & + & a_{mn}x_n & = & 0 \end{array}$$

is also called a **homogeneous system of linear equations**.

Proposition 1.3.14. *The set of $(x_1, x_2, \ldots, x_n) \in F^n$ satisfying the homogeneous system of equations in (1.6) is a subspace of F^n.*

Proof. For $i = 1, 2, \ldots, m$, let

$$W_i = \{(x_1, x_2, \ldots, x_n) \in F^n : a_{i1}x_1 + a_{i2}x_2 + \cdots + a_{in}x_n = 0\}.$$

The set of all solutions of the *homogeneous* system of linear equations in (1.6) is $\bigcap_{i=1}^m W_i$. If we can show that each W_i is a subspace, by Theorem 1.3.13 the intersection of the W_i's is also a subspace. Now we can reduce the problem to the case of one linear equation

$$a_1 x_1 + a_2 x_2 + \cdots + a_n x_n = 0.$$

Clearly, the zero vector is a solution. (It is essential that the system must be *homogeneous!*) Suppose $(\alpha_1, \ldots, \alpha_n)$ and $(\beta_1, \ldots, \beta_n)$ are both solutions of the equation above. Then

$$a_1(\alpha_1 + \beta_1) + a_2(\alpha_2 + \beta_2) + \cdots + a_n(\alpha_n + \beta_n)$$
$$= (a_1(\alpha_1) + a_2(\alpha_2) + \cdots + a_n(\alpha_n))$$
$$+ (a_1(\beta_1) + a_2(\beta_2) + \cdots + a_n(\beta_n))$$
$$= 0 + 0 = 0$$

and

$$a_1(r\alpha_1) + a_2(r\alpha_2) + \cdots + a_n(r\alpha_n)$$
$$= r(a_1\alpha_1 + a_2\alpha_2 + \cdots a_n\alpha_n) = 0$$

for all $r \in \mathbb{R}$. Thus, the solution set of one linear equation is indeed a subspace of F^n by the subspace test. $\square$

On the other hand, if we consider the solution set of some non-linear equation, say $W = \{(x, y) \in \mathbb{R}^2 : 2x^2 - 3y = 0\}$ of $\mathbb{R}^2$. Note that the vector $(3, 6)$ belongs in W. However, the vector $2(3, 6) = (6, 12)$ is not a solution of $2x^2 - 3y = 0$. This shows that W is not an R-subspace of $\mathbb{R}^2$. The solution sets of non-linear equations are never vector spaces. This is the reason that we will not discuss solutions of non-linear equations in this book.

Exercises 1.3

1. Let W be a *nonempty* subset of an F-vector space V such that $av + bw \in W$ for all $a, b \in F$ and $v, w \in W$. Show that W is a subspace of V.

2. Which of the following sets are subspaces of $\mathbb{R}^3$?

   (a) $V = \{(x_1, x_2, x_3) \in \mathbb{R}^3 : x_1 + x_2 + x_3 = 0\}$.

   (b) $V = \{(x_1, x_2, x_3) \in \mathbb{R}^3 : x_1 + x_2 + x_3 \geq 0\}$.

(c) $V = \{(x_1, x_2, x_3) \in \mathbb{R}^3 : x_1 + x_2 = 0\}$.

3. Which of the following sets are subspaces of $\mathscr{F}(\mathbb{R}, \mathbb{R})$?

   (a) $\{f \in \mathscr{F}(\mathbb{R}, \mathbb{R}) : f(0) = 2\}$.

   (b) $\{f \in \mathscr{F}(\mathbb{R}, \mathbb{R}) : f(2) = 0\}$.

4. Is $\mathbb{R}$ a subspace of $\mathbb{C}$ as a complex vector space?

5. Show that $\mathscr{C}([0,1], \mathbb{R})$ is a subspace of $\mathscr{F}([0,1], \mathbb{R})$ and $\mathscr{D}([0,1], \mathbb{R})$ is a subspace of $\mathscr{C}([0,1], \mathbb{R})$.

6. Let W be a subspace of U and U a subspace of V. Show that W is a subspace of V.

7. Let m and $n \in \mathbb{Z}$. Show that $\mathscr{P}_m$ is a subspace of $\mathscr{P}_n$ if and only if $m \leq n$.

8. Does the set of symmetric matrices in $M_n(\mathbb{R})$ form a subspace over $\mathbb{R}$?

9. Suppose given two matrices $A_{m \times n}$ and $B_{n \times r}$. Show that $(AB)^t = B^t A^t$.

1.4 How to solve a system of linear equations

From now on for the sake of convenience, we assume that $F = \mathbb{R}$ if we do not specify F. In other words, the scalars will be real numbers unless we say otherwise. Hence when we say V is a vector space, we actually mean that V is a real vector space. Please bear in mind that our discussions will be valid no matter what F is.

Reduced row echelon form.   Students of Linear Algebra have had some experiences in solving linear equations using different methods. No matter what method one may use, the principle is to decrease the number of variables appearing in a subsystem so that eventually we either reach a contradiction or an equation with exactly one variable. To make this process straightforward, we introduce a useful tool. We will associate a matrix with a system of linear equations. We will then simplify this matrix until the solutions to the system becomes obvious. For this purpose we will introduce the concept of **reduced row echelon form**.

We say a matrix is in reduced row echelon form if it satisfies the following three conditions.

- The nonzero rows precede the zero rows (if any).

- The first nonzero entry in each nonzero row is the only nonzero entry in that column.

- The first nonzero entry in each nonzero row is 1 and it occurs in a column to the right of the first nonzero entry in the preceding row. This entry is called the **pivot** of the corresponding row. In other words, each time we move downward, we must move at least one column to the right to find the pivot of the next row.

For example, the following matrices are in reduced row echelon form:

$$\begin{pmatrix} \mathbf{1} & 0 \\ 0 & \mathbf{1} \\ 0 & 0 \end{pmatrix}, \quad \begin{pmatrix} 0 & \mathbf{1} & 3 & 0 \\ 0 & 0 & 0 & \mathbf{1} \end{pmatrix}, \quad \begin{pmatrix} \mathbf{1} & -5.3 & 0 & 0 \\ 0 & 0 & \mathbf{1} & 0 \\ 0 & 0 & 0 & \mathbf{1} \end{pmatrix}.$$

The entries in boldface are the pivots.

To summarize, the three operations that we may perform on the system without affecting the solution set translate as the three on the augmented matrix given below.

When we are given an arbitrary matrix, we are allowed to perform three types of **elementary row operations**:

- One can permute two rows in an augmented matrix.

- One can multiply a row by a *nonzero* constant.

- One can add to a row in an augmented matrix a constant multiple of a different row.

Using these row operations, it is always possible to transform a matrix to a reduced row echelon form.

Example 1.4.1. Find a reduced row echelon form for the matrix

$$A = \begin{pmatrix} -2 & 1 & 1 \\ 2 & -1 & 1 \\ 2 & 1 & -1 \end{pmatrix}.$$

Solution. First, add the first row to the second and the third rows to eliminate all other entries on column one:

$$A \rightsquigarrow \begin{pmatrix} -2 & 1 & 1 \\ 0 & 0 & 2 \\ 0 & 2 & 0 \end{pmatrix}.$$

Next, permute the second row and the third row so that the second entry in the third row becomes 0:

$$A \rightsquigarrow \begin{pmatrix} -2 & 1 & 1 \\ 0 & 2 & 0 \\ 0 & 0 & 2 \end{pmatrix}.$$

Next, multiply the first, the second, and the third row by $-1/2$, $1/2$, $1/2$ respectively:

$$A \rightsquigarrow \begin{pmatrix} 1 & -1/2 & -1/2 \\ 0 & 1 & 0 \\ 0 & 0 & 1 \end{pmatrix}.$$

We now have the pivots in place. Finally, add $(1/2)\times$the second row and $(1/2)\times$the third row to the first row:

$$A \rightsquigarrow \begin{pmatrix} 1 & 0 & 0 \\ 0 & 1 & 0 \\ 0 & 0 & 1 \end{pmatrix}.$$

We have reached the reduced row echelon form of A.   ◇

Gaussian elimination.   We will describe a systematic method called **Gaussian elimination** to solve any system of linear equations. This method can be best understood by looking at examples.

Now let's try solving this system of linear equations

$$\begin{array}{rl}
x_1 + x_2 + x_3 + x_4 = 0 & \text{(i)} \\
2x_1 + 2x_2 + x_3 - x_4 = 0 & \text{(ii)} \\
3x_1 + 3x_2 + 2x_3 = 0 & \text{(iii)}
\end{array}$$

One can perform three operations to a system without affecting the solution set since all these operations are reversible.

- One may permute two equations in this system.

- One may multiply an equation by a *nonzero* constant. That is, one may replace one of the $f_i = b_i$ by $cf_i = cb_i$ for $c \neq 0$.

- One may multiply one of the equations by a constant and then add it to a different equation in the system. That is, one may replace one of the $f_i = b_i$ by $f_i + cf_j = b_i + cb_j$ for $c \in \mathbb{R}$ and $i \neq j$.

The goal is to use the three types of operations given above to simplify the original system. For example, The system is equivalent to

(1.7)
$$\begin{array}{rl}
x_1 + x_2 + x_3 + x_4 = 0 & \text{(i)} \\
-x_3 - 3x_4 = 0 & \text{(ii)}' = \text{(ii)} - 2 \times \text{(i)} \\
-x_3 - 3x_4 = 0 & \text{(iii)}' = \text{(iii)} - 3 \times \text{(i)}
\end{array}$$

Here, the number of variables is reduced in the second and the third equations in the system. One can see that (iii)$'$ is redundant and we will simply drop it from the system. By performing a simple operation, the system becomes

$$\begin{array}{rl}
x_1 + x_2 + x_3 + x_4 = 0 & \text{(i)} \\
x_3 + 3x_4 = 0 & \text{(ii)}'' = -1 \times \text{(ii)}
\end{array}$$

Just to make the system even simpler, we will take one further step:

$$\begin{array}{rl}
x_1 + x_2 - 2x_4 = 0 & \text{(i)}' = \text{(i)} - \text{(ii)}'' \\
x_3 + 3x_4 = 0 & \text{(ii)}''
\end{array}$$

We now conclude that the solution is

$$\begin{array}{l}
x_3 = -3x_4, \\
x_1 = -x_2 + 2x_4
\end{array}$$

where x_2 and x_4 are arbitrary real numbers.

We will rework this problem by using matrices as an aid. The matrix

$$A = \begin{pmatrix} 1 & 1 & 1 & 1 \\ 2 & 2 & 1 & -1 \\ 3 & 3 & 2 & 0 \end{pmatrix}$$

is called the **coefficient matrix** of the system (1.7). One may observe that the entries are taken from the coefficients in the given system. If one is given a matrix as a coefficient matrix, one may also recover the left hand part of the system. When solving systems of linear equations, it is more often to use the **augmented matrix** which also records the constant terms of the system. The augmented matrix for the system (1.7) is

$$B = \left(\begin{array}{cccc|c} 1 & 1 & 1 & 1 & 0 \\ 2 & 2 & 1 & -1 & 0 \\ 3 & 3 & 2 & 0 & 0 \end{array} \right).$$

The augmented matrix is a convenient and efficient way to record a system of linear equations.

During the process of solving a system, one performs certain operations on the system. The corresponding augmented matrices also receive the same treatment. As in the example just mentioned, the augmented matrix B underwent the transformations:

$$B = \left(\begin{array}{cccc|c} 1 & 1 & 1 & 1 & 0 \\ 2 & 2 & 1 & -1 & 0 \\ 3 & 3 & 2 & 0 & 0 \end{array} \right) \rightsquigarrow \left(\begin{array}{cccc|c} 1 & 1 & 1 & 1 & 0 \\ 0 & 0 & -1 & -3 & 0 \\ 0 & 0 & -1 & -3 & 0 \end{array} \right)$$

$$\rightsquigarrow \left(\begin{array}{cccc|c} 1 & 1 & 1 & 1 & 0 \\ 0 & 0 & 1 & 3 & 0 \\ 0 & 0 & 0 & 0 & 0 \end{array} \right) \rightsquigarrow \left(\begin{array}{cccc|c} 1 & 1 & 0 & -2 & 0 \\ 0 & 0 & 1 & 3 & 0 \\ 0 & 0 & 0 & 0 & 0 \end{array} \right)$$

When we reach the last matrix, there is practically no further simplification. We translate the augmented matrix into its corresponding system and give the solution from there.

After we transform an augmented matrix of a system of linear equations to be in reduced echelon form, if one of the pivots appears in the last column (the column after the dividing line), we have encountered an *inconsistent* system. Suppose this does not happen. The columns on which a pivot lies (the i_1-th, i_2-th, ..., and the i_k-th columns) correspond to variables $(x_{i_1}, x_{i_2}, ..., x_{i_k})$ which are not free (they depend on the other variables) in our solution set. These variables are called the **pivot variables** of the given system. The other variables are called the **free parameters**. The free parameters give us an idea how large the solution space is.

Example 1.4.2. Solve the following system of linear equations over $\mathbb{R}$:

$$\begin{aligned} x + 3y - 3z \qquad\quad + 11u &= 11, \\ x + y + z \qquad\quad + 7u &= 5, \\ 3x + y + 7z + 3w + 20u &= 18. \end{aligned}$$

Solution. The corresponding augmented matrix is

$$B = \left(\begin{array}{ccccc|c} 1 & 3 & -3 & 0 & 11 & 11 \\ 1 & 1 & 1 & 0 & 7 & 5 \\ 3 & 1 & 7 & 3 & 20 & 18 \end{array} \right).$$

Clearly, the fist entry on row one may act as the pivot, and we use it to eliminate the other entries on column one:

$$B \rightsquigarrow \left(\begin{array}{ccccc|c} 1 & 3 & -3 & 0 & 11 & 11 \\ 0 & -2 & 4 & 0 & -4 & -6 \\ 0 & -8 & 16 & 3 & -13 & -15 \end{array} \right) \quad \begin{array}{l} \text{row(ii)} - \text{row(i)} \\ \text{row(iii)} - 3 \times \text{row(i)} \end{array}.$$

We next adjust the first nonzero entry on row two into a pivot:

$$B \rightsquigarrow \left(\begin{array}{ccccc|c} 1 & 3 & -3 & 0 & 11 & 11 \\ 0 & 1 & -2 & 0 & 2 & 3 \\ 0 & -8 & 16 & 3 & -13 & -15 \end{array} \right) \quad (-1/2) \times \text{row(ii)} \ .$$

Use the pivot on the second row to eliminate all other entries on column two:

$$B \rightsquigarrow \left(\begin{array}{ccccc|c} 1 & 0 & 3 & 0 & 5 & 2 \\ 0 & 1 & -2 & 0 & 2 & 3 \\ 0 & 0 & 0 & 3 & 3 & 9 \end{array} \right) \quad \begin{array}{l} \text{row(i)} - 3 \times \text{row(ii)} \\ \text{row(iii)} + 8 \times \text{row(i)} \end{array}.$$

Adjust the first nonzero entry on row three into a pivot:

$$B \rightsquigarrow \left(\begin{array}{ccccc|c} 1 & 0 & 3 & 0 & 5 & 2 \\ 0 & 1 & -2 & 0 & 2 & 3 \\ 0 & 0 & 0 & 1 & 1 & 3 \end{array} \right) \quad (1/3) \times \text{row(iii)}.$$

At this point, we have already reached the reduced row echelon form of B. This final form tells us that

$$
\begin{aligned}
x \quad + 3z \quad + 5u &= 2 \\
y - 2z \quad + 2u &= 3 \\
w + \quad u &= 3
\end{aligned}
$$

We now conclude that

$$
\begin{aligned}
x &= 2 - 3z - 5u \\
y &= 3 + 2z - 2u \\
w &= 3 - u
\end{aligned}
$$

where z and w are arbitrary real numbers. Note that x, y and w are the pivot variables while the remaining two variables z and u are the free parameters.   ◇

Example 1.4.3. Solve

$$
\begin{aligned}
x_1 + \quad x_2 + \quad x_3 + x_4 &= 1 \\
2x_1 + 2x_2 + \quad x_3 - x_4 &= 2. \\
3x_1 + 3x_2 + 2x_3 \quad &= 1
\end{aligned}
$$

Solution. The corresponding augmented matrix is

$$
B = \left(\begin{array}{cccc|c}
1 & 1 & 1 & 1 & 1 \\
2 & 2 & 1 & -1 & 2 \\
3 & 3 & 2 & 0 & 1
\end{array} \right).
$$

We use elementary row operations to simplify B:

$$
B \rightsquigarrow \left(\begin{array}{cccc|c}
1 & 1 & 1 & 1 & 1 \\
0 & 0 & -1 & -3 & 0 \\
0 & 0 & -1 & -3 & -2
\end{array} \right) \quad
\begin{array}{l}
\text{row(ii)} - 2 \times \text{row(i)} \\
\text{row(iii)} - 3 \times \text{row(i)}
\end{array}
$$

$$
\rightsquigarrow \left(\begin{array}{cccc|c}
1 & 1 & 1 & 1 & 1 \\
0 & 0 & -1 & -3 & 0 \\
0 & 0 & 0 & 0 & -2
\end{array} \right) \quad
\text{row(iii)} - \text{row(ii)}
$$

However, the last row is translated as $0 = -2$, a contradiction! This is an **inconsistent** system of linear equations.   ◇

Example 1.4.4. Solve the system of linear equations

$$x_1 + x_2 + x_3 + x_4 = 1$$
$$2x_1 + 2x_2 + x_3 - x_4 = 3.$$
$$3x_1 + 3x_2 + 2x_3 \quad\quad = 4$$

Solution. The corresponding augmented matrix is

$$B = \left(\begin{array}{cccc|c} 1 & 1 & 1 & 1 & 1 \\ 2 & 2 & 1 & -1 & 3 \\ 3 & 3 & 2 & 0 & 4 \end{array} \right).$$

We use elementary row operations to simplify B:

$$B \rightsquigarrow \left(\begin{array}{cccc|c} 1 & 1 & 1 & 1 & 1 \\ 0 & 0 & -1 & -3 & 1 \\ 0 & 0 & -1 & -3 & 1 \end{array} \right) \quad \begin{array}{l} \text{row(ii)} - 2\times\text{row(i)} \\ \text{row(iii)} - 3\times\text{row(i)} \end{array}$$

$$\rightsquigarrow \left(\begin{array}{cccc|c} 1 & 1 & 1 & 1 & 1 \\ 0 & 0 & -1 & -3 & 1 \\ 0 & 0 & 0 & 0 & 0 \end{array} \right) \quad \text{row(iii)} - \text{row(ii)}$$

$$\rightsquigarrow \left(\begin{array}{cccc|c} 1 & 1 & 1 & 1 & 1 \\ 0 & 0 & 1 & 3 & -1 \\ 0 & 0 & 0 & 0 & 0 \end{array} \right) \quad (-1) \times \text{row(ii)}$$

$$\rightsquigarrow \left(\begin{array}{cccc|c} 1 & 1 & 0 & -2 & 2 \\ 0 & 0 & 1 & 3 & -1 \\ 0 & 0 & 0 & 0 & 0 \end{array} \right) \quad \text{row(i)} - \text{row(ii)}$$

We translate the last matrix back to a system of linear equations

$$x_1 + x_2 \quad\quad - 2x_4 = 2$$
$$x_3 + 3x_4 = -1$$

We now conclude that the solution is

$$x_1 = 2 - x_2 + 2x_4$$
$$x_3 = -1 - 3x_4$$

where x_2, x_4 are arbitrary real numbers.   ◇

Next, let's try to solve linear equations over a different field.

Example 1.4.5. Solve the system of linear equations

$$x_1 + x_2 + x_3 + x_4 = 1 \qquad \text{(i)}$$
$$x_2 + x_3 + x_4 = 0 \qquad \text{(ii)}$$
$$x_1 + x_2 \qquad\qquad = 1 \qquad \text{(iii)}$$

over $\mathbb{Z}_2$.

Solution. The arithmetic in $\mathbb{Z}_2$ cannot be simpler! In $\mathbb{Z}_2$, $1 = -1$: there is no distinction between $+$ and $-$! How convenient!

The corresponding augmented matrix is

$$B = \left(\begin{array}{cccc|c} 1 & 1 & 1 & 1 & 1 \\ 0 & 1 & 1 & 1 & 0 \\ 1 & 1 & 0 & 0 & 1 \end{array} \right)$$

$$\rightsquigarrow \left(\begin{array}{cccc|c} 1 & 1 & 1 & 1 & 1 \\ 0 & 1 & 1 & 1 & 0 \\ 0 & 0 & 1 & 1 & 0 \end{array} \right) \quad \text{row(iii)} + \text{row(i)}$$

$$\rightsquigarrow \left(\begin{array}{cccc|c} 1 & 0 & 0 & 0 & 1 \\ 0 & 1 & 1 & 1 & 0 \\ 0 & 0 & 1 & 1 & 0 \end{array} \right) \quad \text{row(i)} + \text{row(ii)}$$

$$\rightsquigarrow \left(\begin{array}{cccc|c} 1 & 0 & 0 & 0 & 1 \\ 0 & 1 & 0 & 0 & 0 \\ 0 & 0 & 1 & 1 & 0 \end{array} \right) \quad \text{row(ii)} + \text{row(iii)}$$

This gives the system

$$x_1 = 1$$
$$x_2 = 0$$
$$x_3 + x_4 = 0$$

Thus $x_1 = 1$, $x_2 = 0$ and $x_3 = x_4$. More precisely, $(x_1, x_2, x_3, x_4) = (1, 0, 0, 0)$ or $(1, 0, 1, 1)$.   ◇

Remark. A non-homogeneous system of linear equations may be inconsistent, but a homogeneous system of linear equations always has at least one solution: the trivial solution $(x_1, x_2, \ldots, x_n) = (0, 0, \ldots, 0)$.

The solution space. Remember that by Proposition 1.3.14 the solution set W of

(1.8)
$$
\begin{array}{rcccccccl}
f_1 = & a_{11}x_1 & + & a_{12}x_2 & + & \cdots & + & a_{1n}x_n & = & 0 \\
f_2 = & a_{21}x_1 & + & a_{22}x_2 & + & \cdots & + & a_{2n}x_n & = & 0 \\
\multicolumn{10}{c}{\dotfill} \\
f_m = & a_{m1}x_1 & + & a_{m2}x_2 & + & \cdots & + & a_{mn}x_n & = & 0
\end{array}
$$

is a subspace of F^n. If instead, we have a nonhomogeneous system

(1.9)
$$
\begin{array}{rcccccccl}
f_1 = & a_{11}x_1 & + & a_{12}x_2 & + & \cdots & + & a_{1n}x_n & = & b_1 \\
f_2 = & a_{21}x_1 & + & a_{22}x_2 & + & \cdots & + & a_{2n}x_n & = & b_2 \\
\multicolumn{10}{c}{\dotfill} \\
f_m = & a_{m1}x_1 & + & a_{m2}x_2 & + & \cdots & + & a_{mn}x_n & = & b_m
\end{array}
$$

what is the solution set?

Let v_0 and v_1 both be solutions of (1.9), then a simple observation shows that $v_1 - v_0$ is a solution of the homogeneous system (1.8). In other words, any solution of the system (1.9) is of the form

$$ v_0 + w, \qquad w \in W. $$

We usually use $v_0 + W$ to denote the set

$$ \{v_0 + w : w \in W\}. $$

We formally state this result as a proposition.

Proposition 1.4.6. *Suppose given a* consistent *system of linear equations as in* (1.9). *Let v_0 be a particular solution of the system* (1.9) *and let W be the solution space of the corresponding homogeneous system* (1.8). *Then the solution set of the system* (1.9) *is $v_0 + W$.*

Next we use a simple example to demonstrate our point in Proposition 1.4.6.

Example 1.4.7. Consider the solution space of the homogeneous system of *one* linear equation over $\mathbb{R}$

$$ x - 2y = 0. $$

The solution space is

$$W = \{(2y, y) : y \in \mathbb{R}\},$$

which is the line L through the origin with slope $1/2$ in $\mathbb{R}^2$. It is a linear subspace of $\mathbb{R}^2$. On the other hand, consider the non-homogeneous system of one linear equation

$$x - 2y = 1.$$

The point $(1, 0)$ is *one* solution. The solution set of $x - 2y = 1$ is

$$(1, 0) + W = \{(1 + 2y, y) : y \in \mathbb{R}\}.$$

The solution set is also a line. It is a line parallel to W and it passes through $(1, 0)$. It is the translation of W towards $(1, 0)$.

Exercises 1.4

1. Find the pivot variables and the solution set for the system of linear equations.

(a)
$$\begin{cases} x + y + z = 0 \\ y + z = 0 \\ 3z = 5 \end{cases}$$

(b)
$$\begin{cases} 2x + y \phantom{+ z } = 0 \\ x + 2y + z = 0 \\ y + 2z + t = 0 \\ z + 2t = 5 \end{cases}$$

(c)
$$\begin{cases} 2x - y \phantom{+ z } = 0 \\ -x + 2y - z = 0 \\ -y + 2z - t = 0 \\ -z + 2t = 5 \end{cases}$$

2. Find a reduced row echelon form for the matrix

$$\begin{pmatrix} 0 & 1 & 3 & 1 & -1 & 2 \\ 1 & -1 & 3 & -4 & 2 & 6 \\ 1 & 1 & -1 & 2 & 1 & 1 \\ 1 & 0 & -1 & 0 & 1 & 1 \end{pmatrix} \in M_{4\times6}(\mathbb{R}).$$

3. Find a reduced row echelon form for the matrix

$$\begin{pmatrix} 0 & 1 & 1 & 1 & 1 & 0 \\ 1 & 1 & 1 & 0 & 0 & 0 \\ 1 & 1 & 1 & 0 & 1 & 1 \\ 1 & 0 & 1 & 0 & 1 & 1 \end{pmatrix} \in M_{4\times6}(\mathbb{Z}_2).$$

4. For which values of a does the following system of equations have a solution?

$$y + z = 2,$$
$$x + y + z = a,$$
$$x + y \quad\; = 2.$$

5. For which values of a does the following system of equations have a solution?

$$y + z = 2,$$
$$x + ay + z = 2,$$
$$x + y \quad\; = 2.$$

Review Exercises for Chapter 1

1. Let V be a real vector space containing two distinct vectors. Show that V is an infinite set.

2. An infinite sequence of real numbers is a sequence such as

$$a_1, a_2, a_3, \ldots, a_n, \ldots$$

where $a_n \in \mathbb{R}$ for all $n \in \mathbb{Z}_+$. The term a_n is called the **general term** of the sequence above. The sequence is generally denoted as $(a_n)_{n=1}^{\infty}$ or simply denoted as (a_n).

Let V be the set of all infinite sequences of real numbers. Define

$$(a_n) + (b_n) = (a_n + b_n), \quad \text{and}$$
$$r(a_n) = (ra_n), \quad \text{for } r \in \mathbb{R}.$$

Show that V is a real vector space.

3. Let the reduced echelon form of A be

$$\begin{pmatrix} 1 & 0 & 2 & 0 & -2 \\ 0 & 1 & -5 & 0 & -3 \\ 0 & 0 & 0 & 1 & 6 \end{pmatrix}.$$

Determine A if the first, second and fourth columns of A are

$$\begin{pmatrix} 1 \\ -1 \\ 3 \end{pmatrix}, \quad \begin{pmatrix} 0 \\ -1 \\ 1 \end{pmatrix} \quad \text{and} \quad \begin{pmatrix} 1 \\ -2 \\ 0 \end{pmatrix}$$

respectively.

4. Let A and B be matrices in $M_n(\mathbb{R})$ such that $AB = I_n$. Can you find the reduced row-echelon form of A?

5. Let $A_{n \times n}$ be a symmetric matrix.

   (a) Suppose A has entries in $\mathbb{R}$. Show that $A = \mathbf{0}$ if $A^m = \mathbf{0}$ for some positive integer m. (Hint: Experiment with a 2×2 matrix A such that $A^2 = \mathbf{0}$.)

   (b) Is the assertion in (a) still true if A is assumed to have entries in $\mathbb{C}$?

CHAPTER 2

Bases and Dimension

Dimension is a feature which defines a vector space. In this chapter we will learn what dimension is.

The letters V and W will always be used to denote $\mathbb{R}$-vector spaces unless otherwise noted. However, the techniques we use and the results we derive will also apply to vector spaces over an arbitrary field F. In later chapters, we will occasionally give examples in rational vector spaces or in complex vector spaces.

2.1 Observations in $\mathbb{R}^2$ and in $\mathbb{R}^3$

Subspaces in $\mathbb{R}^2$. We now consider the nonzero vector $(1,1)$ in $\mathbb{R}^2$. A subspace must contain all scalar products of $(1,1)$, that is, it must contain the subset

$$L = \{(t,t) \in \mathbb{R}^2 : t \in \mathbb{R}\}$$

which is the line through O and $u = (1,1)$. In our previous discussion, we have seen that this line is indeed a subspace in $\mathbb{R}^2$. We conclude that L is the *smallest* subspace in $\mathbb{R}$ containing $(1,1)$. We say that the line L is **spanned** (or $\mathbb{R}$-**spanned** to be exact) by $(1,1)$.

If the subspace contains both $u = (1,1)$ and $v = (1,-2)$, then the subspace must contain the subset

$$W = \{a(1,1) + b(1,-2) \in \mathbb{R}^2 : a, b \in \mathbb{R}\}.$$

The subset W obviously contains the line spanned by u and the line spanned by v. The element $(2,5)$ is on neither line. To test how big W is, we start by testing whether $(2,5)$ is in W. To do this we must solve the system of linear equations

$$\begin{cases} a + b = 2, \\ a - 2b = 5. \end{cases}$$

It is easy to see that $(2,5) = 3(1,1) - (1,-2)$ indeed falls in the subspace W. We say that $(2,5)$ is **spanned** (or $\mathbb{R}$-**spanned** to be exact) by u and v.

Furthermore, it can be observed that u and v can span any vector $w = (\alpha, \beta)$ in $\mathbb{R}^2$. To see this, we need to solve the following system of linear equations

$$\begin{cases} a + b = \alpha, \\ a - 2b = \beta \end{cases}$$

for any pair (α, β) in $\mathbb{R}^2$. The augmented matrix of this system is

$$\left(\begin{array}{cc|c} 1 & 1 & \alpha \\ 1 & -2 & \beta \end{array} \right)$$

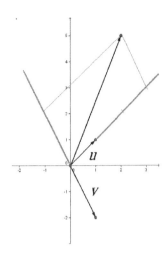

Figure 2.1: The vector $(2,5)$ is spanned by $(1,1)$ and $(1,-2)$.

and its reduced row echelon form is

$$\left(\begin{array}{cc|c} 1 & 0 & \dfrac{2\alpha + \beta}{3} \\ 0 & 1 & \dfrac{\alpha - \beta}{3} \end{array} \right).$$

It follows that $(\alpha, \beta) = \dfrac{2\alpha + \beta}{3}(1,1) + \dfrac{\alpha - \beta}{3}(1,-2)$ is in W. We now conclude that $W = \mathbb{R}^2$. The smallest subspace containing both $(1,1)$ and $(1,-2)$ is $\mathbb{R}^2$ itself.

From Fig. 2.2 one can deduce that the reason u and v can span $\mathbb{R}^2$ is that these two vectors are *not* colinear! If u and v happen to be colinear, the subspace they span will be the line on which they lie. We expect any pair of non-colinear vectors will span $\mathbb{R}^2$.

To summarize, the following list is a complete list of subspaces in $\mathbb{R}^2$:

- $\{O\}$,

- lines through O, and

- $\mathbb{R}^2$

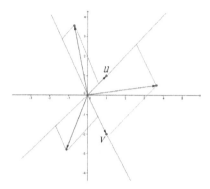

Figure 2.2: Any vector in $\mathbb{R}^2$ can be spanned by $(1, 1)$ and $(1, -2)$.

Subspaces in $\mathbb{R}^3$. A set of colinear vectors will span the line on which these vectors lie. A set of coplanar but non-colinear vectors will span the plane on which these vectors lie. If we are given a set of non-coplanar vectors, these vectors will span $\mathbb{R}^3$. (See Fig. 2.3.)

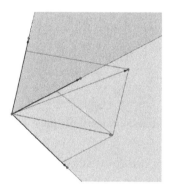

Figure 2.3: Any vector in $\mathbb{R}^3$ can be spanned by three non-coplanar vectors.

To summarize, this is a complete list of subspaces in $\mathbb{R}^3$:

- $\{O\}$,
- lines through O,
- planes through O, and

- $\mathbb{R}^3$.

Example 2.1.1. Show that $u = (1,0,0)$, $v = (1,1,0)$ and $w = (1,1,1)$ are non-coplanar in $\mathbb{R}^3$ and determine how they span $(7,5,-3)$.

Solution. If u, v, w are coplanar, we should be able to find a, $b \in \mathbb{R}$ such that

$$(1,1,1) = a(1,0,0) + b(1,1,0).$$

This means that the system of linear equations

$$a + b = 1$$
$$b = 1$$

should have solutions. A quick observation shows otherwise. Hence the given three vectors are *not* coplanar! We now expect they should span the vector $(7,5,-3)$. In other words, we need to find a, b, $c \in \mathbb{R}$ such that

$$(7,5,-3) = a(1,0,0) + b(1,1,0) + c(1,1,1).$$

For this, we need to solve

$$a + b + c = 7$$
$$b + c = 5$$
$$c = -3$$

It is easy to see that $(7,5,-3) = 2(1,0,0) + 8(1,1,0) - 3(1,1,1)$. ◇

Observations. We often heard sayings such as $\mathbb{R}^2$ is a two-dimensional space or that $\mathbb{R}^3$ is three-dimensional space. What does *dimension* mean exactly? In this section, we can observe that the most economic way to span $\mathbb{R}^2$ is to use two non-colinear vectors, while the most economic way to span $\mathbb{R}^3$ is to use three non-coplanar vectors. The "efficient" way to span a vector space will lead to the discussion of the concept of *bases*, while the number of elements in a base constitutes the concept of *dimension*. We will leave it to the reader to figure out why $\mathbb{R}$ is considered a one-dimensional space and why the trivial space is considered zero-dimensional.

We may also observe that the two-dimensional vector space $\mathbb{R}^2$ contains a zero-dimensional subspace (the origin) and many one-dimensional proper subspaces (lines through the origin). Similarly, the three-dimensional vector space $\mathbb{R}^3$ contains one zero-dimensional-, and plenty of one-dimensional- and two-dimensional proper subspaces (the origin and lines and planes through the origin). Later in this book, we will see that this is a common phenomenon for vector spaces in general.

Exercises 2.1

1. Show that only subspaces in $\mathbb{R}$ are the trivial subspace and $\mathbb{R}$ itself.

2. Find defining equations for the subspace spanned by $(2, -3)$ in $\mathbb{R}^2$.

3. Find defining equations for the subspace spanned by $(1, 3, 5)$ in $\mathbb{R}^3$.

4. Find defining equations for the subspace spanned by $(1, 3, 5)$ and $(-1, -1, -1)$ in $\mathbb{R}^3$.

5. Let
$$\begin{cases} S = \{(1, 2, 3),\ (4, 5, 6),\ (7, 8, 9)\} \\ T = \{(1, 2, 3),\ (4, 5, 6),\ (7, -8, 9)\} \end{cases}$$
be two subsets in $\mathbb{R}^3$.

   (a) Determine if S spans $\mathbb{R}^3$? Do the same thing for T.

   (b) Find defining equations for the subspace spanned by the set which does not span $\mathbb{R}^3$.

2.2   Linear combinations

Throughout this section V stands for a real vector space.

Linear combinations. Let $\mathscr{S} = \{a_1, a_2, \ldots, a_n\}$. It is customary to use $\sum_{a \in \mathscr{S}} a$ or $\sum_{i=1}^n a_i$ to denote the **finite sum** over the elements in $\mathscr{S}$. By convention, the sum over the empty set (the **empty sum**) is defined to be 0!!!

Definition 2.2.1. We say $v \in V$ is a **linear combination** of $v_1, v_2, \ldots, v_n$ (over $\mathbb{R}$) if

$$v = a_1 v_1 + a_2 v_2 + \cdots + a_n v_n$$

for some $a_1, a_2, \ldots, a_n \in \mathbb{R}$. In this case we also say that v is **spanned** by $v_1, v_2, \ldots, v_n$ over $\mathbb{R}$.

In general, let S be a (possibly infinite) subset of V. We say v is a linear combination of elements in S if v is a linear combination of finitely many elements in S. In this case we also say that v is **spanned** by S.

Example 2.2.2. Which of $(0,0,0)$ and $(2,6,8)$ is a linear combination of $u_1 = (1,2,1)$, $u_2 = (-2,-4,-2)$, $u_3 = (0,2,3)$, $u_4 = (2,0,-3)$ and $u_5 = (-3,8,16)$ over $\mathbb{R}$?

Solution. The trivial vector $(0,0,0) = 0u_1 + 0u_2 + 0u_3 + 0u_4 + 0u_5$ is a linear combination of $u_1, \ldots, u_5$.

To see whether the vector $(2,6,8)$ is a linear combination of $u_1, \ldots, u_5$ requires some work. We need to determine whether we can find real numbers $a_1, \ldots, a_5$ such that

$$(2,6,8) = a_1(1,2,1) + a_2(-2,-4,-2) + a_3(0,2,3) + a_4(2,0,-3) + a_5(-3,8,16).$$

This is equivalent to solving the following system of linear equations

$$
\begin{array}{rrrrrrr}
a_1 & - & 2a_2 & & & + & 2a_4 & - & 3a_5 & = & 2 \\
2a_1 & - & 4a_2 & + & 2a_3 & & & + & 8a_5 & = & 6 \\
a_1 & - & 2a_2 & + & 3a_3 & - & 3a_4 & + & 16a_5 & = & 8
\end{array}
$$

It turns out there are infinitely many solutions for this system, and we need only one! For example,

$$(2,6,8) = -4u_1 + 7u_3 + 3u_4$$

is indeed a linear combination of the given vectors. $\qquad \diamond$

Example 2.2.3. Is the polynomial $x^2 + 2x + 3$ in $\mathscr{P}$ a linear combination of $x^2 + 1$, 1 and $x^2 + 3$ over $\mathbb{R}$?

Solution. To answer this question, we need to determine whether there are real numbers a, b and c such that

$$x^2 + 2x + 3 = a(x^2 + 1) + b(1) + c(x^2 + 3).$$

This is impossible since the coefficient of the linear term on the left is 2, while the coefficient of the linear term on the right is 0.   ◇

Example 2.2.4. Is the polynomial $x^2 + 2x + 3$ in $\mathscr{P}$ a linear combination of elements in $\{1, 1 + x, 1 + x^2, 1 + x^3, \dots\}$ over $\mathbb{R}$?

Solution. Yes! The polynomial $x^2 + 2x + 3 = 2(1 + x) + (1 + x^2)$. To find the linear combination there is no need to look for polynomials of degree higher than 2 in the given set.   ◇

Definition 2.2.5. Let S be a subset of V. The **span** or the **linear span** of S, denoted $\mathrm{Sp}\,(S)$, is the set consisting of all linear combinations (including the empty sum) of the vectors in S.

Remarks. (1) $\mathrm{Sp}\,(\phi) = \{O\}$.

(2) If $S = \{v_1, v_2, \dots, v_n\}$ is a finite set of V. Then we also use $\mathrm{Sp}\,(v_1, v_2, \dots, v_n)$ for $\mathrm{Sp}\,(S)$.

(3) $S_1 \subseteq S_2 \Rightarrow \mathrm{Sp}\,(S_1) \subseteq \mathrm{Sp}\,(S_2)$.

Let $v_1, \dots, v_n \in V$. By definition we have that

$$\mathrm{Sp}\,(v_1, \dots, v_n) = \{a_1 v_1 + \dots + a_n v_n : a_1, \dots, a_n \in \mathbb{R}\}.$$

Thus, we often write

(2.1) $$\mathrm{Sp}\,(v_1, \dots, v_n) = \mathbb{R}v_1 + \dots + \mathbb{R}v_n.$$

In particular, $\mathrm{Sp}\,(v) = \{av : a \in \mathbb{R}\} = \mathbb{R}v$.

Example 2.2.6. Does

$$\begin{pmatrix} 1 & 0 \\ 0 & 1 \end{pmatrix} \in \mathrm{Sp}\left(\begin{pmatrix} 1 & 0 \\ -1 & 0 \end{pmatrix}, \begin{pmatrix} 0 & 1 \\ 0 & 1 \end{pmatrix}, \begin{pmatrix} 1 & 1 \\ 0 & 0 \end{pmatrix} \right)$$

in $M_2(\mathbb{R})$?

Solution. We need to determine whether there are a, b, $c \in \mathbb{R}$ such that

$$\begin{pmatrix} 1 & 0 \\ 0 & 1 \end{pmatrix} = a \begin{pmatrix} 1 & 0 \\ -1 & 0 \end{pmatrix} + b \begin{pmatrix} 0 & 1 \\ 0 & 1 \end{pmatrix} + c \begin{pmatrix} 1 & 1 \\ 0 & 0 \end{pmatrix}.$$

For this we need to solve

$$a + c = 1$$
$$b + c = 0$$
$$-a = 0$$
$$b = 1.$$

However, the fact that $a = 0$ and $b = 1$ leads to both $c = 1$ and $c + 1 = 0$, a contradiction. Thus, the given matrix is not in the span. ◇

Example 2.2.7. In $\mathbb{R}^2$, Sp $((1,2))$ is the line through $(0,0)$ and $(1,2)$, and Sp $((1,0),(2,0))$ is the line through $(0,0)$ and $(1,0)$. On the other hand, Sp $((1,0),(1,1)) = \mathbb{R}^2$ since $(1,0)$ and $(1,1)$ are non-colinear vectors.

The linear span as a subspace. Now we discuss various properties of the span of a subset in a vector space.

Proposition 2.2.8. *Let S be a subset of V. Then Sp (S) is a subspace of F.*

Proof. If $S = \varnothing$, Sp (S) is the trivial space. We may assume $S \neq \varnothing$.

The trivial vector $O = 0v$ for any $v \in S$. Hence $O \in$ Sp (S). Let v and $w \in$ Sp (S). We have that

$$v = a_1 v_1 + a_2 v_2 + \cdots + a_m v_m,$$
$$w = b_1 w_1 + b_2 w_2 + \cdots + b_n w_n$$

where v_i, $w_j \in S$. Let a, $b \in \mathbb{R}$. Then

$$av + bw = aa_1 v_1 + aa_2 v_2 + \cdots + aa_m v_m + bb_1 w_1 + bb_2 w_2 + \cdots + bb_n w_n$$

is in Sp (S). Thus by the subspace test, Sp (S) is a subspace of V. □

Corollary 2.2.9. *The span Sp (S) is the smallest subspace of V containing S. Specifically, if W is a subspace of V and $S \subseteq W$ then Sp $(S) \subseteq W$.*

Proof. Since any vector v in Sp (S) is also a linear combination of vectors in W, v is in W by the subspace test. □

We may summarize §2.1 as follows. The span of nothing is the trivial space. The span of a nonzero vector is the line on which it lies. The span of two non-colinear vectors is the plane on which they lie. The span of three non-coplanar vectors is the (3-dimensional) space on which they lie.

Corollary 2.2.10. *A subset W of V is a subspace if and only if $\text{Sp}(W) = W$.*

Proof. The "if" part follows from Proposition 2.2.8. The "only if" part follows from Corollary 2.2.9. $\square$

Corollary 2.2.11. *Let S be a subset of V and let $v \in V$. Then the following are true:*

(a) *If $v \in \text{Sp}(S)$ then $\text{Sp}(v) \subseteq \text{Sp}(S)$.*

(b) $\text{Sp}(S \cup \{v\}) = \text{Sp}(S)$ *if and only if $v \in \text{Sp}(S)$.*

Proof. (a) Since $\{v\} \subseteq S$, it is clear that $\text{Sp}(v) \subseteq \text{Sp}(S)$.

(b) The "if" part: If $v \in \text{Sp}(S)$, then $S \cup \{v\} \subseteq \text{Sp}(S)$. By Corollary 2.2.9, $\text{Sp}(S \cup \{v\}) \subseteq \text{Sp}(S)$. Hence $\text{Sp}(S \cup \{v\}) = \text{Sp}(S)$.

The "only if" part: The vector v is in $\text{Sp}(S \cup \{v\}) = \text{Sp}(S)$. $\square$

Generators. Next we discuss the concept of *generators*.

Definition 2.2.12. Let S be a subset of V. We say S **generates** or **spans** V (over $\mathbb{R}$) if $\text{Sp}(S) = V$. In this case we also say that S is a **generating set** or a **set of generators** for V (over $\mathbb{R}$).

Definition 2.2.13. In $\mathbb{R}^n$ we write

$$e_i^{(n)} = (\quad 0, \quad 0, \quad \ldots, \quad 1, \quad 0, \quad \ldots, \quad 0, \quad)$$
$$\uparrow$$
$$\text{the } i\text{-th place}$$

We will simply write e_i for $e_i^{(n)}$ when n is understood.

Definition 2.2.14. Fix i_0 and j_0. In $M_{m \times n}(\mathbb{R})$ we define $e_{i_0 j_0}^{(m,n)} = \left(a_{ij}\right)_{m \times n}$ where

$$a_{ij} = \begin{cases} 1, & \text{if } i = i_0 \text{ and } j = j_0, \\ 0, & \text{otherwise.} \end{cases}$$

We usually simply write $e_{i_0 j_0}$ for $e_{i_0 j_0}^{(m,n)}$ when m and n are understood.

Proposition 2.2.15.   *a. The set $\{e_i\}_{i=1}^n$ spans $\mathbb{R}^n$ over $\mathbb{R}$.*

   b. The set $\{1, x, x^2, \ldots, x^n\}$ spans $\mathscr{P}_n$ over $\mathbb{R}$.

   c. The set $\{e_{ij} : i = 1, \ldots, m, \ j = 1, \ldots, n\}$ spans $M_{m \times n}(\mathbb{R})$ over $\mathbb{R}$.

Proof. (a) Any element $v = (a_1, \ldots, a_n)$ or $\mathbb{R}^n$ is a linear combination of the e_i's since $v = a_1 e_1 + \cdots + a_n e_n$.

   (b) Any polynomial $f = a_0 + a_1 x + a_2 x^2 + \cdots + a_n x^n$ in $\mathscr{P}_n$ is a linear combination of $\{1, x, x^2, \ldots, x^n\}$ since $f = a_0 \cdot 1 + a_1 \cdot x + a_2 \cdot x^2 + \cdots + a_n \cdot x^n$.

   (c) Any matrix $A = \left(a_{ij}\right)_{m \times n}$ in $M_{m \times n}(\mathbb{R})$ is a linear combination of the e_{ij}'s since $A = \sum_{i,j} a_{ij} e_{ij}$. $\square$

Example 2.2.16. Does the set $\{(1,1,1,1), (1,1,1,0), (1,1,0,0), (1,0,0,0)\}$ span $\mathbb{R}^4$ over $\mathbb{R}$? To see this, we need to see if

$$(a, b, c, d) = x(1,1,1,1) + y(1,1,1,0) + z(1,1,0,0) + w(1,0,0,0)$$

is solvable for any choice of $(a, b, c, d) \in \mathbb{R}^4$. In other words, we need to see whether the system of equations

$$
\begin{array}{ccccccccc}
x & + & y & + & z & + & w & = & a \\
x & + & y & + & z & & & = & b \\
x & + & y & & & & & = & c \\
x & & & & & & & = & d
\end{array}
$$

is consistent. This is indeed so since $(x, y, z, w) = (d, c - d, b - c, a - b)$.

Example 2.2.17. Do $x^2 + 2x + 1$, $x^2 + 3x$ and $x^2 + 2x + 2$ span $\mathscr{P}_2$ over $\mathbb{R}$? To see this we need to determine whether we can find α, β and $\gamma \in \mathbb{R}$ such that

$$ax^2 + bx + c = \alpha(x^2 + 2x + 1) + \beta(x^2 + 3x) + \gamma(x^2 + 2x + 2)$$

for any choice of (a, b, c). Equivalently, we need to solve the system of linear equations

$$
\begin{array}{ccccccc}
\alpha & + & \beta & + & \gamma & = & a \\
2\alpha & + & 3\beta & + & 2\gamma & = & b \\
\alpha & & & + & 2\gamma & = & c
\end{array}
$$

Use the method in the previous chapter, we can see that $\alpha = 6a - 2b - c$, $\beta = b - 2a$, $\gamma = -3a + b + c$. We conclude that the three given polynomials span $\mathscr{P}_2$.

Example 2.2.18. Find a set of generators for the solution space of

$$\begin{cases} 2x + 2y - 10z = 0, \\ 5x + 7y - 21z = 0, \\ 3x + 4y - 13z = 0. \end{cases}$$

Solution. The solution set is

$$\{(7z, -2z, z) \in \mathbb{R}^3 : z \in \mathbb{R}\} = \{z(7, -2, 1) \in \mathbb{R}^3 : z \in \mathbb{R}\}.$$

The solution space is generated by $\{(7, -2, 1)\}$.   ◇

Definition 2.2.19. If V can be spanned by a finite set over $\mathbb{R}$ we say V is a **finite dimensional** vector space over $\mathbb{R}$ (or whatever field). Otherwise we say V is **infinite dimensional**.

From Proposition 2.2.15 we see that $\mathbb{R}^n$, $\mathscr{P}_n$ and $M_{m \times n}(\mathbb{R})$ are finite dimensional over $\mathbb{R}$. The focus of this course will be on finite dimensional vector spaces.

The intersection and the sum of two subspaces. Remember that the intersection of an arbitrary family of subspaces remains a subspace. Using Theorem 2.2.9, we have the following immediate corollary.

Corollary 2.2.20. *Let S_1, S_2 be subsets of V. Then*

$$\mathrm{Sp}\,(S_1 \cap S_2) \subseteq \mathrm{Sp}\,(S_1) \cap \mathrm{Sp}\,(S_2).$$

We now give an example to show that equality may not hold in the previous corollary.

Example 2.2.21. Let $S_1 = \{(1, 0)\}$ and $S_2 = \{(2, 0)\}$. Check that $\mathrm{Sp}\,(S_1 \cap S_2) \subsetneq \mathrm{Sp}\,(S_1) \cap \mathrm{Sp}\,(S_2)$.

Even though the intersection of two subspaces is always a subspace, the union of two subspaces is seldom a subspace.

Example 2.2.22. Let $W = \mathrm{Sp}\,((1,0))$ and $W' = \mathrm{Sp}\,((0,1))$. Then $W \cup W'$ is the union of the x-axis and the y-axis. It is clear that $W \cup W'$ is not a subspace of $\mathbb{R}^2$ by the discussion in the previous space.

We now proceed to find out what is the smallest subspace containing both W and W'.

Definition 2.2.23. Let W and W' be subspaces of V. We define

$$W + W' = \{w + w' : w \in W, w' \in W'\}$$

and $W + W'$ is called the **sum** of W and W'.

Proposition 2.2.24. *Let W and W' be subspaces of V. Then $W + W'$ is a subspace of V. In fact, $W + W'$ is the smallest subspace in V containing both W and W'.*

Proof. The trivial vector O belongs to both W and W'. Hence $O = O + O \in W + W'$. Let $v_i = w_i + w_i'$ for $i = 1, 2$ such that $w_i \in W$ and $w_i' \in W'$. Then $v_1 + v_2 = (w_1 + w_2) + (w_1' + w_2') \in W + W'$. For any $r \in \mathbb{R}$, $rv_1 = rw_1 + rw_1' \in W + W'$. By the subspace test, $W + W'$ is a subspace of V.

Furthermore, Let X be a subspace of V containing both W and W'. Then for all $w \in W$ and $w \in W'$, we have $w + w' \in X$. This implies that $W + W' \subseteq X$. $\qquad\square$

Using this proposition, we have the following immediate results.

Corollary 2.2.25. *Let W and W' be subspaces of V such that $W \subseteq W'$. Then $W + W' = W'$.*

Corollary 2.2.26. *Let S_1 and S_2 be subsets of V. Then*

$$\mathrm{Sp}\,(S_1 \cup S_2) = \mathrm{Sp}\,(S_1) + \mathrm{Sp}\,(S_2)\,.$$

Proof. Since $S_i \subseteq \mathrm{Sp}\,(S_i) \subseteq \mathrm{Sp}\,(S_1) + \mathrm{Sp}\,(S_2)$ for $i = 1, 2$, we have that $\mathrm{Sp}\,(S_1 \cup S_2) \subseteq \mathrm{Sp}\,(S_1) + \mathrm{Sp}\,(S_2)$ by Corollary 2.2.9. On the other hand, since $S_i \subseteq S_1 \cup S_2$, we have $\mathrm{Sp}\,(S_i) \subseteq \mathrm{Sp}\,(S_1 \cup S_2)$ for $i = 1, 2$. This implies that $\mathrm{Sp}\,(S_1) + \mathrm{Sp}\,(S_2) \subseteq \mathrm{Sp}\,(S_1 \cup S_2)$ by Proposition 2.2.24. $\qquad\square$

Corollary 2.2.27. *Let $v_1, v_2, \ldots, v_n \in V$. Then*

$$\mathrm{Sp}\,(v_1, v_2, \ldots, v_n) = \mathrm{Sp}\,(v_1) + \mathrm{Sp}\,(v_2) + \cdots + \mathrm{Sp}\,(v_n).$$

Remember that $\mathrm{Sp}\,(v_i) = \mathbb{R}v_i$. Thanks to Corollary 2.2.27, we have that

$$\mathrm{Sp}\,(v_1, v_2, \ldots, v_n) = \mathbb{R}v_1 + \mathbb{R}v_2 + \cdots + \mathbb{R}v_n.$$

Fortunately this does not contradict to the expression in (2.1).

Exercises 2.2

1. What is the least number of vectors needed to span the trivial space $\{O\}$?

2. Is $\begin{pmatrix} 1 & 0 \\ 0 & 1 \\ 0 & 0 \end{pmatrix}$ a linear combination of $\begin{pmatrix} 1 & 0 \\ -1 & 0 \\ -1 & 0 \end{pmatrix}$, $\begin{pmatrix} 0 & 1 \\ 0 & 1 \\ 0 & 1 \end{pmatrix}$, $\begin{pmatrix} 1 & 1 \\ 0 & 0 \\ 0 & 0 \end{pmatrix}$ in $M_{3\times 2}(\mathbb{R})$?

3. Is $x^3 + x^2 - 3x - 2$ a linear combination of $-x^3 + 2x^2 + 5x - 3$ and $-x^2 - 4x + 4$?

4. Find a generating set for the solution space of

$$\begin{cases} 3x + 2y - 4z + 2u - 5v = 0, \\ 5x - 2y + 3z - 2u + 3v = 0. \end{cases}$$

5. Find a generating set for $\mathscr{P}$. Is $\mathscr{P}$ a finite dimensional vector space over $\mathbb{R}$?

6. Is $\mathrm{Sp}\,(\{(t, t+1) \in \mathbb{R}^2 : t = 0, 1, 2, 3, \ldots\})$ a finite dimensional real vector space?

7. Let W and W' be two subspaces of V. Show that $W \cup W'$ is a subspace of V if and only if $W \subseteq W'$ or $W' \subseteq W$.

8. Prove Corollary 2.2.25.

9. Is it true that $\mathrm{Sp}\left(\begin{pmatrix} 3 \\ 0 \\ 4 \end{pmatrix}\right) + \mathrm{Sp}\left(\begin{pmatrix} 3 \\ -1 \\ 5 \end{pmatrix}, \begin{pmatrix} -6 \\ 5 \\ -13 \end{pmatrix}\right) = \mathbb{R}^3$?

2.3 Linear dependence and independence

Linear dependence and independence. Linear (in)dependence is one of the most important features concerning a set of vectors. We cannot delete any element from an linearly independent set without sacrificing the integrity of its span.

Definition 2.3.1. The vectors $v_1, v_2, \ldots, v_n$ in V are said to be **linearly dependent** (over $\mathbb{R}$) if there exist $a_1, a_2, \ldots, a_n \in \mathbb{R}$, not all zero, such that

$$a_1 v_1 + a_2 v_2 + \cdots + a_n v_n = O.$$

We say $v_1, v_2, \ldots, v_n$ are **linearly independent** (over $\mathbb{R}$) if $v_1, \ldots, v_n$ are *not* linearly dependent, that is, if

$$a_1 v_1 + a_2 v_2 + \cdots + a_n v_n = O \implies a_1 = a_2 = \cdots = a_n = 0.$$

A *finite* set is said to be linearly independent (or linearly dependent respectively) if the vectors in it are linearly independent (or dependent respectively). The *empty* set is assumed to be linearly independent by default.

An *infinite* set is linearly dependent if it contains a linearly dependent finite subset. An *infinite* set is linearly independent if every finite subset is linearly independent.

Remarks. (1) If a finite set contains O then it is automatically linearly dependent. If a sequence of vectors has repeated elements then they are linearly dependent.

(2) Let $S_1 \subseteq S_2$ be two subsets of V. If S_1 is linearly dependent then S_2 is linearly dependent. If S_2 is linearly independent then S_1 is linearly independent.

Now, let's look at a few examples.

Example 2.3.2. In $\mathbb{R}^3$ is the set $\{(1, 2, 3), \ (-1, 0, 2), \ (-1, -2, -2)\}$ linearly independent over $\mathbb{R}$?

To check this, we need to solve

$$a(1, 2, 3) + b(-1, 0, 2) + c(-1, -2, -2) = (0, 0, 0).$$

That is, we need to determine whether the system of linear equations

$$
\begin{aligned}
a - \ b - \ c &= 0 \\
2a \quad\ \ - 2c &= 0 \\
3a + 2b - 2c &= 0
\end{aligned}
$$

has a nontrivial solution. This answer is no! Thus, the given set is linearly independent over $\mathbb{R}$.

Example 2.3.3. In $M_{2\times3}(\mathbb{R})$ is the set

$$\left\{ \begin{pmatrix} 1 & -3 & 2 \\ -4 & 0 & 5 \end{pmatrix}, \begin{pmatrix} -3 & 7 & 4 \\ 6 & -2 & -7 \end{pmatrix}, \begin{pmatrix} -2 & 3 & 11 \\ -1 & -3 & 2 \end{pmatrix} \right\}$$

linearly independent over $\mathbb{R}$? For this, we need to see whether there are nontrivial solutions for

$$
a \begin{pmatrix} 1 & -3 & 2 \\ -4 & 0 & 5 \end{pmatrix} + b \begin{pmatrix} -3 & 7 & 4 \\ 6 & -2 & -7 \end{pmatrix} + c \begin{pmatrix} -2 & 3 & 11 \\ -1 & -3 & 2 \end{pmatrix}
$$
$$
= \begin{pmatrix} 0 & 0 & 0 \\ 0 & 0 & 0 \end{pmatrix}.
$$

This is equivalent to solving the system

$$
\begin{aligned}
a - 3b - \ 2c &= 0 \\
-3a + 7b + \ 3c &= 0 \\
2a + 4b + 11c &= 0 \\
-4a + 6b - \ \ c &= 0 \\
- 2b - \ 3c &= 0 \\
5a - 7b + \ 2c &= 0
\end{aligned}
$$

over $\mathbb{R}$. It turns out this only solution is $(a, b, c) = (0, 0, 0)$. The three given matrices are linearly independent over $\mathbb{R}$.

Example 2.3.4. Let $S = \{(1,1,0),(1,0,1),(0,1,1)\}$ be a subset of the F-vector space F^3.

When $F = \mathbb{R}$, S is a linearly independent set. However, if instead $F = \mathbb{Z}_2$, S is linearly dependent since

$$(1,1,0) + (1,0,1) + (0,1,1) = (2,2,2) = (0,0,0)$$

over $\mathbb{Z}_2$.

Example 2.3.5. Consider the vector space $\mathbb{C}$. The set $S = \{1,i\}$ is linearly independent over $\mathbb{R}$, whereas it is linearly dependent over $\mathbb{C}$:

$$\underset{\text{scalar}}{1} \quad \cdot \quad \underset{\text{vector}}{1} \quad + \quad \underset{\text{scalar}}{i} \quad \cdot \quad \underset{\text{vector}}{i} \quad = \quad 0.$$

Example 2.3.6. Is the subset $S = \{(1-i,2),\ (3,5+i)\}$ of $\mathbb{C}^2$ linearly independent over $\mathbb{R}$? Over $\mathbb{C}$?

Over $\mathbb{R}$, we need to solve

$$a(1-i,2) + b(3,5+i) = (0,0)$$

for $a,\ b \in \mathbb{R}$. This implies that

$$a(1-i) + 3b = (a+3b) - ai = 0$$
$$2a + b(5+i) = (2a+5b) + bi = 0.$$

Or equivalently,

$$a + 3b = 0, \quad a = 0,$$
$$2a + 5b = 0, \quad b = 0.$$

Clearly $a = b = 0$, and we conclude that S is linearly independent over $\mathbb{R}$.

Now consider the same problem over $\mathbb{C}$. We need to solve

$$\alpha(1-i,2) + \beta(3,5+i) = (0,0)$$

for $\alpha,\ \beta \in \mathbb{C}$. Let $\alpha = a + bi$ and $\beta = c + di$, where $a,\ b,\ c,\ d \in \mathbb{R}$. Hence we need to solve

$$(a+bi)(1-i) + 3(c+di) = (a+b+3c) + (-a+b+3d)i = 0$$
$$2(a+bi) + (c+di)(5+i) = (2a+5c-d) + (2b+c+5d)i = 0.$$

This translates as the system of linear equations

$$a + b + 3c = 0, \quad -a + b + 3d = 0,$$
$$2a + 5c - d = 0, \quad 2b + c + 5d = 0.$$

Again, the solution for this system is $a = b = c = d = 0$, or $\alpha = \beta = 0$. The given set is linearly independent over $\mathbb{C}$.

Example 2.3.7. Let u, v and w be linearly independent vectors in the real vector space V. Are the vectors

$$u + v, \quad v + w, \quad w + u$$

linearly independent? Let

$$a(u + v) + b(v + w) + c(w + u) = O$$

where a, b, $c \in \mathbb{R}$. This implies that

$$(a + c)u + (a + b)v + (b + c)w = O.$$

Since u, v and w are linearly independent over $\mathbb{R}$, we should have $a + c = a + b = b + c = 0$. A straightforward calculation shows that $a = b = c = 0$. Hence the three given vectors are linearly independent over $\mathbb{R}$.

Proposition 2.3.8.   (a) *In $\mathbb{R}^n$ the set $\{e_i\}_{i=1}^n$ is linearly independent over $\mathbb{R}$.*

   (b) *In $\mathscr{P}$ or in $\mathscr{P}_n$ the set $\{1, x, x^2, \ldots, x^n\}$ is linearly independent over $\mathbb{R}$.*

   (c) *In $M_{m \times n}(\mathbb{R})$ the set $\{e_{ij} : i = 1, \ldots, m, \quad j = 1, \ldots, n\}$ is linearly independent over $\mathbb{R}$.*

Proof. (a) Note that $\sum_{i=1}^n a_i e_i = (a_1, \ldots, a_n)$. If $\sum_{i=1}^n a_i e_i = (0, \ldots, 0)$, we must have $a_1 = \cdots = a_n = 0$.

   (b) If $a_0 \cdot 1 + a_1 \cdot x + \cdots + a_n \cdot x^n$ is the zero polynomial in $\mathscr{P}_n$, the coefficients $a_1, a_2, \ldots, a_n$ must all be 0.

   (c) The linear combination $\sum a_{ij} e_{ij} = (a_{ij})_{m \times n}$. Hence, if $\sum a_{ij} e_{ij}$ is the zero matrix, the entries a_{ij}'s are all 0.   □

Example 2.3.9. The set $S = \{1, x, x^2, x^3, \dots\}$ is linearly independent over $\mathbb{R}$. This is clear since any finite subset T of S is contained in $\{1, x, x^2, \dots, x^n\}$ for a sufficiently large n. Hence T is linearly independent over R

To express a vector as the linear combination of a linearly independent set. The following result demonstrates the advantage of having a linearly independent set.

Proposition 2.3.10. *Let v_1, v_2, $\dots$, v_n be vectors in a vector space V. Then $v_1, \dots, v_n$ are linearly independent if and only if for each $v \in \mathrm{Sp}\,(v_1, \dots, v_n)$ there exists a unique expression*

$$v = a_1 v_1 + a_2 v_2 + \cdots + a_n v_n,$$

where $a_1, a_2, \dots, a_n \in \mathbb{R}$.

Proof. The "if" part: Clearly,

$$O = 0v_1 + 0v_2 + \cdots + 0v_n.$$

Thus when $O = a_1 v_1 + a_2 v_2 + \cdots + a_n v_n$, we must have $a_1 = a_2 = \cdots = a_n = 0$ by assumption.

The "only if" part: Assume

$$\begin{aligned}
v &= a_1 v_1 + a_2 v_2 + \cdots + a_n v_n \\
&= b_1 v_1 + b_2 v_2 + \cdots + b_n v_n.
\end{aligned}$$

This implies that

$$(a_1 - b_1)v_1 + (a_2 - b_2)v_2 + \cdots + (a_n - b_n)v_n = O.$$

Since $v_1, v_2, \dots, v_n$ are linearly independent over $\mathbb{R}$, we must have $a_i - b_i = 0$ for $i = 1, 2, \dots, n$. In other words, $a_i = b_i$ for $i = 1, 2, \dots, n$. $\qquad \square$

Example 2.3.11. The vectors e_1, e_2, e_3 and $v = (3, 2, 4)$ are obviously dependent over $\mathbb{R}$ since they satisfy the nontrivial linear relation

$$v - 3e_1 - 2e_2 - 4e_3 = O.$$

Now

$$(-1, -2, 3) = -e_1 - 2e_2 + 3e_3 = v - 4e_1 - 4e_2 - e_3$$

are simply two of the many (infinitely many, in fact) different linear combinations expressing $(-1, -2, 3)$ using e_1, e_2, e_3 and v. It is quite an inconvenience.

Example 2.3.12. In $\mathscr{P}$ is the set $S = \{1 + x,\ 2 - x - x^2,\ 3 - x/2 - x^3\}$ linearly independent over $\mathbb{R}$? Determine whether $f = -9 - x - 2x^2 + 4x^3$ is spanned by S. If yes, express f as the linear combination of elements in S.

Solution. Let

$$a(1 + x) + b(2 - x - x^2) + c\left(3 - \frac{x}{2} - x^3\right) = 0$$

where $a, b, c \in \mathbb{R}$. By comparing the coefficients we need to solve

$$a + 2b + 3c = 0,$$
$$a - b - \frac{c}{2} = 0,$$
$$-b = 0,$$
$$-c = 0.$$

Hence $a = b = c = 0$. The set S is linearly independent over $\mathbb{R}$.

Next we verify whether f is spanned by S. For this, we need to find $a, b, c \in \mathbb{R}$ such that

$$a(1 + x) + b(2 - x - x^2) + c\left(3 - \frac{x}{2} - x^3\right)$$
$$= -9 - x - 2x^2 + 4x^3.$$

Comparing the coefficients we need to solve

(1) $$a + 2b + 3c = -9,$$
(2) $$a - b - \frac{c}{2} = -1,$$
$$-b = -2,$$
$$-c = 4.$$

Clearly $b = 2$ and $c = -4$. Substituting the values of b and c into (1) and (2), we have

$$a + 4 - 12 = a - 8 = -9,$$
$$a - 2 + 2 = a = -1.$$

This is a consistent system of equations. We conclude that

$$f = -(1 + x) + 2(2 - x - x^2) - 4\left(3 - \frac{x}{2} - x^3\right).$$

$\diamond$

Linear independence and linear spans. Next, we provide a new point of view on linear independence. More importantly we provide guidelines how to construct a linearly independent set.

Lemma 2.3.13. *The vectors v_1, v_2, ..., v_n in V are linearly independent if and only if the set $\{v_1, v_2, \ldots, v_k\}$ is linearly independent for each k.*

Proof. See Remark (2) to Definition 2.3.1. □

Lemma 2.3.14. *Let S be a subset of V and let $v \in V$. If $v \in \mathrm{Sp}\,(S)$ then $S \cup \{v\}$ is linearly dependent.*

Proof. Since $v \in \mathrm{Sp}\,(S)$,

$$v = a_1 v_1 + a_2 v_2 + \cdots + a_n v_n$$

for some $v_1, \ldots, v_n \in S$ and $a_1, \ldots, a_n \in \mathbb{R}$. This gives a non-trivial relation

$$1 \cdot v - a_1 v_1 - a_2 v_2 - \cdots - a_n v_n = 0.$$

Thus $S \cup \{v\}$ is linearly dependent. □

By virtue of Lemma 2.3.14, we sometimes say that v is *linearly dependent on* S if $v \in \mathrm{Sp}\,(S)$. Thanks to Corollary 2.2.11, $\mathrm{Sp}\,(S \cup \{v\}) = \mathrm{Sp}\,(S)$ if v is linearly dependent on S. Hence, adding a linearly dependent element does not enlarge a span.

Proposition 2.3.15. *Let S be a linearly independent subset of V and let $v \in V$. Then $S \cup \{v\}$ is linearly independent if and only if $v \notin \mathrm{Sp}\,(S)$.*

Proof. The "only if" part is simply Lemma 2.3.14. Conversely, suppose $S \cup \{v\}$ is linearly dependent over the underlying field. There is a non-trivial relation

$$(2.2) \qquad av + a_1 v_1 + a_2 v_2 + \cdots + a_n v_n = O$$

where $v_1, \ldots, v_n \in S$ and scalars $a, a_1, \ldots, a_n$. If $a = 0$, then the relation (2.2) gives a non-trivial relation for the vectors $v_1, \ldots, v_n$ in S, a contradiction. Hence $a \neq 0$. We now have

$$v = a^{-1}(-a_1 v_1 + a_2 v_2 + \cdots + a_n v_n) \in \text{Sp}\,(S).$$

This proves the "if" part. $\qquad\qquad\qquad\qquad\qquad\qquad\qquad\qquad\qquad$ $\square$

Theorem 2.3.16. *Let $v_1, \ldots, v_n$ be vectors in V. The following are true:*

(a) *The vectors v_1, v_2, $\ldots$, v_n are linearly dependent if and only if $v_k \in \text{Sp}\,(v_1, \ldots, v_{k-1})$ for some k with $1 \leq k \leq n$.*

(b) *The vectors v_1, v_2, $\ldots$, v_n are linearly independent if and only if $v_k \notin \text{Sp}\,(v_1, v_2, \ldots, v_{k-1})$ for each k.*

Proof. The "if" part of (a): By Lemma 2.3.14 we have that $\{v_1, v_2, \ldots, v_k\}$ is linearly dependent. Hence $\{v_1, v_2, \ldots, v_n\}$ is also linearly dependent.

The "only if" part of (a): By Lemma 2.3.13, if $\{v_1, v_2, \ldots, v_n\}$ is linearly independent, then $\{v_1, v_2, \ldots, v_k\}$ is also linearly independent for $1 \leq k \leq n$. Proposition 2.3.15 tells us that $v_k \notin \text{Sp}\,(v_1, v_2, \ldots, v_{k-1})$ for each k.

Part (b) is simply a restatement of part (a). $\qquad\qquad\qquad\qquad\qquad$ $\square$

The following are two immediate results.

Corollary 2.3.17. *Let u and v be vectors in V. The following statements are true:*

(a) *The vector u is linearly independent if and only if $u \neq O$.*

(b) *The vectors u, v are linearly independent if and only if $u \neq O$ and v is not a scalar multiple of u, that is, there are no $a \in \mathbb{R}$ such that $v = au$.*

If one delete elements from a linearly independent set, one ends up with a smaller span.

Corollary 2.3.18. *Let $v_1, \ldots, v_n \in V$. Then $v_1, \ldots, v_n$ are linearly independent if and only if*

$$\{O\} \subsetneqq \operatorname{Sp}(v_1) \subsetneqq \operatorname{Sp}(v_1, v_2) \subsetneqq \cdots \subsetneqq \operatorname{Sp}(v_1, \ldots, v_n).$$

If we are to construct a linearly independent set from scratch, we first pick a nonzero element v_1. Then we choose an element $v_2 \notin \operatorname{Sp}(v_1)$ and a third element $v_3 \notin \operatorname{Sp}(v_1, v_2)$ and so on. Remember in §2.1 how we constructed a linearly independent set in $\mathbb{R}^3$? First, we randomly pick a nonzero vector. We know that a nonzero vector spans a line. Then we pick a vector outside that line so that the two chosen vectors are non-colinear. The two non-colinear vectors span a plane. Then we pick a vector outside that plane. Then we have three non-coplanar vectors which span the space $\mathbb{R}^3$. In this manner, we obtain a set of generators for $\mathbb{R}^3$ which is also linearly independent. Hopefully, this process can be done for more general vector spaces. By Choosing linearly independent elements, we steadily expand the subspaces and eventually we expect to have enough generators to span any given vector space.

Example 2.3.19. The vectors

$$(2, 5, 0, 0), \ (-3, 4, 0, 0), \ (1, 1, 3, 0) \text{ and } (1, 1, 3, 1)$$

in $\mathbb{R}^4$ are linearly independent over $\mathbb{R}$.

Solution. We start with the non-zero vector $(2, 5, 0, 0)$. Since $(-3, 4, 0, 0)$ is not a real multiple of $(2, 5, 0, 0)$, $(2, 5, 0, 0) \notin \operatorname{Sp}((-3, 4, 0, 0))$. By observing the third entry, we can see that $(1, 1, 3, 0) \notin \operatorname{Sp}((2, 5, 0, 0), (-3, 4, 0, 0))$. Similarly, by observing the fourth entry, we have

$$(1, 1, 3, 1) \notin \operatorname{Sp}((2, 5, 0, 0), (-3, 4, 0, 0), (1, 1, 3, 0)).$$

Thus, these vectors are linearly independent by Proposition 2.3.16(b). ◇

Example 2.3.20. In the real vector space $\mathscr{P}$, the polynomial $1 + x \notin \operatorname{Sp}(1)$ since the constant polynomial can only span other constant polynomials.

Similarly, $1 + x^2 \notin \mathrm{Sp}\,(1, 1 + x)$ since the first two polynomials can only span polynomials of degree ≤ 1. Following this argument we can see that the set $\{1,\ 1 + x,\ 1 + x^2,\ 1 + x^3, \ldots,\ 1 + x^n\}$ is linearly independent over $\mathbb{R}$?

The previous example inspires the following corollary to Theorem 2.3.16.

Corollary 2.3.21. *A set consisting of polynomials of different degrees in $\mathscr{P}$ or in $\mathscr{P}_n$ forms a linearly independent subset.*

Exercises 2.3

1. Is the set $\{1,\ 1+x,\ 1+x+x^2,\ 1+x+x^2+x^3, \ldots\}$ linearly independent over $\mathbb{R}$?

2. Which of the following sets in $\mathscr{P}$ are linearly independent over $\mathbb{R}$?

$$A = \{1,\ x,\ x^2\}.$$
$$B = \{1 + x,\ 1 - x,\ x^2,\ 1\}.$$
$$C = \{x^2 - 1,\ x + 1,\ x^2 - x,\ x^2 + x\}.$$
$$D = \{x - x^2,\ x^2 - x\}.$$
$$E = \{1,\ 1 - x,\ 1 - x^2\}.$$

3. Is the set
$$\left\{ \cos x,\ \sin x,\ \sin\left(x + \frac{\pi}{2}\right) \right\}$$
linearly dependent in $\mathscr{C}(\mathbb{R},\ \mathbb{R})$?

4. Let u, v, w and x be linearly independent vectors in the real vector space V. Are the vectors

$$u + v + w + x,\ 2u + 2v + w - x,\ u - v + w,\ u - w + x$$

linearly independent?

5. Consider the vector space $\mathbb{C}^2$. Is the set $\{(1,0),\ (0,i),\ (1+i,0)\}$ linearly independent over $\mathbb{R}$? How about over $\mathbb{C}$?

6. Let $f,\ g,\ h \in \mathscr{C}(\mathbb{R},\ \mathbb{R})$ be such that

$$f(0) = 1, \quad f(1) = 0, \quad f(2) = 0,$$
$$g(0) = 0, \quad g(1) = 1, \quad g(2) = 0,$$
$$h(0) = 0, \quad h(1) = 0, \quad h(2) = 1.$$

Show that $f,\ g,\ h$ are linearly independent over $\mathbb{R}$.

2.4 Bases and dimension

In this section we are finally ready to present the two central ideas in Linear Algebra.

Bases. It is not enough to simply find a generating set for a given vector space. Often we need the generating set to be something better!

Definition 2.4.1. A **basis** or **base** of a vector space V is a linearly independent subset of V which also spans V.

From previous discussion, we have already encountered several examples of bases.

Example 2.4.2. (1) The empty set is a basis for the trivial vector space $\{O\}$.

(2) The vectors $e_1,\ e_2,\ \ldots,\ e_n$ form a basis for $\mathbb{R}^n$ over $\mathbb{R}$.

(3) The vectors $1,\ x,\ \ldots,\ x^n$ form a basis for $\mathscr{P}_n$ over $\mathbb{R}$. In fact, the infinite set

$$\{1, x, x^2, x^3, x^4, \ldots\}$$

is a basis for $\mathscr{P}$ over $\mathbb{R}$.

(4) The matrices e_{ij}, $1 \leq i \leq m$ and $1 \leq j \leq n$, form a basis for $M_{m \times n}(\mathbb{R})$ over $\mathbb{R}$.

The bases mentioned in Example 2.4.2 are so useful we call them the **standard bases** for their respecting vector spaces.

Example 2.4.3. The set $\{1\}$ is a basis for $\mathbb{C}$ over $\mathbb{C}$. The set $\{1, i\}$ is a basis for $\mathbb{C}$ over $\mathbb{R}$.

The following theorem gives us the advantage of having bases.

Theorem 2.4.4. *Let* $\mathscr{B} = \{u_1, u_2, \ldots, u_n\}$ *be a subset of* V. *Then* $\mathscr{B}$ *is a basis for* V *over* $\mathbb{R}$ *if and only if for every vector* $v \in V$ *there exists a unique choice of* $(a_1, a_2, \ldots, a_n) \in \mathbb{R}^n$ *such that*

$$v = a_1 u_1 + a_2 u_2 + \cdots + a_n u_n.$$

Proof. This theorem follows from Proposition 2.3.10 and the definition of bases. $\qquad\square$

The uniquely determined n-tuple $(a_1, a_2, \ldots, a_n)$ in the previous theorem is usually called the **coefficients** or **coordinates** of v with respect to the **ordered basis** $(u_1, u_2, \ldots, u_n)$.

Example 2.4.5. Consider the set $\mathscr{B} = \{1, 1+x, 1+x+x^2\}$ in $\mathscr{P}_2$. This is a linearly independent subset by Corollary 2.3.21. For any choice of a, b, $c \in \mathbb{R}$,

$$a + bx + cx^2 = (a - b) \cdot 1 + (b - c)(1 + x) + c(1 + x + x^2).$$

This shows that $\mathscr{B}$ spans $\mathscr{P}_2$. Thus $\mathscr{B}$ forms a basis for $\mathscr{P}_2$ over $\mathbb{R}$. The coefficients of $a + bx + cx^2$ with respect to the ordered basis $(1, 1+x, 1+x+x^2)$ is $(a - b, b - c, c)$.

Next comes the first "heart" of Linear Algebra (for finite dimensional vector spaces at least).

Theorem 2.4.6. *Let* V *be generated by a finite subset* S. *Then some subset of* S *is a basis for* V. *Hence any finite dimensional vector space has a finite base.*

Proof. If S is already linearly independent, it itself is a basis for V. Otherwise, find distinct vectors $v_1, \ldots, v_n$ in S and nonzero scalars $a_1, \ldots, a_n$ such that

$$a_1 v_1 + a_2 v_2 + \cdots + a_n v_n = O$$
$$\implies v_1 = a_1^{-1}(-a_2 v_2 - \cdots - a_n v_n) \in \mathrm{Sp}\left(S \setminus \{v_1\}\right).$$

In other words, v_1 is dependent on $S \setminus \{v_1\}$. Hence

$$\mathrm{Sp}\,(S \setminus \{v_1\}) = \mathrm{Sp}\,(S) = V.$$

After dropping one element from S we still have a generating set for V. If this smaller set is linearly independent we will have reached a basis for V. Otherwise we may drop one more element from the generating set. Since S is finite, this process must eventually stop and we will obtain a basis for V. □

Example 2.4.7. Find a basis for $V = \mathrm{Sp}\,((1,2,3),\ (1,1,1),\ (2,-3,-8))$.

Solution. First we solve the equation

$$a(1,2,3) + b(1,1,1) + c(2,-3,-8) = (0,0,0)$$

for $a, b, c \in \mathbb{R}$. This is equivalent to solving the system of linear equations

$$a + b + 2c = 0,$$
$$2a + b - 3c = 0,$$
$$3a + b - 8c = 0.$$

After a straightforward computation, the solution is

$$a = 5c, \quad b = -7c$$

where c is an arbitrary real number. Letting $c = 1$, we have a non-trivial relation

$$5(1,2,3) - 7(1,1,1) + (2,-3,-8) = (0,0,0).$$

Hence $(2,-3,-8) = -5(1,2,3) + 7(1,1,1)$ is spanned by $(1,2,3)$ and $(1,1,1)$. We have $V = \mathrm{Sp}\,((1,2,3),(1,1,1))$. The vector $(1,2,3)$ is not a real multiple of $(1,1,1)$. Thus these two vectors are linearly independent over $\mathbb{R}$. We conclude that $\{(1,2,3),\ (1,1,1)\}$ is a basis for V over $\mathbb{R}$. ◇

Example 2.4.8. Let

$$S = \{(1,1,0,0,1),\ (1,1,0,1,1),\ (0,1,1,1,1),\ (2,1,-1,1,1)\}$$

be a subset of $\mathbb{R}^5$. Find a subset T of S such that T is a basis for $\mathrm{Sp}\,(S)$.

Solution. First we verify whether S is linearly independent over $\mathbb{R}$. For this, we need to solve

$$a(1,1,0,0,1) + b(1,1,0,1,1) + c(0,1,1,1,1) + d(2,1,-1,1,1)$$
$$= (0,0,0,0,0).$$

This is equivalent to solving the system of equations

$$a + b + 2d = 0,$$
$$a + b + c + d = 0,$$
$$c - d = 0,$$
$$b + c + d = 0,$$
$$a + b + c + d = 0.$$

A straightforward calculation may give us a non-trivial relation, say

$$-2(1,1,0,1,1) + (0,1,1,1,1) + (2,1,-1,1,1) = (0,0,0,0,0).$$

Thus $(2,1,-1,1,1)$ is linearly dependent on the rest of the elements in S. We may delete it without hurting $\mathrm{Sp}\,(S)$.

Next we solve

$$a(1,1,0,0,1) + b(1,1,0,1,1) + c(0,1,1,1,1) = (0,0,0,0,0).$$

This time it turns out $a = b = c = 0$. Hence

$$T = \{(1,1,0,0,1), (1,1,0,1,1), (0,1,1,1,1)\}$$

forms a basis for $\mathrm{Sp}\,(S)$ over $\mathbb{R}$. Note that this is not the only choice for T. We may also drop either $(1,1,0,1,1)$ or $(0,1,1,1,1)$ (certainly not both) to form a basis for $\mathrm{Sp}\,(S)$.   ◇

Dimension.   The following is an important technical lemma which is essential in proving Theorem 2.4.12.

Theorem 2.4.9 (Replacement Theorem). *Let $\mathscr{B}$ be a basis for V over $\mathbb{R}$ and let $\{v_1, v_2, \ldots, v_m\}$ be a linearly independent subset in V. Then there exist m distinct vectors $u_1, u_2, \ldots, u_m \in \mathscr{B}$ such that*

$$(\mathscr{B} \setminus \{u_1, u_2, \ldots, u_m\}) \cup \{v_1, v_2, \ldots, v_m\}$$

remains a basis for V over $\mathbb{R}$.

This theorem tells us that it is possible to replace m elements in a basis by any set of m linearly independent vectors in its span.

Proof. We prove this theorem by induction on m.

Step 1. To prove the case for $m = 1$, we first describe how we choose u_1.

If $v_1 \in \mathcal{B}$, we simply choose $u_1 = v_1$. We now assume that $v_1 \notin \mathcal{B}$. Since $v_1 \in V = \text{Sp}\,(\mathcal{B})$, we may write

$$(2.3) \qquad\qquad v_1 = a_1 w_1 + a_2 w_2 + \cdots + a_k w_k,$$

where the a_i's are scalars and the u_j's are chosen from $\mathcal{B}$. Since no linearly independent sets contain the zero vector, the vector $v \neq O$. Hence, one of the a_i's must be nonzero. Without loss of generality we may assume $a_1 \neq 0$. Let $u_1 = w_1$.

Step 2. Let $\mathcal{B}_1 = (\mathcal{B} \setminus \{u_1\}) \cup \{v_1\}$. We will show that $\mathcal{B}_1$ is a basis for V.

We first check that $V = \text{Sp}\,(\mathcal{B}_1)$. Clearly, $\mathcal{B} \setminus \{u_1\} \subseteq \mathcal{B}_1 \subseteq \text{Sp}\,(\mathcal{B}_1)$. As for u_1 we have that

$$u_1 = a_1{}^{-1}(v_1 - a_2 w_2 - \cdots - a_k w_k) \in \text{Sp}\,(\mathcal{B}_1)$$

using the identity (2.3). Thus $\mathcal{B} \subseteq \text{Sp}\,(\mathcal{B}_1)$. It follows that

$$V = \text{Sp}\,(\mathcal{B}) \subseteq \text{Sp}\,(\mathcal{B}_1) \subseteq V.$$

We conclude that $V = \text{Sp}\,(\mathcal{B}_1)$.

Next we verify that $\mathcal{B}_1$ is linearly independent. Let

$$(2.4) \qquad\qquad b v_1 + c_1 x_1 + c_2 x_2 + \cdots + c_\ell x_\ell = O$$

where $b, c_1, \ldots, c_\ell$ are scalars and $x_1, \ldots, x_\ell$ are distinct vectors in $\mathcal{B} \setminus \{u_1\}$. We need to show that $b_1 = c_1 = \cdots = c_\ell = 0$.

Suppose $b \neq 0$. Then

$$v_1 = b^{-1}(-c_1 x_1 - c_2 x_2 - \cdots - c_\ell x_\ell) \in \text{Sp}\,(\mathcal{B} \setminus \{u_1\})$$

using the identity (2.4). It follows that $\mathcal{B}_1 \subseteq \text{Sp}\,(\mathcal{B} \setminus \{u_1\})$ and thus $V = \text{Sp}\,(\mathcal{B} \setminus \{u_1\})$. However, this implies that $u_1 \in \text{Sp}\,(\mathcal{B} \setminus \{u_1\})$ and $\mathcal{B}$

is linearly dependent, a contradiction. We conclude that $b = 0$. Now from the identity (2.4) we have that

$$c_1 x_1 + \cdots + c_\ell x_\ell = O.$$

Since this is a relation in the linearly independent set $\mathscr{B}$, we have that $c_1 = \cdots = c_\ell = 0$.

Step 3. We now assume the induction hypothesis for $m - 1$. More precisely, we assume that there are distinct vectors $u_1, \ldots, u_{m-1}$ in $\mathscr{B}$ such that

$$\mathscr{B}_{m-1} = (\mathscr{B} \setminus \{u_1, \ldots, u_{m-1}\}) \cup \{v_1, \ldots, v_{m-1}\}$$

remains a basis for V over $\mathbb{R}$.

Step 4. Next we describe how to pick u_m.

If $v_m \in \mathscr{B} \setminus \{u_1, \ldots, u_{m-1}\}$, we simply choose $u_m = v_m$. Suppose not. Since $v_m \in V = \mathrm{Sp}\,(\mathscr{B}_{m-1})$, we can find scalars $a_1, \ldots, a_k$ and distinct vectors $w_m, \ldots, w_k \in \mathscr{B} \setminus \{u_1, \ldots, u_{m-1}\}$ such that

$$(2.5) \qquad v_m = a_1 v_1 + \cdots + a_{m-1} v_{m-1} + a_m w_m + \cdots + a_k w_k.$$

If $a_m = \cdots = a_k = 0$ then $v_m \in \mathrm{Sp}\,(v_1, \ldots, v_{m-1})$, which means that $\{v_1, \ldots, v_m\}$ is linearly dependent, a contradiction. Hence, one of the a_i, $i = m, \ldots, k$, must be nonzero. Without loss of generality we may assume that $a_m \neq 0$. Choose $u_m = w_m$. Our choice of u_m guarantees that $u_m \notin \mathscr{B} \setminus \{u_1, \ldots, u_m\}$. Thus $u_1, \ldots, u_m$ are m distinct vectors in $\mathscr{B}$.

Step 5. Let $\mathscr{B}_m = (\mathscr{B} \setminus \{u_1, \ldots, u_m\}) \cup \{v_1, \ldots, v_m\}$. Then $\mathscr{B}_m$ is a basis for V. This completes the induction steps.

First we verify that $V = \mathrm{Sp}\,(\mathscr{B}_m)$. Note that

$$\mathscr{B}_{m-1} = (\mathscr{B}_m \setminus \{v_m\}) \cup \{u_m\}.$$

Clearly, $\mathscr{B}_m \setminus \{v_m\} \subseteq \mathscr{B}_m \subseteq \mathrm{Sp}\,(\mathscr{B}_m)$. As for u_m we have that

$$u_m = a_m^{-1}(v_m - a_1 v_1 - \cdots - a_{m-1} v_{m-1} - a_{m+1} w_{m+1} - \cdots - a_k w_k)$$
$$\in \mathrm{Sp}\,(\mathscr{B}_m)$$

using the identity (2.5). Thus $\mathrm{Sp}\,(\mathscr{B}_m)$ contains the generating set $\mathscr{B}_{m-1}$ for V. It follows that $V = \mathrm{Sp}\,(\mathscr{B}_1)$.

Next we verify that $\mathscr{B}_m$ is linearly independent. Note that

$$\mathscr{B}_m = (\mathscr{B}_{m-1} \setminus \{u_m\}) \cup \{v_m\}.$$

Let

$$(2.6) \quad b_1 v_1 + \cdots + b_{m-1} v_{m-1} + b_m v_m + c_1 x_1 + c_2 x_2 + \cdots + c_\ell x_\ell = O$$

where $b_1, \ldots, b_m, c_1, \ldots, c_\ell$ are scalars and $x_1, \ldots, x_\ell$ are distinct vectors in $\mathscr{B} \setminus \{u_1, \ldots, u_m\}$. We need to show that $b_1 = \cdots = b_m = c_1 = \cdots = c_\ell = 0$. Suppose $b_m \neq 0$. Then

$$\begin{aligned} v_m &= -b_m{}^{-1}(b_1 v_1 + \cdots + b_{m-1} v_{m-1} + c_1 x_1 + c_2 x_2 + \cdots + c_\ell x_\ell) \\ &\in \operatorname{Sp}(\mathscr{B}_{m-1} \setminus \{u_m\}) \end{aligned}$$

using identity (2.6). It follows that $\mathscr{B}_m \subseteq \operatorname{Sp}(\mathscr{B}_{m-1} \setminus \{u_m\})$ and thus $\mathscr{B}_{m-1} \setminus \{u_m\}$ spans V. However, this implies that $u_m \in \operatorname{Sp}(\mathscr{B}_{m-1} \setminus \{u_m\})$ and $\mathscr{B}_{m-1}$ is linearly dependent, a contradiction to our assumption in Step 3. We conclude that $b_m = 0$. Now from the identity (2.6) we have that

$$b_1 v_1 + \cdots + b_{m-1} v_{m-1} + c_1 x_1 + \cdots + c_\ell x_\ell = O.$$

Since this is a relation in the linearly independent set $\mathscr{B}_{m-1}$, we also have that $b_1 = \cdots = b_{m-1} = c_1 = \cdots = c_\ell = 0$. $\qquad \square$

Corollary 2.4.10. *Let V be a vector space with a basis of n elements and let $\mathscr{L}$ be any linearly independent subset of V. Then $|\mathscr{L}| \leq n$.*

Corollary 2.4.11. *If a vector space V contains an infinite subset which is also linearly independent then V is infinite dimensional.*

Proof. If V is finite dimension, V contains a finite basis by Theorem 2.4.6. Now the results follows from Corollary 2.4.10. $\qquad \square$

Here comes the second "heart" of Linear Algebra for finite dimensional vector spaces.

Theorem 2.4.12. *Let V be a finite dimensional vector space. Then any two bases of V have the same number of vectors in them.*

Proof. By Theorem 2.4.6, V contains a finite basis $\mathscr{B}$. If $\mathscr{B}'$ is another basis for V, then $\mathscr{B}'$ is certainly linearly independent. By Corollary 2.4.10, $|\mathscr{B}'| \leq |\mathscr{B}|$. However, $\mathscr{B}$ is also linearly independent, we also have $|\mathscr{B}| \leq |\mathscr{B}'|$. We conclude that $|\mathscr{B}| = |\mathscr{B}'|$. $\qquad\square$

Definition 2.4.13. We define the **dimension** of V over $\mathbb{R}$, denoted $\dim_{\mathbb{R}} V$ (or simply $\dim V$) as follows. If V is infinite dimensional then we define $\dim_{\mathbb{R}} V = \infty$. If V is finite dimensional we define $\dim_{\mathbb{R}} V$ to be the number of elements in a basis for V over $\mathbb{R}$.

From Example 2.4.2 we have the following results regarding dimension.

Corollary 2.4.14.   (a) $\dim_{\mathbb{R}}\{O\} = 0$.

(b) $\dim_{\mathbb{R}} \mathbb{R}^n = n$.

(c) $\dim_{\mathbb{R}} \mathscr{P}_n = n + 1$ *and* $\dim_{\mathbb{R}} \mathscr{P} = \infty$.

(d) $\dim_{\mathbb{R}} M_{m \times n}(\mathbb{R}) = mn$.

(e) $\dim_{\mathbb{R}} \mathbb{C} = 2$ *and* $\dim_{\mathbb{C}} \mathbb{C} = 1$.

We summarize two related and important ideas from Theorem 2.4.6 and from Corollary 2.4.10 here.

Corollary 2.4.15. *Let* $\dim_{\mathbb{R}} V = n$. *Let* $\mathscr{G}$ *be a generating set for* V *and let* $\mathscr{L}$ *be a linearly independent subset in* V. *Then*

$$|\mathscr{G}| \geq n \qquad and \qquad |\mathscr{L}| \leq n.$$

How to construct a basis for a finite dimensional vector space. How do we construct from scratch a basis for a finite dimensional vector space? There are two ways to do it. According to Theorem 2.4.6, we may start with any finite generating set. If it is linearly dependent, one vector in the set is linearly dependent on the rest of the vectors in the generating set by Theorem 2.3.16(1). We eliminate this vector from the original generating set and we still have a generating set by Lemma 2.2.11. If the new set is linearly independent then we have a basis. Otherwise we may repeat this process. Since we have at most n elements to eliminate from the original generating set, sooner or later we must come upon a linearly independent generating set and voilà, there is a basis for V.

There is an alternative method to construct a basis if we don't have a generating set to start with. In fact, we described it in §2.3. We first start with any nonzero vector u_1. Corollary 2.3.17 tells us that u_1 is linearly independent. If $V = \mathrm{Sp}\,(u_1)$ then we have obtained a basis for v. Otherwise we find a $u_2 \in V \setminus \{u_1\}$. By Lemma 2.3.14, u_1, u_2 are linearly independent. If $V = \mathrm{Sp}\,(u_1, u_2)$ then we have reached a basis for V. Otherwise we repeat the process described. If this process does not stop, we can find an infinite set which is linearly independent. In this case we conclude that V is infinite dimensional using Corollary 2.4.10. If V is finite dimensional the process must stop and we reach a linearly independent set which also spans V. We now have successfully constructed a basis for V. We have just explained following results.

Theorem 2.4.16. *Let V be a finite dimensional vector space. The following statements are true.*

(a) *Any generating set can be reduced to a basis. Any linearly independent set can be expanded to a basis.*

(b) *A basis is a minimal generating set. A basis is a maximal linearly independent set.*

Remark. Theorem 2.4.6, Replacement Theorem, Theorem 2.4.12 and Theorem 2.4.16 are true even for infinite dimensional vector spaces, although we will need more advanced Set Theory to prove them.

Corollary 2.4.17. *Let $\dim_{\mathbb{R}} V = n$. Then a subset $\mathscr{B}$ of V that satisfies any of the following three conditions is a basis of V:*

(i) *$\mathscr{B}$ is both a linearly independent and a generating set;*

(ii) *$\mathscr{B}$ is a linearly independent set consisting of n elements;*

(iii) *$\mathscr{B}$ is a generating set consisting of n elements.*

Proof. Condition (i) obviously implies both (b) and (c).

Assume (ii). If $\mathscr{B}$ is not a generating set, $\mathscr{B}$ can be enlarged to a basis by Theorem 2.4.16(a). This basis will contain more than n elements, contradicting Theorem 2.4.12. This gives us (i).

Assume (iii). If $\mathscr{B}$ is not linearly independent, $\mathscr{B}$ can be reduced to a basis by Theorem 2.4.16(a). This basis will contain less than n elements, contradicting Theorem 2.4.12. Hence $\mathscr{B}$ is linearly independent. This gives us (i). $\square$

The following Corollary is a simple observation of Proposition 2.4.16.

Corollary 2.4.18. *If W is a* proper *subspace of a finite dimensional vector space V. Then*

$$\dim_\mathbb{R} W < \dim_\mathbb{R} V.$$

Proof. A basis for W is a linearly independent subset in V. Since it does not span V, it can be enlarged to a basis for V, which must contain more elements than the given basis for W. $\square$

Example 2.4.19. Let

$$W = \{(a_1, a_2, a_3, a_4, a_5) \in \mathbb{R}^5 : a_1 + a_3 + a_5 = 0, \ a_2 = a_4\}.$$

Find a basis for W and find $\dim_\mathbb{R} W$.

Solution. The set W is the solution space for the system of linear equations

$$a_1 + a_3 + a_5 = 0,$$
$$a_2 - a_4 = 0.$$

The coefficient matrix to the system is

$$\begin{pmatrix} 1 & 0 & 1 & 0 & 1 \\ 0 & 1 & 0 & -1 & 0 \end{pmatrix}$$

which is already in reduced row-Echelon form. Hence the solution is

$$a_1 = -a_3 - a_5,$$
$$a_2 = a_4$$

for arbitrary real numbers a_3, a_4, a_5. Thus

$$W = \{(-a_3 - a_5, a_4, a_3, a_4, a_5) : a_3, a_4, a_5 \in \mathbb{R}\}.$$

The vectors in W may be written as

$$a_3(-1, 0, 1, 0, 0) + a_4(0, 1, 0, 1, 0) + a_5(-1, 0, 0, 0, 1).$$

This shows that $W = \text{Sp}(S)$ where

$$S = \{(-1, 0, 1, 0, 0), (0, 1, 0, 1, 0), (-1, 0, 0, 0, 1)\}.$$

It is straightforward to check that S is linearly independent over $\mathbb{R}$. Hence S is a basis for W and $\dim_{\mathbb{R}} W = 3$. $\diamond$

Exercises 2.4

1. Expand $\{(1, 2, 1)\}$ to a basis for $\mathbb{R}^3$.

2. Expand $\{x^2 + x + 1\}$ to a basis for $\mathscr{P}_2$.

3. In $M_2(\mathbb{R})$ is the set $\{e_{11} + e_{12}, \ e_{11} + 2e_{12}\}$ linear transformation over $\mathbb{R}$? If yes, expand it to a basis for $M_2(\mathbb{R})$.

4. Which of the following subsets in $\mathbb{R}^3$ are bases for $\mathbb{R}^3$?

$$A = \{(1, 1, 1), \ (1, 1, 0), \ (1, 0, 0)\}.$$
$$B = \{(1, 0, 1), \ (0, 1, 0), \ (1, 1, 1)\}.$$
$$C = \{(-1, 1, 1), \ (1, -1, 1), \ (1, 1, -1)\}.$$
$$D = \{(1, -1, 0), \ (1, 0, -1), \ (-1, 0, 1)\}.$$
$$E = \{(1, 2, 3), \ (4, 5, 6), \ (7, 8, 9)\}.$$

5. Which of the following sets in $\mathscr{P}_2$ are bases for $\mathscr{P}_2$?

$$A = \{1, \ x, \ x^2\}.$$
$$B = \{1 + x, \ 1 - x, \ x^2, \ 1\}.$$
$$C = \{x^2 - 1, \ x + 1, \ x^2 - x, \ x^2 + x\}.$$
$$D = \{x - x^2, \ x^2 - x\}.$$
$$E = \{1, \ 1 - x, \ 1 - x^2\}.$$
$$F = \{1, \ x - 1, \ (x - 1)(x - 2)\}.$$
$$G = \{x^2 + x + 1, \ x^2 - x + 1, \ x^2 - x - 1\}.$$

6. Find a basis and determine the dimension for each of the following subspaces of $\mathscr{P}_3$:

$$S = \{p(x) \in \mathscr{P}_3 : p(0) = 0\}.$$
$$T = \{p(x) \in \mathscr{P}_3 : p(1) = 0\}.$$
$$U = \{p(x) \in \mathscr{P}_3 : \frac{d}{dx}p(x) = 0\}.$$
$$V = \{p(x) \in \mathscr{P}_3 : p(x) = a_0 + a_1x + a_3x^3, \text{ where } a_0, a_1, a_3 \in \mathbb{R}\}.$$
$$W = \{p(x) \in \mathscr{P}_3 : p(-x) = p(x)\}.$$

7. Find a basis and determine the dimension for each of the following vector spaces:

$$U = \{(x, y, z) \in \mathbb{R}^3 : x - z = 0\}.$$
$$V = \{(x, y, z) \in \mathbb{R}^3 : x + y = 0 \text{ and } y + z = 0\}.$$
$$W = \{(x, y, z, w) \in \mathbb{R}^4 : x - z = 0 \text{ and } x + y + z + w = 0\}.$$

8. Let V be the subspace of $\mathbb{R}^3$ generated by $(1, 2, 3)$, $(2, 3, 4)$, $(3, 0, 1)$ and $(1, 0, -1)$. Which of the generators are redundant?

9. Let $\mathscr{B} = \{(1, 0, 1), \ (1, 1, 1), \ (0, 0, 1)\}$.

   (a) Show that $\mathscr{B}$ is a basis for $\mathbb{R}^3$.

   (b) Replace elements in $\mathscr{B}$ by $(1, 2, 3)$ to obtain another basis for $\mathbb{R}^3$.

10. Let $a_1, \ldots, a_n$ be a sequence of real numbers which are not all zero. Let V be the subspace

$$\{(x_1, \ldots, x_n) \in \mathbb{R}^n : a_1 x_1 + a_2 x_2 + \cdots + a_n x_n = 0\}$$

of $\mathbb{R}^n$. Show that $\dim V = n - 1$.

11. Let $W \subseteq V$ be finite dimensional vector spaces. Show that $V = W$ if $\dim V = \dim W$.

12. Let V be an n-dimensional vector space and let k be an integer such that $0 \le i \le n$. Show that V contains a k-dimensional subspace.

2.5 Applications

In this section we gather what we know about finite dimensional vector spaces to obtain some interesting mathematical results.

Dimension and reduced row echelon forms. So far, we have not been looking for bases with great efficiency. We now provide an easier method to find bases for a linear span of $\mathbb{R}^n$.

Suppose we are given a set of generators for a linear span. To delete the redundant generators is one way of obtaining a basis. However, it is not always required that our chosen basis be a subset of the original generating set, especially when you simply want to determine the dimension of a certain space. Hence, it is convenient to replace our original generators with a new set which is readily seen as linearly independent. Next, we provide some guidelines on how to transform generating sets.

Proposition 2.5.1. *Suppose* $V = \mathrm{Sp}\,(v_1, \ldots, v_n)$. *The followings sets also generate* V:

(i) $\{v_1, \ \ldots, \ v_{i-1}, \ v_j, \ v_{i+1}, \ldots, \ v_{j-1}, \ v_i, \ v_{j+1}, \ldots, v_n\}$ *where* $1 \le i < j \le n$;

(ii) $\{v_1, \ \ldots, \ v_{i-1}, \ u v_i, \ v_{i+1}, \ldots, \ v_n\}$ *where* $1 \le i \le n$ *and* $u \in \mathbb{R} \setminus \{0\}$;

(iii) $\{v_1, \ \ldots, \ v_{i-1}, \ v_i + a v_j, \ v_{i+1}, \ldots, \ v_n\}$ *where* $1 \le i, \ j \le n$, $i \ne j$ *and* $a \in \mathbb{R}$.

Proof. The set in (i) in in fact the original generating set. The two sets in (ii) and (iii) contain $\{v_1, \ldots, v_n\} \setminus \{v_i\}$. It remains to show v_i is also in the span of the two given sets. This is true since in case (ii) we have $v_i = u^{-1}(uv_i)$, and in case (iii) we have $v_i = (v_i + av_j) - av_j$. $\square$

Definition 2.5.2. Let $A \in M_{m \times n}(\mathbb{R})$. Define the **column space** of A to be the subspace of $\mathbb{R}^m$ generated by the columns of A. Similarly, the **row space** of A is the subspace of $\mathbb{R}^n$ generated by the rows of A.

For example, the linear span $\mathrm{Sp}\,((1,0,0),\ (2,3,5))$ is the row space of the matrix $\begin{pmatrix} 1 & 0 & 0 \\ 2 & 3 & 5 \end{pmatrix}$. Proposition 2.5.1 can now be restated as follows.

Proposition 2.5.3. *Let A be an $m \times n$ matrix and B be another matrix obtained by performing various elementary row (column, respectively) operations to A. Then the row (column, respectively) space of A is also the row (column, respectively) space of B.*

In particular, the row space of A is the same as the row space of its reduced row echelon form.

When dealing with a linear span in $\mathbb{R}^n$, say, generated by m generators, we may use the generators to form the rows of an $m \times n$ matrix. Then we perform elementary row operations to obtain a new set of generators which is more to our liking.

Lemma 2.5.4. *The nonzero rows of a reduced row echelon form are linearly independent.*

Proof. Let $w_1, \ldots, w_r$ be the nonzero rows of a reduced row echelon form. Consider the linear combination $v = a_1 w_1 + a_2 w_2 + \ldots a_r w_r$. If the pivot of the i-th row fall in the j-th column, then a_i is the j-th entry of v. In case the linear combination $v = O$, we have $a_1 = a_2 = \cdots = a_r = 0$. $\square$

Definition 2.5.5. Let $A \in M_{m \times n}(\mathbb{R})$. Define the **column rank** of A, denoted $\mathrm{col\,rk}\ A$, to be the dimension of the column space of A. Define the **row rank** of A, denoted $\mathrm{row\,rk}\ A$, to be the dimension of the row space of A.

Proposition 2.5.6. *The nonzero rows of the reduced row echelon form of a matrix form a basis for its row space. The row rank of a matrix is the number of nonzero rows in its reduced row echelon form.*

Proof. The result follows from Proposition 2.5.3 and Lemma 2.5.4.   □

In fact, to compute the row rank of a matrix, it is not necessary to go all the way to reduced echelon forms. We may perform a series of elementary row operations to a matrix so that it satisfies the two conditions given in Proposition 2.5.7. Once we reach this stage, we may multiply each row by the inverse of the first nonzero entry in that row. Then we have the pivots sitting where it belongs. Even if we proceed further to produce the reduced row echelon form, the number of nonzero rows will not change. We now have the following result.

Proposition 2.5.7. *Perform elementary row operations to a matrix A to obtain a matrix B such that*

(i) *the nonzero rows precede the zero rows (if any), and*

(ii) *the first nonzero entry in each nonzero row occurs in a column to the right of the first nonzero entry in the preceding row.*

Then the nonzero rows of B form a basis for the row space of A. The number of the nonzero rows of B is the row rank of A and the dimension of row space of A.

Example 2.5.8. Find a basis for

$$\mathrm{Sp}\left((1,1,0,0,1),\ (1,1,0,1,1),\ (0,1,1,1,1),\ (2,1,-1,1,1)\right).$$

Determine its dimension.

Solution. The given span is the row space of

$$A = \begin{pmatrix} 1 & 1 & 0 & 0 & 1 \\ 1 & 1 & 0 & 1 & 1 \\ 0 & 1 & 1 & 1 & 1 \\ 2 & 1 & -1 & 1 & 1 \end{pmatrix}$$

We now perform elementary row operations on A:

$$A \rightsquigarrow \begin{pmatrix} 1 & 1 & 0 & 0 & 1 \\ 0 & 0 & 0 & 1 & 0 \\ 0 & 1 & 1 & 1 & 1 \\ 0 & -1 & -1 & 1 & -1 \end{pmatrix} \begin{array}{l} \text{row(ii)} - \text{row(i)} \\ \\ \text{row(iv)} - 2 \times \text{row(i)} \end{array}$$

$$\rightsquigarrow \begin{pmatrix} 1 & 1 & 0 & 0 & 1 \\ 0 & 1 & 1 & 1 & 1 \\ 0 & 0 & 0 & 2 & 0 \\ 0 & 0 & 0 & 1 & 0 \end{pmatrix} \begin{array}{l} \\ \text{row(iii)} \\ \text{row(iv)} + \text{row(iii)} \\ \text{row(ii)} \end{array}$$

$$\rightsquigarrow \begin{pmatrix} 1 & 1 & 0 & 0 & 1 \\ 0 & 1 & 1 & 1 & 1 \\ 0 & 0 & 0 & 1 & 0 \\ 0 & 0 & 0 & 0 & 0 \end{pmatrix} \begin{array}{l} \\ \\ \text{row(iv)} \\ \text{row(iii)} - 2 \times \text{row(iv)} \end{array}.$$

We conclude that the dimension of the row space of A is 3.   ◇

A matrix satisfying the requirements as in Proposition 2.5.7 is called a **row echelon form**. The final form for the matrix A in Example 2.5.8 is in row echelon form. The row echelon form is more relaxed than the *reduced* row echelon form in the sense that the entries above a pivot does not have to be zero! In the following example, we will see that we may determine the row rank of a matrix even before we reach the row echelon form.

Example 2.5.9. Let

$$S = \{(2, -3, 1),\ (1, 4, -2),\ (-8, 12, -4),\ (1, 37, -17),\ (-3, -5, 8)\}$$

be a subset in $\mathbb{R}^3$. Does S generate $\mathbb{R}^3$? Find a basis for $\mathrm{Sp}\,(S)$ which is a subset of S.

Solution. The span of S is the row space of

$$A = \begin{pmatrix} 2 & -3 & 1 \\ 1 & 4 & -2 \\ -8 & 12 & -4 \\ 1 & 37 & -17 \\ -3 & -5 & 8 \end{pmatrix}.$$

Perform elementary row operation on A to simplify the matrix:

$$A \rightsquigarrow \begin{pmatrix} 1 & 4 & -2 \\ 2 & -3 & 1 \\ 2 & -3 & 1 \\ 1 & 37 & -17 \\ -3 & -5 & 8 \end{pmatrix} \begin{matrix} \text{row(ii)} \\ \text{rowi} \\ (-1/4) \times \text{row(iii)} \\ \\ \\ \end{matrix}$$

$$\rightsquigarrow \begin{pmatrix} 1 & 4 & -2 \\ 0 & -11 & 5 \\ 0 & 0 & 0 \\ 0 & 33 & -15 \\ 0 & 7 & 2 \end{pmatrix} \begin{matrix} \\ \text{row(ii)} - 2 \times \text{row(i)} \\ \text{row(iii)} - \text{row(ii)} \\ \text{row(iv)} - \text{row(i)} \\ \text{row(v)} + 3 \times \text{row(i)} \end{matrix}$$

$$\rightsquigarrow \begin{pmatrix} 1 & 4 & -2 \\ 0 & -11 & 5 \\ 0 & 0 & 0 \\ 0 & 0 & 0 \\ 0 & 7 & 2 \end{pmatrix} \begin{matrix} \\ \\ \\ \text{row(iv)} + 3 \times \text{row(ii)} \\ \end{matrix} .$$

Even though this matrix is not in row echelon form yet, it is obvious that the three nonzero rows are linearly independent. Hence $\dim_\mathbb{R} \mathrm{Sp}\,(S) = 3$. It follows that $\mathrm{Sp}\,(S) = \mathbb{R}^3$.

From the work above, a safe bet is to choose

$$\{(2, -3, 1), (1, 4, -2), (-3, -5, 8)\},$$

for a basis of $\mathbb{R}^3$. Just to be doubly sure, we perform elementary row operations on

$$\begin{pmatrix} 2 & -3 & 1 \\ 1 & 4 & -2 \\ -3 & -5 & 8 \end{pmatrix} \rightsquigarrow \begin{pmatrix} 1 & 4 & -2 \\ 2 & -3 & 1 \\ -3 & -5 & 8 \end{pmatrix} \rightsquigarrow \begin{pmatrix} 1 & 4 & -2 \\ 0 & -11 & 5 \\ 0 & 7 & 2 \end{pmatrix} .$$

The row rank is 3. Hence $\{(2, -3, 1), (1, 4, -2), (-3, -5, 8)\}$ is a basis for $\mathbb{R}^3$.

$\diamond$

Dimension formula.   Before giving a basis, it is often convenient to know its dimension first. The following theorem gives a dimension formula.

Theorem 2.5.10. *Let V and W be finite dimensional subspaces of U. Then*

$$\dim(V + W) + \dim(V \cap W) = \dim V + \dim W.$$

This theorem explains the simple fact why two planes in $\mathbb{R}^3$ intersect at last along a line if they intersect at all. Suppose two planes P_1 and P_2 in $\mathbb{R}^3$ intersect, and we may assume that they intersect at the origin without loss of generality. Then by the dimension formula,

$$\dim(P_1 \cap P_2) = 2 + 2 - \dim(P_1 + P_2) \geq 4 - 3 = 1.$$

Hence P_1 and P_2 must intersect at a subspace of dimension ≥ 1. To be more precise, if $P_1 = P_2$, then $\dim(P_1 + P_2) = \dim P_1 = 2$. We have that $\dim(P_1 \cap P_2) = 2$, a triviality. Otherwise, $P_1 + P_2 = \mathbb{R}^3$, and we have $\dim(P_1 \cap P_2) = 1$. In this case P_1 and P_2 intersect exactly at a line.

Proof. Let $\dim(V \cap W) = t$. Find a basis $\{u_1, \dots, u_t\}$ for $V \cap W$. This is a linearly independent subset in V and in W. By Theorem 2.4.16, it can expanded to a basis

$$\{u_1, \dots, u_t, v_1, \dots, v_r\}$$

for V and a basis

$$\{u_1, \dots, u_t, w_1, \dots, w_s\}$$

for W. This says that $\dim V = r + t$ and $\dim W = s + t$. We are done if we can show that $\dim(V + W) = r + s + t$. We claim that

$$\mathscr{B} = \{u_1, \dots, u_t, v_1, \dots, v_r, w_1, \dots, w_s\}$$

is a basis for $V + W$, and this will complete the proof.

A vector in $V + W$ is in the form of $v = w$ where $v \in V$ and $w \in W$. Write

$$v = a_1 u_1 + \cdots + a_t u_t + b_1 v_1 + \cdots + b_r v_r$$
$$w = c_1 u_1 + \cdots + c_t u_t + d_1 w_1 + \cdots + d_s w_s$$

where a_i, b_j, c_k, d_l are all scalars. Then $v + w$ is the linear combination

$$(a_1 + c_1)u_1 + \cdots + (a_t + c_t)u_t + b_1 v_1 + \cdots + b_r v_r + d_1 w_1 + \cdots + d_s w_s.$$

Hence $V + W$ is spanned by $\mathscr{B}$.

Suppose

$$a_1 u_1 + \cdots + a_t u_t + b_1 v_1 + \cdots + b_r v_r + c_1 w_1 + \cdots + c_s w_s = O.$$

Then

$$b_1 v_1 + \cdots + b_r v_r = -a_1 u_1 + \cdots - a_t u_t - c_1 w_1 - \cdots - c_s w_s$$

is a vector in $V \cap W$. Thus

$$b_1 v_1 + \cdots + b_r v_r = d_1 u_1 + \cdots + d_t u_t.$$

However, $\{u_1, \ldots, u_t, v_1, \ldots, v_r\}$ is a basis for V. This implies that

$$b_1 = \cdots = b_r = d_1 = \cdots = d_t = 0.$$

But then it implies that

$$c_1 w_1 + \cdots + c_s w_s = O.$$

Since $w_1, \ldots, w_s$ are linearly independent, we also have

$$c_1 = \cdots = c_s = 0.$$

We have just shown that $\mathscr{B}$ is linearly independent. We conclude that $\mathscr{B}$ is indeed a basis for $V + W$. $\qquad\square$

Example 2.5.11. Give a basis for

$$\mathrm{Sp}\left((1,0,0,1),\ (1,1,0,2)\right) \cap \mathrm{Sp}\left((-1,0,1,-1),\ (1,2,0,3)\right).$$

Solution. Let

$$V = \mathrm{Sp}\left((1,0,0,1),(1,1,0,2)\right) \text{ and } W = \mathrm{Sp}\left((-1,0,1,-1),(1,2,0,3)\right).$$

Clearly, $\dim V = \dim W = 2$. To find $\dim(V + W)$, we need to calculate the row rank of

$$\begin{pmatrix} 1 & 0 & 0 & 1 \\ 1 & 1 & 0 & 2 \\ -1 & 0 & 1 & -1 \\ 1 & 2 & 0 & 3 \end{pmatrix}.$$

We skip the process and just tell you that $\dim(V + W) = 3$. Thus $\dim(V \cap W) = 1$. To find a basis for $V \cap W$, we need to find a nonzero vector in $V \cap W$. Let

$$a(1, 0, 0, 1) + b(1, 1, 0, 2) = c(-1, 0, 1, -1) + d(1, 2, 0, 3).$$

We need to find a special solution for

$$a + b = -c + d,$$
$$b = c + 2d,$$
$$0 = c,$$
$$a + 2b = -c + 3d.$$

After a straightforward calculation,

$$a = -1, \quad b = 2, \quad c = 0, \quad d = 1$$

is such a solution. Hence $(1, 2, 0, 3) \in V \cap W$. The set $\{(1, 2, 0, 3)\}$ is a basis for $V \cap W$.   ◇

Example 2.5.12. Let $V = \{(a, b, c) \in \mathbb{R}^3 : b - c = 0\}$. Find a subspace W of $\mathbb{R}^3$ such that $V \cap W = \{O\}$ and $V + W = \mathbb{R}^3$.

Solution. Since

$$V = \{(a, b, b) : a, b \in \mathbb{R}\}$$
$$= \{a(1, 0, 0) + b(0, 1, 1) : a, b \in \mathbb{R}\}$$
$$= \mathrm{Sp}\,((1, 0, 0), (0, 1, 1)),$$

we have that $\{(1, 0, 0), (0, 1, 1)\}$ is a basis for V. This says that $\dim V = 2$. From dimension formula, $\dim W = 1$. We need to find a vector not in V. The vector $(0, 0, 1)$ is such a choice. We choose $W = \mathrm{Sp}\,((0, 0, 1))$. Of course, this is not the only choice.   ◇

The Lagrange Interpolation Formula.   Polynomials in $\mathscr{P}_n$ can be characterized by its coefficients. Or they may also be characterized by their values at $n + 1$ distinct points of $\mathbb{R}$. Suppose we now want to find a polynomial $g \in \mathscr{P}_n$ so that its values at the $n+1$ distinct real numbers c_0, $c_1, \ldots, c_n$ are $a_0, a_1, \ldots, a_n$ respectively. If we use the relations $g(c_i) = a_i$ to obtain a *non-homogeneous* system of $n+1$ linear equations in $n+1$ unknowns (the coefficients in g), this is an inefficient way to find g.

Definition 2.5.13. We define the **Kronecker delta**

$$\delta_{ij} = \begin{cases} 1, & \text{if } i = j, \\ 0, & \text{if } i \neq j. \end{cases}$$

The Kronecker delta is a useful notation that students will encounter in many mathematical courses. For example, the identity matrix can be denoted as $I_n = (\delta_{ij})_{n \times n}$.

Definition 2.5.14. The polynomials $f_0, f_1, \ldots, f_n$ defined by

$$f_i(x) = \frac{(x - c_0) \ldots (x - c_{i-1})(x - c_{i+1}) \ldots (x - c_n)}{(c_i - c_0) \ldots (c_i - c_{i-1})(c_i - c_{i+1}) \ldots (c_i - c_n)}$$

are called the **Lagrange polynomials associated with** $c_0, c_1, \ldots, c_n$.

Remark. The polynomials $f_i \in \mathscr{P}_n$ is a polynomial function from $\mathbb{R}$ to $\mathbb{R}$ such that

$$f_i(c_j) = \delta_{ij}, \qquad \text{for } i, j = 0, 1, \ldots, n.$$

Lemma 2.5.15. *The set $\{f_0, f_1, \ldots, f_n\}$ is linearly independent over $\mathbb{R}$.*

Proof. Let $a_0, a_1, \ldots, a_n$ be scalars such that

$$a_0 f_0 + a_1 f_1 + \cdots + a_n f_n = 0.$$

After substituting x by c_i, we obtain that $a_i = 0$ for each i. □

Theorem 2.5.16. *The set $\{f_0, f_1, \ldots, f_n\}$ is a basis for $\mathscr{P}_n$ over $\mathbb{R}$.*

Proof. Remember that $\dim \mathscr{P}_n = n+1$ and use Lemma 2.5.15. □

Any polynomial g in $\mathscr{P}_n$ can be uniquely represented as

$$g = \sum_{i=0}^{n} a_i f_i$$

for a suitable choice of $(a_0, a_1, \ldots, a_n)$ in $\mathbb{R}^n$. Note that $g(c_i) = a_i$ for $i = 0, 1, \ldots, n$. Thus we obtain the **Lagrange Interpolation Formula**

$$g = \sum_{i=0}^{n} g(c_i) f_i.$$

We conclude that for each choice of n-tuples $(a_0, a_1, \ldots, a_n)$ of real numbers, there exists one and only one real polynomial g of degree at most n so that $g(c_i) = a_i$ for all $i = 0, 1, \ldots, n$. In fact, $g = \sum_{i=0}^{n} a_i f_i$.

Example 2.5.17. There is exactly one real polynomial of degree at most 2 whose graph passes through the points $(1, 8)$, $(2, 5)$ and $(3, -4)$. It is

$$g = \frac{8(x-2)(x-3)}{(-1)(-2)} + \frac{5(x-1)(x-3)}{(1)(-1)} - \frac{4(x-1)(x-2)}{(2)(1)}.$$

Exercises 2.5

1. Compute the dimension of

$$\text{Sp}\left((5, 3, 0, 2, 1),\ (2, -1, -1, 3, 5),\ (3, 5, -3, 6, 7),\ (-2, 4, 1, -4, 3)\right).$$

2. Under what conditions on a will the set

$$\{(a, 1, 0),\ (1, a, 1),\ (0, 1, a)\}$$

be a basis for $\mathbb{R}^3$?

3. Let a, b and c be real numbers.

   (a) Under what condition on a and b will the set $\{(1, a),\ (1, b)\}$ be linearly independent over $\mathbb{R}$?

   (b) Under what condition on a, b and c will the set

   $$\{(1, a, a^2),\ (1, b, b^2),\ (1, c, c^2)\}$$

   be linearly independent over $\mathbb{R}$?

   (c) Guess and prove a generalization of (a) and (b) to $\mathbb{R}^n$.

4. Can you repeat Problem (3) for $\mathbb{C}^2$, $\mathbb{C}^3$ and $\mathbb{C}^n$ over $\mathbb{C}$?

5. Show that the dimension of the solution space of m homogeneous linear equations in n variables is at least $n - m$.

6. Let V be a finite dimensional vector space and let W be a subspace. Show that V contains a subspace U such that $U + W = V$ and $U \cap W = \{O\}$.

2.6   Further examples

In this section we will take a look at more examples of interesting vector spaces: the function space, the direct sum and the quotient space. In this section, we will also work on scalars other than real numbers.

Function Spaces.   After the discussion of the Lagrange Interpolation formula, I think it is appropriate now to introduce the function space.

Definition 2.6.1. Let S be an arbitrary set and let F be a field. We use $\mathscr{F}(S, F)$ to denote *all* functions from S to F.

Example 2.6.2. Let $S = \{\spadesuit, \heartsuit, \clubsuit\}$ There are exactly 8 elements in $\mathscr{F}(S, \mathbb{Z}_2)$. For example,

$$
\begin{array}{ccc}
f_1: & S & \longrightarrow & \mathbb{Z}_2 \\
& \spadesuit & \longmapsto & 0 \\
& \heartsuit & \longmapsto & 1 \\
& \clubsuit & \longmapsto & 1
\end{array}
\quad \text{and} \quad
\begin{array}{ccc}
f_1: & S & \longrightarrow & \mathbb{Z}_2 \\
& \spadesuit & \longmapsto & 1 \\
& \heartsuit & \longmapsto & 0 \\
& \clubsuit & \longmapsto & 1
\end{array}
$$

are two of them. For the value of $\spadesuit$, there are two choices 0 or 1. The same is true for the value of $\heartsuit$ and of $\clubsuit$. Hence, there are altogether $2 \cdot 2 \cdot = 8$ such functions.

Remark. In general we have the formula

$$|\mathscr{F}(S, F)| = |F|^{|S|}$$

if both S and F are finite sets.

The function that sends s to $0 \in F$ for all $s \in S$ is called the **zero function**. We now want to define vector addition and scalar multiplication on $\mathscr{F}(S, F)$ to make it into a vector space. Take any two functions f and g from $\mathscr{F}(S, F)$. For vector addition we define $f + g \colon S \to F$ by letting

$$(f + g)(s) \stackrel{\text{def}}{=} f(s) + g(s)$$

for all $s \in S$. For any $a \in F$ we define the scalar multiplication $af \colon S \to F$ by letting

$$(af)(s) \stackrel{\text{def}}{=} af(s)$$

for all $s \in S$. It is now easy to check that $\mathscr{F}(S, F)$ is a vector space over F with the zero function as the zero vector space. (See Exercise 1.) This type of vector spaces and their subspaces are called **function spaces**. In fact, we have encountered function spaces before. For example, $\mathscr{C}(\mathbb{R}, \mathbb{R})$ is an $\mathbb{R}$-subspace of $\mathscr{F}(\mathbb{R}, \mathbb{R})$, and hence both are function spaces. $\mathscr{P}$ and $\mathscr{P}_n$ may also be considered as function spaces.

Let's make the problem a little bit simpler by considering the case when S is finite. Now we construct a "standard" basis for $\mathscr{F}(S, F)$ when S is finite. For $s \in S$, we define $\chi_s \colon S \to F$ so that

$$\chi_s(t) = \begin{cases} 1, & \text{if } t = s, \\ 0, & \text{otherwise.} \end{cases}$$

Let $s_1, \ldots, s_n \in S$. For each i assign a value a_i in F. Let

$$f = \sum_{i=1}^{n} a_i \chi_i.$$

It is easy to see that $f(s_i) = a_i$ for each i.

Lemma 2.6.3. *The set (finite or infinite)*

$$\{\chi_s : s \in S\}$$

is linearly independent over F.

Proof. Let $s_1, \ldots, s_n$ be distinct elements in S and $a_1, \ldots, a_n$ be scalars such that $f = \sum_{i=1}^{n} a_i \chi_{s_i}$ is the zero function. Then $a_i = f(s_i) = a_i$ for all i. $\square$

Proposition 2.6.4. *When* $S = \{s_1, s_2, \ldots, s_n\}$, *the set*

$$\mathcal{B} = \{\chi_{s_1}, \chi_{s_2}, \ldots, \chi_{s_n}\}$$

forms a basis for $\mathscr{F}(S, F)$ *over* F. *Hence we conclude that*

$$\dim_F \mathscr{F}(S, F) = |S|.$$

Proof. By Virtue of Lemma 2, it remains to show that $\mathcal{B}$ spans $\mathscr{F}(S, F)$. It is easy to see that

$$f = \sum_{i=1}^{n} f(s_i)\chi_{s_i}$$

for all $f \in \mathscr{F}(S, F)$. $\qquad\qquad\qquad\qquad\qquad\qquad\qquad\qquad\square$

The (external) direct sum. We may "add" two vector spaces together to form a new vector space. Let V and W be two F-vector spaces. We define a new F-vector space called the **(external) direct sum of V and W**, denoted $V \oplus W$. Set-wise, let

$$V \oplus W = \{(v, w) : v \in V, w \in W\}.$$

Take any two elements (v, w) and $(v', w') \in V \oplus W$. We define the vector addition by letting

$$(v, w) + (v', w') = (v + v', w + w').$$

Let $a \in F$. We define the scalar multiplication by letting

$$a(v, w) = (av, aw).$$

It is easy to check that $V \oplus W$ is an F-vector space. (See Exercise 2.)

Example 2.6.5. The direct sum $\mathbb{R} \oplus \mathbb{R}$ is $\mathbb{R}^2$. The direct sum of n copies of $\mathbb{R}$ is $\mathbb{R}^n$.

Proposition 2.6.6. *Let V and W be two finite dimensional F-vector spaces. Let $\{v_1, \ldots, v_n\}$ and $\{w_1, \ldots, w_m\}$ be F-bases of V and W respectively. Then*

$$\mathcal{B} = \{(v_1, O), (v_2, O), \ldots, (v_n, O), (O, w_1), (O, w_2), \ldots, (O, w_m)\}$$

is an F-basis for $V \oplus W$. Furthermore, we have that

$$\dim_F V \oplus W = \dim_F V + \dim_F W.$$

Proof. Linear independence: Let

$$\sum_{i=1}^{n} a_i(v_i, O) + \sum_{j=1}^{m} b_j(O, w_j) = (O, O)$$

where a_i, b_j are scalars. The left-hand side of the sum is

$$\left(\sum_{i=1}^{n} a_i v_i, \; \sum_{j=1}^{m} b_j w_j \right).$$

Hence $\sum_{i=1}^{n} a_i v_i = O$ and $\sum_{j=1}^{m} b_j w_j = O$. It implies that $a_i = b_j = 0$ for all i and j.

The set $\mathcal{B}$ spans $V \oplus W$: Let $(v, w) \in V \oplus W$. Find scalars a_i and b_j so that $v = \sum_{i=1}^{n} a_i v_i$ and $w = \sum_{j=1}^{m} b_j w_j$. Then

$$(v, w) = \left(\sum_{i=1}^{n} a_i v_i, \; \sum_{j=1}^{m} b_j w_j \right).$$

We have now shown that $\mathcal{B}$ is a base for V over F. It also implies that $\dim V \oplus W = n + m$. □

If one understands V and W, then it is really not difficult to understand $V \oplus W$.

The internal direct sum. There is a different type of direct sum, called the **internal direct sum** opposed to the *external* direct sum we described above.

Definition 2.6.7. Let W_1 and W_2 be subspaces of V. An F-vector space V is called the **(internal) direct sum** of it subspaces W_1 and W_2 if

$$V = W_1 + W_2 \qquad \text{and} \qquad W_1 \cap W_2 = \phi.$$

In this case we also write $W_1 \oplus W_2$.

Remark. The external direct sum "add" together two (probably unrelated) vector spaces while the internal direct sum "add" together two subspaces in the same vector space. Let W_1 and W_2 be subspaces of V such that V is the *internal* direct sum of $W_1 \oplus W_2$. The *external* direct sum of W_1 and W_2 uses the same notation $W_1 \oplus W_2$. This causes confusion. However, later the reader will be asked to check in an exercise that the two types of direct sums are basically the same in this case.

It is easy to see that $\mathbb{R}^2 = \mathrm{Sp}\,((1,0)) \oplus \mathrm{Sp}\,((0,1)) = \mathrm{Sp}\,((1,0)) \oplus \mathrm{Sp}\,((1,1))$. Usually there are infinitely many different ways to split a vector space into a direct sum of subspaces (when its dimension is greater than one and the scalar field is infinite).

Proposition 2.6.8. *Let $V = W_1 \oplus W_2$ for the two subspaces W_1 and W_2 of V. For any $v \in V$ there is a unique expression*

$$v = w_1 + w_2, \qquad \text{for } w_1 \in W_1 \text{ and } w_2 \in W_2.$$

Proof. Since $V = W_1 + W_2$, we can find $w_1 \in W_1$ and $w_2 \in W_2$ such that $v = w_1 + w_2$ for any $v \in V$.

Now suppose $v = w_1 + w_2 = w_1' + w_2'$ for $w_1, w_1' \in W_1$ and $w_2, w_2' \in W_2$. Note that

$$w_1 - w_1' = w_2' - w_2 \in W_1 \cap W_2 = \{O\}.$$

Hence $w_1 = w_1'$ and $w_2 = w_2'$.   □

From the dimension formula (Proposition 2.5.10) we have the following result on $\dim V \oplus W$. (*Cf.* Proposition 2.6.6.)

Proposition 2.6.9. *Let W_1 and W_2 be the subspaces of an F-vector space V. If $V = W_1 \oplus W_2$ then*

$$\dim V = \dim W_1 + \dim W_2.$$

Example 2.6.10. Let $W_1 = \mathrm{Sp}\,((1,0,0),\ (1,1,0))$ and $W_2 = \mathrm{Sp}\,((1,1,1))$.

(a) Show that $\mathbb{R}^3 = W_1 \oplus W_2$.

(b) Find $w_1 \in W_1$ and $v_2 \in W_2$ such that $(1,2,3) = w_1 + v_2$.

Solution. (a) Clearly $W_1 + W_2 = \mathrm{Sp}\,((1,0,0),\ (1,1,0),\ (1,1,1))$. It is easy to verify that $\{(1,0,0),\ (1,1,0),\ (1,1,1)\}$ is linearly independent, it is a basis for $\mathbb{R}^3$. Form Proposition 2.5.10, we know that $\dim(V \cap W) = 0$. Thus $\mathbb{R}^3 = W_1 \oplus W_2$.

(b) First, we find that

$$(1,2,3) = -(1,0,0) - (1,1,0) + 3(1,1,1).$$

We choose $w_1 = -(1,0,0) - (1,1,0)$ and $w_2 = 3(1,1,1)$.   ◇

The Quotient Space. In contrast to the direct sum, there is a way to "subtract" a subspace from a vector space.

Definition 2.6.11. Let V be an F-vector space and let W be a subspace of V. For any $v \in V$ the set

$$v + W = \{v + w : w \in W\}$$

is called the **coset** of v modulo W.

Remark. The coset $v+W$ is a set containing v. Remember that the solution set of a system of linear equations is a coset of a special solution modulo the solution set of its corresponding homogeneous system (see Proposition 1.4.6).

Lemma 2.6.12. *Let W be a subspace of an F-vector space V and let v, $v' \in V$. Then either $(v + W) \cap (v' + W) = \varnothing$ or $v + W = v' + W$.*

Proof. If $(v + W) \cap (v' + W) \neq \varnothing$, find a vector u in the intersection. We may find w, $w' \in W$ such that $u = v + w$ and $u = v' + w'$.

We now show that $v + W \subseteq v' + w$. For any vector $w'' \in W$, we have that

$$v + w'' = u - w + w'' = v' + w' - w + w'' \in v' + W.$$

By symmetry, we also have $v' + W \subseteq v + W$. To conclude, $v + W = v' + W$. $\qquad\square$

Lemma 2.6.13. *Let W be a subspace of an F-vector space V. The following are true for v, $v' \in V$:*

(a) *The coset $v + W = v' + W$ if and only if $v - v' \in W$.*

(b) *If $v' \in v + W$, then $v + W = v' + W$.*

(c) *The coset $v + W = W$ if and only if $v \in W$.*

Proof. (a) The "only if" part: The element $v \in v + W = v' + W$. So $v = v' + w$ for some $w \in W$. Thus $v - v' = w \in W$.

The "if" part: The vector $v = v' + (v - v') \in v' + W$. Since v is also in $v + W$, $v + W = v' + W$ by Lemma 2.6.12.

(b) If $v' \in v + W$, then $v' = v + w$ for some $w \in W$. This implies that $v - v' \in W$. The result follows form (a).

Part (c) also follows form (b). □

Remark. The element v is called a **representative** of the coset $v + W$. Lemma 2.6.13 says that any element in a coset can be used as a representative.

Set-wise, we define the **quotient space of V modulo W**, denoted V/W, to be

$$\{v + W : v \in V\}$$

The coset $v + W$ is now viewed as an element. For the sake of simplicity, $v + W$ is also often written as $\bar{v}$. Take any two elements $v_1 + W$ and $v_2 + W$ in V/W we define the vector addition:

$$(v_1 + W) + (v_2 + W) = (v_1 + v_2) + W,$$

and the scalar multiplication for $a \in F$:

$$a(v + W) = av + W.$$

We leave to the Reader to check that the vector addition and scalar multiplication are well defined (not depending on the representatives we choose) and that V/W is an F-vector space (see Exercises 7).

Proposition 2.6.14. *Let W be a subspace of an F-vector space V. Then*

$$\dim V/W = \dim V - \dim W.$$

We leave this as an exercise. (See Exercise 8.)

Example 2.6.15. Let $V = \mathbb{R}^2$ and let $W = \text{Sp}((1,1))$. For any vector $(a,b) \in \mathbb{R}^2$, the coset $(a,b) + W = (a - b, 0) + W$. We may choose the representative so that the second coordinate is 0. Thus

$$V/W = \{(a,0) + W : a \in \mathbb{R}\}.$$

By Lemma 2.6.13, $(a,0) + W$ represents a distinct element in V/W for each $a \in F$. Note that W is the line passing through the origin and $(1,1)$ on $\mathbb{R}^2$ and the cosets of W are the parallel lines to W. Now in the quotient space each line is considered as a vector and the line though the origin is the zero vector.

Exercises 2.6

1. Verify that $\mathscr{F}(S, F)$ is an F-vector space.

2. Let V and W be two vector spaces. Verify that $V \oplus W$ is also a vector space.

3. Show that $\{\chi_s : s \in \mathbb{Z}\}$ is not a generating set for $\mathscr{F}(\mathbb{Z}, \mathbb{R})$.

4. Show that $\mathbb{R}^2 = \mathrm{Sp}\,((1, 2)) \oplus \mathrm{Sp}\,((2, 3))$.

5. Let $\{v_1, \ldots, v_n\}$ be a basis for V over $\mathbb{R}$. Show that $V = \mathrm{Sp}\,(v_1, \ldots, v_i) \oplus \mathrm{Sp}\,(v_{i+1}, \ldots, v_n)$.

6. Let W_1 and W_2 be two subspaces of V. Let $\{v_1, \ldots, v_k\}$ be a basis for W_1 and $\{w_1, \ldots, w_m\}$ be a basis for W_2. Show that $V = W_1 \oplus W_2$ if and only if $\{v_1, \ldots, v_k, w_1, \ldots, w_m\}$ is a basis for V.

7. Let W be a subspace of a vector space V over F. Check that the vector addition and scalar multiplication defined for V/W are well defined. To be precise, if $v_1 + W = v_1' + W$ and $v_2 + W = v_2' + W$, show that

$$(v_1 + W) + (v_2 + W) = (v_1' + W) + (v_2' + W)$$

and

$$a(v_1 + W) = a(v_1' + W)$$

for all $a \in F$.

Show that V/W is an F-vector space.

8. Prove Proposition 2.6.14.

9. Find a basis for $\mathbb{R}^3 / \mathrm{Sp}\,((1, 1, 0))$ over $\mathbb{R}$.

Review Exercises for Chapter 2

1. We say $A = \left(a_{ij}\right)$ is an **upper triangular** matrix if $a_{ij} = 0$ for $i > j$. Similarly, we say A is **lower triangular** if $a_{ij} = 0$ for $i < j$.

   Let A be a square upper triangular matrix in $M_n(\mathbb{R})$ with no nonzero diagonal entries. Show that the columns of A are linearly independent over $\mathbb{R}$.

2. Let S be a set of nonzero polynomials in $\mathscr{P}$ such that no two polynomials in S have the same degree. Show that S is a linearly independent set in $\mathscr{P}$.

3. Prove Theorem 2.4.4 for infinite subsets. To be precise, let $\mathscr{B}$ be a subset of a vector space V. Show that $\mathscr{B}$ is a basis of V if and only if for any element $v \in V$, there exist a unique choice of distinct elements $u_1, \ldots, u_n \in \mathscr{B}$ and a unique choice of *nonzero* scalars $a_1, \ldots, a_n$ such that
$$v = a_1 u_1 + a_2 u_2 + \cdots a_n u_n.$$

4. Let V be an n-dimensional vector space over $\mathbb{C}$. Find $\dim_{\mathbb{R}} V$.

5. Let V be the real vector space of infinite sequences of real numbers. For $i \in \mathbb{Z}_+$, let $s_i = (a_{in})$ be the infinite sequence with
$$a_{in} = \begin{cases} 1, & \text{if } n = i; \\ 0, & \text{if } n \neq i. \end{cases}$$

   Show that $\{s_i : i \in \mathbb{Z}_+\}$ is linearly independent over $\mathbb{R}$.

6. For each $k \in \mathbb{Z}$, let
$$f_k = \begin{cases} 0, & \text{if } x \leq k; \\ x - k, & \text{if } l \leq x \leq k + 1; \\ 1, & \text{if } k + 1 \leq x. \end{cases}$$

   Show that $\{f_k : k \in \mathbb{Z}\}$ is a linearly independent subset of $\mathscr{C}(\mathbb{R}, \mathbb{R})$ over $\mathbb{R}$.

7. Let V be the space of all functions on $\mathbb{R}$.

   (a) Find the dimension of the subspace generated by $\{\cos(x + k) : k = 0, 1, 2, 3, \dots\}$.

   (b) Find the dimension of the subspace generated by $\{\sin^k x : k = 0, 1, 2, 3, \dots\}$.

CHAPTER 3

Linear Transformations and Matrices

A linear transformation is a function between two vector spaces. But it is more than a function. It has to respect the structure of vector spaces.

Throughout the whole chapter V and W denote real vector spaces unless otherwise noted.

3.1   Linear transformations

In this section, we describe what a linear transformation is and investigate some of the basic properties.

Definition and examples.   To be precise, a linear transformation between two vector spaces is a function preserving the vector addition and the scalar multiplication.

Definition 3.1.1. We call a function (map, or mapping) $T \colon V \to W$ a **$\mathbb{R}$-linear transformation** (or a **linear transformation over** $\mathbb{R}$) from V to W if for all v, $v' \in V$ and $a \in \mathbb{R}$ we have

(i) $T(v + v') = T(v) + T(v')$, and

(ii) $T(av) = aT(v)$.

Proposition 3.1.2. *Let* $T \colon V \to W$ *be a linear transformation. Then*

$$T(O_V) = O_W \qquad and \qquad T(-v) = -T(v)$$

for all $v \in V$.

Proof. By definition of linear transformations, we have

$$T(O_V) = T(O_V + O_V) = T(O_V) + T(O_V)$$
$$\implies T(O_V) = O_W, \qquad \text{by Cancellation Law.}$$

For $v \in V$,

$$T(v) + T(-v) = T(v + (-v)) = T(O_V) = O_W.$$

So $T(-v) = -T(v)$. $\qquad\qquad\qquad\qquad\qquad\qquad\qquad\qquad$ □

The following is another way to check whether a function is a linear transformation.

Proposition 3.1.3. *The function* $T \colon V \to W$ *is a linear transformation if and only if*

$$T(av + a'v') = aT(v) + a'T(v')$$

for all v, $v' \in V$ *and* a, $a' \in \mathbb{R}$.

Proof. The "only if" part: Since T is a linear transformation,

$$T(av + a'v') = T(av) + T(a'v') = aT(v) + a'T(v').$$

The "if" part: For $a \in \mathbb{R}$ and $v, v' \in V$,

$$T(v + v') = T(1v + 1v') = 1T(v) + 1T(v') = T(v) + T(v')$$
$$T(av) = T(av + 1O_V) = aT(v) + 1T(O_V)$$
$$= aT(v) + O_W = aT(v).$$

This completes the proof.   □

The following are two types of useful linear transformations. Let V and W be vector spaces.

- The **identity transformation**: $\mathbf{1}_V : V \to V$ is the linear transformation defined by $\mathbf{1}_V(v) = v$ for all $v \in V$.

- The **zero transformation**: $\mathbf{0}_V : V \to W$ is the linear transformation sending v to O_W for all $v \in V$.

Example 3.1.4. For a fixed angle α, define $T_\alpha : \mathbb{R}^2 \to \mathbb{R}^2$ by letting

$$T_\alpha(x, y) = (x \cos \alpha - y \sin \alpha, \ x \sin \alpha + y \cos \alpha).$$

This is the mapping sending the vector (x, y) to the vector obtained by rotating (x, y) counterclockwise by the angle α.

For all $x, y \in \mathbb{R}$,

$$T_\alpha((x, y) + (x', y')) = T_\alpha(x + x', y + y'))$$
$$= ((x + x') \cos \alpha - (y + y') \sin \alpha, \ (x + x') \sin \alpha + (y + y') \cos \alpha)$$
$$= (x \cos \alpha - y \sin \alpha, \ x \sin \alpha + y \cos \alpha)$$
$$\qquad + (x' \cos \alpha - y' \sin \alpha, \ x \sin \alpha + y \cos \alpha)$$
$$= T_\alpha(x, y) + T_\alpha(x', y')$$
$$T_\alpha(a(x, y)) = T_\alpha(ax, ay)$$
$$= (ax \cos \alpha - ay \sin \alpha, \ ax \sin \alpha + ay \cos \alpha)$$
$$= a(x \cos \alpha - y \sin \alpha, \ x \sin \alpha + y \cos \alpha) = aT_\alpha(x, y).$$

Thus T_α is a linear transformation. We call this linear transformation the **rotation** by α.

Rotation preserves distance and angles and it does not distort shape. It belongs to one of the most elementary type of transformations: the **rigid transformations**.

Example 3.1.5. The linear transformation $T_1 \colon \mathbb{R}^2 \to \mathbb{R}^2$ sending $(x,\ y)$ to $(x,\ -y)$ is called the **reflection** about the x-axis. The linear transformation $T_2 \colon \mathbb{R}^2 \to \mathbb{R}^2$ sending $(x,\ y)$ to $(x,\ 0)$ is called the **projection** on the x-axis. The linear transformation $T_3 \colon \mathbb{R}^3 \to \mathbb{R}^3$ sending (x,y,z) to $(x,y,0)$ is called the **projection** onto the xy-plane.

Example 3.1.6. Let W_1 and W_2 be two subspaces of V such that $V = W_1 \oplus W_2$. Let $T \colon V \to W_1$ be the function defined so that $T(v) = w_1$ if $v = w_1 + w_2$ with $w_i \in W_i$, $i = 1, 2$. Then T is a linear transformation, and it is called the **projection** of V on W_1 along W_2.

Example 3.1.7. Define $T \colon M_{m \times n}(\mathbb{R}) \to M_{n \times m}(\mathbb{R})$ by sending A to A^{t} for all $A \in M_{m \times n}(\mathbb{R})$. This is a linear transformation since

$$(A + B)^{\mathrm{t}} = A^{\mathrm{t}} + B^{\mathrm{t}}$$
$$(aA)^{\mathrm{t}} = aA^{\mathrm{t}}$$

for all $A \in M_{m \times n}(\mathbb{R})$ and for all $a \in \mathbb{R}$.

Example 3.1.8. Define $T \colon \mathscr{P}_n \to \mathscr{P}_{n-1}$ by sending $f(x)$ to $f'(x)$. This is a linear transformation since

$$(f + g)' = f' + g'$$
$$(af)' = af'$$

for all $f,\ g \in \mathscr{P}$ and $a \in \mathbb{R}$.

Example 3.1.9. Suppose $T \colon \mathbb{R}^2 \to \mathbb{R}^2$ is a linear transformation such that $T(1, 0) = (1, 4)$ and $T(1, 1) = (2, 5)$. What is $T(2, 3)$?

Solution. Since $(2, 3) = -(1, 0) + 3(1, 1)$, we have

$$T(2, 3)$$
$$= T(-(1, 0) + 3(0, 1))$$
$$= -T(1, 0) + 3T(0, 1)$$
$$= -(1, 4) + 3(2, 5) = (5, 11)$$

by definition of linear transformations.   ◇

Proposition 3.1.10. *Let* U, V *and* W *be vector spaces over* $\mathbb{R}$. *Let* $S: U \to V$ *and* $T: V \to W$ *be linear transformations. Then* $T \circ S$ *is a linear transformation from* U *to* W.

Proof. For all scalars a, b and all u, $u' \in U$, we have

$$T \circ S(au + bu') = T(S(au + bu'))$$
$$= T(aS(u) + bS(u')), \qquad S \text{ being a linear transformation,}$$
$$= aT(S(u)) + bT(S(u')), \qquad T \text{ being a linear transformation,}$$
$$= aT \circ S(u) + bT \circ S(u').$$

Thus $T \circ S$ is a linear transformation.   □

The universal property of linear transformations. The following theorem gives the most fundamental property of linear transformations. It basically says that linear transformations can be constructed rather *freely*.

Theorem 3.1.11. *Let* $\{u_1, u_2, \ldots, u_n\}$ *be a basis for* V, *and let* w_1, w_2, $\ldots$, w_n *be arbitrary elements (which may even be repeated) in* W. *Then there exists one and only one linear transformation* $T: V \to W$ *such that* $T(u_i) = w_i$ *for* $i = 1, 2, \ldots, n$. *In this case,*

$$(3.1) \qquad T(a_1u_1 + a_2u_2 + \cdots + a_nu_n) = a_1w_1 + a_2w_2 + \cdots + a_nw_n$$

where a_1, a_2, $\ldots$, $a_n \in \mathbb{R}$.

Proof. By Theorem 2.4.4, the map T defined as in (3.1) is well-defined. It is easy to verify that T is indeed a linear transformation (see Exercise 1). The linear transformation T sends u_i to w_i as required. If S is another linear transformation sending u_i to w_i for each i, then

$$S\left(\sum_{i=1}^{n} a_iu_i\right) = \sum_{i=1}^{n} a_iS(u_i) = \sum_{i=1}^{n} a_iT(u_i) = T\left(\sum_{i=1}^{n} a_iu_i\right).$$

This shows that $S = T$. Hence T is the unique linear transformation satisfying the requirement.   □

Example 3.1.12. (1) Is there a linear transformation $T\colon \mathbb{R}^3 \to \mathbb{R}^2$ sending $(1,0,2)$ to $(1,1)$ and $(-2,0,-4)$ to $(1,2)$?

(2) Is there a linear transformation sending $(1,0,2)$ to $(1,1)$ and $(0,-2,-4)$ to $(1,2)$?

Solution. (1) Note that $(-2,0,-4) = -2(1,0,2)$. By requirement we must have $T(-2,0,-4) = (1,2)$. By definition of linear transformation, $T(-2,0,-4) = -2T(1,0,2) = (2,2)$. There is a contradiction.

(2) The set $\{(1,0,2),\ (0,-2,-4)\}$ is linearly independent. It can be expanded to a basis of $\mathbb{R}^3$. Hence it is possible to find a linear transformation satisfying the requirement by Theorem 3.1.11.   ◇

In fact, it does not matter if you describe a linear transformation in its entirety or if you just give the values for a basis.

Example 3.1.13. Remember that $\{1, x, x^2, \ldots, x^n\}$ is a basis for $\mathscr{P}_n$. By Theorem 3.1.11 there is a linear transformation $T\colon \mathscr{P}_n \to \mathscr{P}_{n-1}$ sending 1 to 0 and x^i to ix^{i-1} for $i = 1, 2, \ldots, n$. This is the same linear transformation as in Example 3.1.8.

There is a linear transformation S from $\mathbb{R}^{n+1}$ to $\mathscr{P}_n$ sending e_i to x^{i-1} for $i = 1, \ldots, n+1$. Clearly,

$$S(a_1, a_2, \ldots, a_n, a_{n+1})$$
$$= S(a_1 e_1 + a_2 e_2 + \cdots + a_{n+1} e_{n+1})$$
$$= a_1 + a_2 x + \cdots + a_n x^{n-1} + a_{n+1} x^n.$$

Conversely, there is also a linear transformation T from $\mathscr{P}_n$ to $\mathbb{R}^{n+1}$ sending x^i to e_{i+1} for $i = 0, 2, \ldots, n$. This map is given such that

$$T(a_1 + a_2 x + \cdots + a_n x^{n-1} + a_{n+1} x^n)$$
$$= a_1 e_1 + a_2 e_2 + \cdots + a_{n+1} e_{n+1}$$
$$= (a_1, a_2, \ldots, a_n, a_{n+1}).$$

It is quite obvious these two linear transformations give a one-to-one correspondence between $\mathbb{R}^{n+1}$ and $\mathscr{P}_n$:

$$(a_1, a_2, \ldots, a_n, a_{n+1}) \quad \rightsquigarrow \quad a_1 + a_2 x + \cdots + a_n x^{n-1} + a_{n+1} x^n.$$

Moreover, the one-to-one correspondence is compatible with the vector addition and scalar multiplication in both vector spaces:

$$(a_1, a_2, \ldots, a_n, a_{n+1}) + (b_1, b_2, \ldots, b_n, b_{n+1})$$
$$= (a_1 + b_1, a_2 + b_2, \ldots, a_n + b_n, a_{n+1} + b_{n+1})$$
$$\updownarrow$$
$$a_1 + a_2 x + \cdots + a_n x^{n-1} + a_{n+1} x^n$$
$$+ b_1 + b_2 x + \cdots + b_n x^{n-1} + b_{n+1} x^n$$
$$= (a_1 + b_1) + (a_2 + b_2)x + \cdots$$
$$+ (a_n + b_n)x^{n-1} + (a_{n+1} + b_{n+1})x^n,$$

and for $r \in \mathbb{R}$,

$$r(a_1, a_2, \ldots, a_n, a_{n+1}) \qquad r(a_1 + a_2 x + \cdots + a_n x^{n-1} + a_{n+1} x^n)$$
$$= (ra_1, ra_2, \ldots, ra_n, ra_{n+1}) \qquad = ra_1 + ra_2 x + \cdots + ra_n x^{n-1} + ra_{n+1} x^n.$$

We might as well consider a polynomial of degree $n + 1$ as a point in $\mathbb{R}^{n+1}$ with the coefficients as the coordinates. The two vector spaces seem "fundamentally" the same! In the following sections, we will further explore this idea.

Exercises 3.1

1. Show that the map T defined as in (3.1) is a linear transformation.

2. Which of the following functions are linear transformations?

   (a) $T \colon \mathbb{R}^2 \to \mathbb{R}^2$ defined by $T(x, y) = (2x - y, x)$.

   (b) $T \colon \mathbb{R}^2 \to \mathbb{R}^3$ defined by $T(x, y) = (2x - y, x, x + y)$.

   (c) $T \colon \mathbb{R}^2 \to \mathbb{R}^2$ defined by $T(x, y) = (2x - y, x + y + 1)$.

   (d) $T \colon \mathbb{R}^2 \to \mathbb{R}^3$ defined by $T(x, y) = (2xy, x, x + y)$.

   (e) $\mathbb{R} \colon \mathbb{R}^2 \to$ defined by $T(x, y) = (1, -1)$.

3. Let $T: \mathscr{P} \to \mathscr{P}$ be the function given by

$$T(f(x)) = \int_0^x f(t)\, dt.$$

(a) Compute $T(1 + x - 2x^3)$ and $T(x^3 - x^5)$.

(b) Show that T is a linear transformation.

(c) Is the set $\{T(1),\ T(x),\ \ldots,\ T(x^n)\}$ linearly independent?

4. Let W_1 be the plane $z = 0$ in $\mathbb{R}^3$ and W_2 be the line $x = y = 0$ in $\mathbb{R}^3$.

(a) Show that $\mathbb{R}^3 = W_1 \oplus W_2$.

(b) Show that the projection T_3 in Example 3.1.5 is the projection of $\mathbb{R}^3$ on W_1 along W_2 in Example 3.1.6.

5. Let $V = \mathrm{Sp}\,((1, 2, 3),\ (3, 5, 7)) \subseteq \mathbb{R}^3$.

(a) Show that $\{(1, 2, 3),\ (3, 5, 7)\}$ form a basis for V.

(b) Show that $(4, 5, 6) \in V$.

(c) Let $T: V \to \mathbb{R}$ be the linear transformation defined by

$$T(1, 2, 3) = 3 \qquad \text{and} \qquad T(3, 5, 7) = -1.$$

Find $T(4, 5, 6)$.

6. Let $\mathscr{L}(V,\ W)$ be the set of all $\mathbb{R}$-linear transformations from V to W. Let T and T' be in $\mathscr{L}(V,\ W)$ and $a \in \mathbb{R}$. Define

$$
\begin{aligned}
T + T'\colon\quad V &\longrightarrow W \\
v &\longmapsto T(v) + T'(v)
\end{aligned}
$$

and

$$
\begin{aligned}
aT\colon\quad V &\longrightarrow W \\
v &\longmapsto aT(v)
\end{aligned}.
$$

(a) Show that $T + T'$ and aT are in $\mathscr{L}(V,\ W)$.

(b) The addition and scalar multiplication defined make $\mathscr{L}(V,\ W)$ into a vector space over $\mathbb{R}$.

(c) If $\dim V = m$ and $\dim W = n$, show that $\dim \mathscr{L}(V,\ W) = mn$.

3.2   Isomorphisms

In order to establish a more precise description of how we can regard two vector spaces to be "fundamentally" the same, we first digress to review some terminology of functions.

A digression to functions.   A function is also called a *map* or a *mapping*. Remember that the function $f\colon A \to B$ is **one-one** (or **injective**) if $f(a) = f(a')$ implies $a = a'$. On the other hand, the function f is **onto** (or **surjective**) if for any $b \in B$ there exists $a \in A$ such that $f(a) = b$. We say f is **bijective** or is a **one-to-one correspondence** if f is both one-one and onto.

Lemma 3.2.1. *Let A, B and C be sets and let $f\colon A \to B$ and $g\colon B \to C$ be functions. Then the following are true.*

(a) *If f and g are one-one then $g \circ f$ is one-one.*

(b) *If f and g are onto then $g \circ f$ is onto.*

(c) *If f and g are bijective then $g \circ f$ is bijective.*

(d) *If $g \circ f$ is one-one then f is one-one.*

(e) *If $g \circ f$ is onto then g is onto.*

We leave the proof of this lemma as an exercise (see Exercise 1).

Now suppose the function $f\colon A \to B$ is bijective. Since f is onto, for all $b \in B$ there is an $a \in A$ such that $f(a) = b$. Moreover, since f is one-one, a is the only element in A such that $f(a) = b$. We can now define a new function $g\colon B \to A$ such that $g(b) = a$. In fact, the relation between f and g can be perfectly given by

$$f(a) = b \quad \text{if and only if} \quad g(b) = a.$$

It can be furthered checked that

$$(3.2) \qquad\qquad g \circ f = 1_A \quad \text{and} \quad f \circ g = 1_B,$$

that is,

$$g(f(a)) = a \quad \text{and} \quad f(g(b)) = b \qquad \text{for all } a \in A \text{ and } b \in B.$$

Functions f and g satisfying (3.2) are called **invertible functions**. The function g is called the **inverse function** of f, and is usually denoted as f^{-1}. Similarly, f is the inverse function of g and we can write $f = g^{-1}$.

Isomorphisms. Back to our discussion on linear transformations, we make the following definition.

Definition 3.2.2. We say a linear transformation T is an **isomorphism** if T is one-one and onto.

An isomorphism between two vector spaces V and W gives a one-to-one correspondence between elements of V and elements of W. As we can see from the discussion at the end of §3.1, an isomorphism corresponds the sum of two vectors to the sum of the corresponding vectors. It also corresponds a scalar product to the same scalar product of the corresponding vector. It shows that we can treat the two vector spaces V and W as the same vector space except that the vectors are labeled differently. This is expressed most eloquently in Shakespeare's words:

> *What's in a name? That which we call a rose*
> *By any other name would smell as sweet.*

The same idea is implemented further by the following proposition.

Proposition 3.2.3. *Let U, V and W be vector spaces over $\mathbb{R}$.*

(a) *The identity transformation 1_V is an isomorphism from V to V.*

(b) *Let $T\colon V \to W$ be an isomorphism. The inverse function of T, denoted T^{-1}, is an isomorphism.*

(c) *Let $S\colon U \to V$ and $T\colon V \to W$ be isomorphisms. Then $T \circ S$ is an isomorphism from U to W.*

Proof. Part (a) is trivial.

(b) For $w, w' \in W$, there exist $v, v' \in V$ such that $T(v) = w$ and $T(v') = w'$. Note that $T(v + v') = T(v) + T(v') = w + w'$. Hence $T^{-1}(w + w') = v + v' = T^{-1}(w) + T^{-1}(w')$. For $a \in \mathbb{R}$, $T(av) = aT(v) = aw$. Thus $T^{-1}(aw) = av = aT^{-1}(w)$. To conclude, T^{-1} is an linear transformation. Since T^{-1} is bijective (see Exercise 3), T is an isomorphism.

Part (c) follows from Proposition 3.1.10 and Lemma 3.2.1(c).   □

Definition 3.2.4. We say V is **isomorphic to** W over $\mathbb{R}$, denoted $V \simeq_{\mathbb{R}} W$ or $V \simeq W$ if $\mathbb{R}$ is understood, if there is an isomorphism over $\mathbb{R}$ from V to W.

We provide an example to show that isomorphisms do preserve properties regarding vector spaces.

Proposition 3.2.5. *Let $T\colon V \to W$ be a linear transformation and let $v_1, \ldots, v_m$ be elements in V.*

(a) *Suppose $T(v_1), T(v_2), \ldots, T(v_m)$ are linearly independent. Then v_1, v_2, $\ldots$, v_m are also linearly independent.*

(b) *Let T be one-one. If $v_1, v_2, \ldots, v_m$ are linearly independent then $T(v_1), T(v_2), \ldots, T(v_m)$ are also linearly independent.*

(c) *Let T be onto. If $v_1, v_2, \ldots, v_m$ span V then $T(v_1), T(v_2), \ldots, T(v_m)$ span W.*

Proof. (a) Let $a_1, a_2, \ldots, a_n$ be scalars such that

$$a_1 v_1 + a_2 v_2 + \cdots + a_n v_n = O.$$

Then

$$O = T(a_1 v_1 + a_2 v_2 + \cdots + a_n v_n)$$
$$= a_1 T(v_1) + a_2 T(v_2) + \cdots + a_n T(v_n).$$

Since $T(v_1), T(v_2), \ldots, T(v_n)$ are assumed to be linearly independent, we have $a_1 = a_2 = \cdots = a_n = 0$.

(b) Let $a_1, a_2, \ldots, a_n$ be scalars such that

$$a_1 T(v_1) + a_2 T(v_2) + \cdots + a_n T(v_n) = O.$$

Since

$$T(a_1 v_1 + a_2 v_2 + \cdots + a_n v_n)$$
$$= a_1 T(v_1) + a_2 T(v_2) + \cdots + a_n T(v_n)$$
$$= O,$$

we have $a_1v_1 + a_2v_2 + \cdots + a_nv_n = O$ by the assumption that T is injective. Now $a_1 = a_2 = \cdots = a_n = 0$ by the assumption that $v_1, \ldots, v_n$ are linearly independent.

(c) Let $w \in W$. Since T is onto, we may find $v \in V$ such that $T(v) = w$. Let

$$v = a_1v_1 + a_2v_2 + \cdots + a_nv_n, \quad a_i \text{ scalars.}$$

Then

$$w = T(v) = T(a_1v_1 + a_2v_2 + \cdots + a_nv_n)$$
$$= a_1T(v_1) + a_2T(v_2) + \cdots + a_nT(v_n).$$

Hence $V = \text{Sp}\,(T(v_1), T(v_2), \ldots, T(v_n))$. $\qquad\qquad\square$

We see that an injective linear transformation maps a linearly independent set to a linearly independent set. A surjective linear transformation maps a generating set to a generating set.

Corollary 3.2.6. *An isomorphism maps a basis onto a basis.*

Corollary 3.2.7. *If $V \cong W$, then $\dim V = \dim W$.*

Exercises 3.2

1. Prove Lemma 3.2.1.

2. (a) Find an example of functions f and g such that $g \circ f$ is one-one while g is not.

   (b) Find an example of functions f and g such that $g \circ f$ is onto while f is not.

3. Suppose the function $f \colon A \to B$ has an inverse function. Show that f is bijective.

4. Determine which of the linear transformations in Examples 3.1.4–3.1.8 are isomorphisms?

5. Show that the linear transformation $T\colon \mathscr{P}_5 \to M_{2\times 3}(\mathbb{R})$ given by

$$1 \longmapsto e_{11}, \quad x \longmapsto e_{12}, \quad x^2 \longmapsto e_{13},$$
$$x^3 \longmapsto e_{21}, \quad x^4 \longmapsto e_{22}, \quad x^5 \longmapsto e_{23}$$

is an isomorphism.

6. Show that $\mathbb{R}^m \oplus \mathbb{R}^n \simeq \mathbb{R}^{m+n}$.

7. Let W_1 and W_2 be two subspaces of V such that $V = W_1 \oplus W_2$ (the internal direct sum). Let U be the external direct sum of W_1 and W_2. Show that

$$
\begin{array}{ccc}
T\colon & U & \longrightarrow & V \\
& (w_1, w_2) & \longmapsto & w_1 + w_2
\end{array}
$$

is an isomorphism. This problem tells us that there is no real distinction between the internal direct sum and the external direct sum.

3.3 Range and null space

In this section we discuss how to check whether a linear transformation is onto or one-one.

Range and null space. Let $T\colon V \to W$ be a linear transformation. First we describe two important subspaces of V and W.

Definition 3.3.1. Let $T\colon V \to W$ be a linear transformation. We define the **range** (or **image**) of T, denoted Range (T), to be the subset of W consisting of the images (under T) of all vectors in V; that is,

$$\text{Range}\,(T) = \{T(v) \in W : v \in V\}.$$

We define the **null space** (or **kernel**) of T, denoted Null (T), to be the set of all vectors in V such that $T(v) = O$; that is,

$$\text{Null}\,(T) = \{v \in V : T(v) = O\}.$$

Proposition 3.3.2. *Let $T\colon V \to W$ be a linear transformation. Then* Range (T) *is a subspace of W and* Null (T) *is a subspace of V.*

Proof. Let $w, w' \in \text{Range}(T)$ and a be a scalar. There exist v and v' such that $T(v) = w$ and $T(v') = w'$. Thus

$$w + w' = T(v) + T(v') = T(v + v') \in \text{Range}(T),$$
$$aw = aT(v) = T(av) \in \text{Range}(T).$$

The set $\text{Range}(T)$ is a subspace.

Let $v, v' \in \text{Null}(T)$ and a be a scalar. Then

$$T(v + v') = T(v) + T(v') = O + O = O,$$
$$T(av) = aT(v) = aO = O.$$

This shows that $v + v'$ and av are both in $\text{Null}(T)$. Thus $\text{Null}(T)$ is a subspace. $\qquad\square$

Next we give an easy criterion for checking whether a linear transformation is one-one.

Proposition 3.3.3. *Let T be a linear transformation. Then T is one-one if and only if* $\text{Null}(T) = \{O\}$.

Proof. The "only if" part: Since T is one-one, the only vector mapped to O_W is O_V. Hence $\text{Null}(T) = \{0\}$.

The "if" part: Suppose $T(v) = T(v')$ for $v, v' \in V$. Then $T(v - v') = T(v) - T(v') = O$. By assumption, $v - v' \in \text{Null}(T) = \{O\}$. This says $v = v'$. Thus T is one-one. $\qquad\square$

Clearly, a linear transformation $T : V \to W$ is onto if and only if $\text{Range}(T) = W$. The following proposition tells us how to compute $\text{Range}(T)$.

Proposition 3.3.4. *Let $T : V \to W$ be a linear transformation where $V = \text{Sp}(u_1, u_2, \ldots, u_n)$. Then*

$$\text{Range}(T) = \text{Sp}(T(u_1), T(u_2), \ldots, T(u_n)).$$

Proof. Since $T(u_i) \in \text{Range}(T)$ for all i, we have

$$\text{Sp}(T(u_1), T(u_2), \ldots, T(u_n)) \subseteq \text{Range}(T).$$

Conversely,

$$w \in \text{Range}\,(T)$$
$$\implies w = T(a_1 u_1 + a_2 u_2 + \cdots + a_n u_n) \quad \text{for some scalars } a_1, \ldots, a_n,$$
$$\implies w = a_1 T(u_1) + a_2 T(u_2) + \cdots + a_n T(u_n)$$
$$\in \text{Sp}\,(T(u_1), T(u_2), \ldots, T(u_n))\,.$$

Hence we also have $\text{Range}\,(T) \subseteq \text{Sp}\,(T(u_1), T(u_2), \ldots, T(u_n))$. $\qquad\square$

Example 3.3.5. Is the linear transformation in Example 3.1.9 an isomorphism?

Solution. Observe that both $\{(1,0),(1,1)\}$ and $\{(1,4),(2,5)\}$ are bases for $\mathbb{R}^2$. By Proposition 3.3.4, Range (T) contains a basis. Hence T is onto. On the other hand, suppose $T(a(1,0) + b(1,1)) = a(1,4) + b(2,5) = (0,0)$. Since $\{(1,4),(2,5)\}$ is linearly independent, we have $a = b = 0$. This says that Null $(T) = \{(0,0)\}$. By Proposition 3.3.8, T is one-one. We conclude that T is an isomorphism. $\qquad\diamond$

Example 3.3.6. Define the function $T\colon \mathscr{P}_2 \to M_2(\mathbb{R})$ by

$$T(f(x)) = \begin{pmatrix} f(1) - f(2) & 0 \\ 0 & f(0) \end{pmatrix}.$$

First, we check that T is a linear transformation:

$$T(f(x) + g(x)) = \begin{pmatrix} f(1) + g(1) - f(2) - g(2) & 0 \\ 0 & f(0) + g(0) \end{pmatrix}$$
$$= \begin{pmatrix} f(1) - f(2) & 0 \\ 0 & f(0) \end{pmatrix} + \begin{pmatrix} g(1) - g(2) & 0 \\ 0 & g(0) \end{pmatrix}$$
$$= T(f(x)) + T(g(x))$$

and

$$T(af(x)) = \begin{pmatrix} af(1) - af(2) & 0 \\ 0 & af(0) \end{pmatrix}$$
$$= a \begin{pmatrix} f(1) - f(2) & 0 \\ 0 & f(0) \end{pmatrix} = aT(f(x))$$

for any $f(x), g(x) \in \mathscr{P}_2$ and $a \in \mathbb{R}$.

The set $\{1, x, x^2\}$ is a basis for $\mathscr{P}_2$.

$$T(1) = \begin{pmatrix} 1-1 & 0 \\ 0 & 1 \end{pmatrix} = \begin{pmatrix} 0 & 0 \\ 0 & 1 \end{pmatrix},$$

$$T(x) = \begin{pmatrix} 1-2 & 0 \\ 0 & 0 \end{pmatrix} = \begin{pmatrix} -1 & 0 \\ 0 & 0 \end{pmatrix},$$

$$T(x^2) = \begin{pmatrix} 1-4 & 0 \\ 0 & 0 \end{pmatrix} = \begin{pmatrix} -3 & 0 \\ 0 & 0 \end{pmatrix},$$

by Proposition 3.3.4,

$$\text{Range}\,(T) = \text{Sp}\left(\begin{pmatrix} 0 & 0 \\ 0 & 1 \end{pmatrix}, \begin{pmatrix} -1 & 0 \\ 0 & 0 \end{pmatrix}, \begin{pmatrix} -3 & 0 \\ 0 & 0 \end{pmatrix} \right)$$

$$= \text{Sp}\left(\begin{pmatrix} 0 & 0 \\ 0 & 1 \end{pmatrix}, \begin{pmatrix} 1 & 0 \\ 0 & 0 \end{pmatrix} \right)$$

$$= \left\{ \begin{pmatrix} a & 0 \\ 0 & b \end{pmatrix} : a, b \in \mathbb{R} \right\}.$$

Next, let

$$T(a + bx + cx^2) = \begin{pmatrix} -b - 3c & 0 \\ 0 & a \end{pmatrix} = \begin{pmatrix} 0 & 0 \\ 0 & 0 \end{pmatrix}.$$

This implies that $a = 0$ and $b = -3c$. Hence

$$\text{Null}\,(T) = \{-3cx + cx^2 : c \in \mathbb{R}\} = \text{Sp}\left(x^2 - 3x\right).$$

We would like to remark that $\dim \text{Range}\,(T) = 2$ and $\dim \text{Null}\,(T) = 1$.

Rank and nullity.   As we can see from the previous examples, the sizes of the range and of the null space of a linear transformation often play important roles. Hence we have the following definition.

Definition 3.3.7. Let T be a linear transformation. Define the **rank** of T, denoted $\text{rk}\,T$ to be the dimension of Range (T). Define the **nullity** of T, denoted nullity T, to be the dimension of Null (T).

If $T\colon V \to W$ is a linear transformation, it is clear that T is one-one if and only if nullity $T = 0$, and T is onto if and only if $\operatorname{rk} T = \dim W$.

The following is the second important dimension formula that we encounter.

Theorem 3.3.8. *Let V and W be finite dimensional vector spaces and let $T\colon V \to W$ be a linear transformation. Then*

$$\text{nullity } T + \operatorname{rk} T = \dim V.$$

Proof. Let $\{w_1, \ldots, w_m\}$ be a basis for Range (T). For each i, find v_i such that $T(v_i) = w_i$. Also find a basis $\{u_1, \ldots, u_n\}$ for the null space of T. We claim that

$$\mathscr{B} = \{v_1, \ldots, v_m, u_1, \ldots, u_n\}$$

is a basis for V. Assuming this claim,

$$\dim V = m + n = \dim \operatorname{Range}(T) + \dim \operatorname{Null}(T)$$

and we are done.

Linear independence: Let

$$a_1 v_1 + \cdots + a_m v_m + b_1 u_1 + \cdots + b_n u_n = O$$

where a_i and b_i are scalars. Then

$$
\begin{aligned}
O &= T(a_1 v_1 + \cdots + a_m v_m + b_1 u_1 + \cdots + b_n u_n) \\
&= a_1 T(v_1) + \cdots + a_m T(v_m) + b_1 T(u_1) + \cdots + b_n T(u_n) \\
&= a_1 T(v_1) + \cdots + a_m T(v_m).
\end{aligned}
$$

Since $\{T(v_1), \ldots, T(v_m)\}$ is linearly independent, we have $a_i = 0$ for all i. It follows that

$$b_1 u_1 + \cdots + b_n u_n = O.$$

Since $\{u_1, \ldots, u_n\}$ is linearly independent, we also have $b_1 = \cdots = b_n = 0$.

The set $\mathscr{B}$ spans V: Let $v \in V$. Suppose

$$T(v) = a_1 w_1 + a_2 w_2 + \cdots + a_m w_m.$$

Then $v - (a_1v_1 + \cdots + a_mv_m) \in \mathrm{Null}\,(T)$. We may find scalars $b_1, \ldots, b_n$ such that

$$v - (a_1v_1 + \cdots + a_mv_m) = b_1u_1 + \cdots + b_nu_n.$$

Thus $v \in \mathrm{Sp}\,(v_1, \ldots, v_m, u_1, \ldots, u_n)$.   □

Both the rank and the nullity of a linear transformation are easy to compute. However, this theorem makes it sufficient to find just one of them.

Corollary 3.3.9. *Let V and W be finite dimensional vector spaces and let $T\colon V \to W$ be a linear transformation. Then T is one-one if and only if $\mathrm{rk}\,T = \dim V$.*

Corollary 3.3.10. *Let $T\colon V \to W$ be a linear transformation and let $\dim V = \dim W < \infty$. The following conditions are equivalent:*

(i) $\mathrm{rk}\,T = \dim V$;

(ii) *T is one-one;*

(iii) *T is onto;*

(iv) *T is an isomorphism.*

Example 3.3.11. Let $T\colon \mathbb{R}^2 \to \mathbb{R}^2$ be the linear transformation defined by

$$T(a, b) = (2b - a, 3a).$$

It is easy to verify that T is an isomorphism. We have $T(1, 0) = (-1, 3)$, and $T(0, 1) = (2, 0)$. The range $\mathrm{Range}\,(T) = \mathrm{Sp}\,((-1, 3), (2, 0))$. Since $\{(-1, 3),\ (2, 0)\}$ is linearly independent, it is a basis for $\mathbb{R}^2$. Thus we have that $\mathrm{Range}\,(T) = \mathbb{R}^2$ and $\mathrm{rk}\,T = 2$. By Theorem 3.3.8, nullity $T = 0$. We conclude that T is an isomorphism.

Example 3.3.12. Let $T\colon \mathscr{P}_2 \to \mathscr{P}_3$ be the function defined by

$$T(f(x)) = 2f'(x) + \int_0^x 6f(t)\,dt.$$

(1) Check that T is a linear transformation.

(2) Find $\mathrm{rk}\,T$ and nullity T.

(3) Is T one-one or onto?

Solution. (1) For any $f(x), g(x) \in \mathscr{P}_2$ and any scalar a,

$$T(f(x) + g(x)) = 2(f(x) + g(x))' + \int_0^x 6(f(t) + g(t)) \, dt$$

$$= 2f'(x) + 2g'(x) + \int_0^x 6f(t) \, dt + \int_0^x 6g(t) \, dt$$

$$= T(f(x)) + T(g(x))$$

and

$$T(af(x)) = 3(af(x))' + \int_0^x 6af(t) \, dt$$

$$= a \left(3f'(x) + \int_0^x 6f(t) \, dt \right)$$

$$= aT(f(x)).$$

Thus T is an linear transformation.

(2) We use the standard basis $\{1, x, x^2\}$ for $\mathscr{P}_2$. Then

$$T(1) = 6x,$$
$$T(x) = 2 + 3x^2,$$
$$T(x^2) = 4x + 2x^3.$$

It follows that Range $(T) = \mathrm{Sp}\left(6x, 2 + 3x^2, 4x + 3x^3\right)$. Hence $\mathrm{rk}\, T = 3$. Since $\dim \mathscr{P}_2 = 3$, we have nullity $T = 0$.

(3) The linear transformation T is one-one. Since $\dim \mathscr{P}_3 = 4 > \mathrm{rk}\, T$, T is not onto.   ◇

Classification of finite dimensional vector spaces.   The process of determining which vector spaces are isomorphic is called the **classification** of vector spaces. Classification of objects (and morphisms) of interest is often the central problem in any discipline within theoretical mathematics.

Theorem 3.3.13. *Let V and W be finite dimensional vector spaces. Then $V \simeq W$ if and only if $\dim V = \dim W$.*

Proof. The "only if" part: This is Corollary 3.2.7.

The "if" part: Suppose $\dim V = \dim W = n$. Find a basis $\{v_1, \ldots, v_n\}$ for V and a basis $\{w_1, \ldots, w_n\}$ for W. By Theorem 3.1.11 we can construct a linear transformation $S\colon V \to W$ sending v_i to w_i for each i. Conversely, we can also construct a linear transformation $T\colon W \to V$ sending w_i to v_i for each i. Note that $T \circ S$ is the unique linear transformation sending v_i to v_i which does the same thing as $\mathbf{1}_V$. Hence $T \circ S = \mathbf{1}_V$. Similarly, we have that $S \circ T = \mathbf{1}_W$. This shows that T and S are inverse to each other and thus S and T are both isomorphisms. $\square$

Corollary 3.3.14. *Let V be a real vector space of dimension n. Then $V \simeq \mathbb{R}^n$.*

Corollary 3.3.15. (a) $\mathscr{P}_n \simeq \mathbb{R}^{n+1}$.

 (b) $M_{m \times n}(\mathbb{R}) \simeq \mathbb{R}^{mn}$.

 (c) $\mathscr{F}(S, \mathbb{R}) \simeq \mathbb{R}^n$ *if* $|S| = n$.

We make an observation from the proof of Theorem 3.3.13. For any arbitrary basis $\{v_1, \ldots, v_n\}$ for V and any arbitrary basis $\{w_1, \ldots, w_n\}$ for W, we may identify the two vector spaces V and W, by making the correspondence (identification)

$$a_1 v_1 + a_2 v_2 + \cdots a_n v_n \quad \rightsquigarrow \quad a_1 w_1 + a_2 w_2 + \cdots a_n w_n.$$

In particular, we may make the following correspondence between elements of V and elements of $\mathbb{R}^n$:

$$v = a_1 v_1 + a_2 v_2 + \cdots + a_n v_n \quad \rightsquigarrow \quad (a_1, a_2, \ldots, a_n)$$

for $a_i \in \mathbb{R}$. Thus, next time you encounter a problem dealing with an n-dimensional vector space V, you can transform it into a problem dealing with $\mathbb{R}^n$.

Example 3.3.16. (1) Using the standard basis $\{1, x, x^2\}$, we may identify the polynomial $f(x) = a_0 + a_1 x + a_2 x^2$ in $\mathscr{P}_2$ with the vector (a_0, a_1, a_2) in $\mathbb{R}^3$.

(2) Using the standard basis $\{e_{11}, e_{12}, e_{21}, e_{22}\}$, we may identify the matrix $A = \begin{pmatrix} a_{11} & a_{12} \\ a_{21} & a_{22} \end{pmatrix}$ in $M_{2\times 2}(\mathbb{R})$ with the vector $(a_{11}, a_{12}, a_{21}, a_{22})$ in $\mathbb{R}^4$.

(3) Let $S = \{1, 2, 3\}$. Using the standard basis $\{\chi_1, \chi_2, \chi_3\}$, we may identify the function

$$
\begin{array}{rcl}
f : & S & \longrightarrow & \mathbb{R} \\
& 1 & \longmapsto & a_1 \\
& 2 & \longmapsto & a_2 \\
& 3 & \longmapsto & a_3
\end{array}
$$

with the vector (a_1, a_2, a_3) in $\mathbb{R}^3$.

Example 3.3.17. Consider the basis $\{(1, 0, 0), (1, 1, 0), (1, 1, 1)\}$ for $\mathbb{R}^3$. For any (a_1, a_2, a_3) in $\mathbb{R}^3$, we have

$$(a_1, a_2, a_3) = (a_1 - a_2)(1, 0, 0) + (a_2 - a_3)(1, 1, 0) + a_3(1, 1, 1).$$

We say (a_1, a_2, a_3) is the coordinate of (a_1, a_2, a_3) with respect to the standard basis, and we say that $(a_1 - a_2, a_2 - a_3, a_3)$ is the coordinate of (a_1, a_2, a_3) with respect to the ordered basis $((1, 0, 0), (1, 1, 0), (1, 1, 1))$.

Example 3.3.18. Let u, v, w and x be linearly independent vectors in the real vector space V. To check whether the vectors

$$(3.3) \qquad u + v + w + x, \ 2u + 2v + w - x, \ u - v + w, \ u - w + x$$

are linearly independent, we can adopt the following approach. Let $W = \mathrm{Sp}\,(u, v, w, x)$. Since $\{u, v, w, x\}$ is linearly independent over $\mathbb{R}$, we can think of it as a basis for W and conclude that $\dim W = 4$. With respect to the ordered basis (u, v, w, x), the vectors in (3.3) are identified as

$$(3.4) \qquad (1, 1, 1, 1), \ (2, 2, 1, -1), \ (1, -1, 1, 0), \ (1, 0, -1, 1)$$

in $\mathbb{R}^4$. To verify that the vectors in (3.3) are linearly independent is equivalent to verifying the vectors in (3.4) are linearly independent. Thus, we may calculate the row rank of

$$
\begin{pmatrix}
1 & 1 & 1 & 1 \\
2 & 2 & 1 & -1 \\
1 & -1 & 1 & 0 \\
1 & 0 & -1 & 1
\end{pmatrix}
$$

which turns out to be 4. We now conclude that the vectors in (3.3) are linearly independent.

Exercises 3.3

In the following problems all vector spaces are assumed to be finite dimensional.

1. Try constructing a linear transformation $T: \mathbb{R}^4 \to \mathbb{R}^4$ such that

$$\text{Range}\,(T) = \text{Sp}\,((1,0,1,0),\ (0,1,0,1)).$$

2. Compute the rank and nullity of the linear transformations in Examples 3.1.4, 3.1.5, 3.1.7–3.1.9 and Problems (2)(a), (b), (3) in §3.1.

3. Determine the rank and nullity of the projection in Example 3.1.6. Give then necessary and sufficient condition such that the projection is an isomorphism.

4. Let $S: U \to V$ and $T: V \to W$ be linear transformations.

   (a) Show that $\text{rk}\,T \le \dim V$ and $\text{rk}\,T \le \dim W$.

   (b) Show that $\text{nullity}\,T \le \dim V$.

   (c) Show that $\text{rk}\,T \circ S \le \text{rk}\,S$ and $\text{rk}\,T \circ S \le \text{rk}\,T$.

   (d) Show that $\text{nullity}\,T \circ S \ge \text{nullity}\,S$.

   (e) Is there any relation between $\text{nullity}\,T \circ S$ and $\text{nullity}\,T$?

5. Let V and W be finite dimensional vector spaces of equal dimension and let $T: V \to W$ be a linear transformation. Suppose $\mathscr{B}$ is a basis for V. Show that T is an isomorphism if and only if

$$T(\mathscr{B}) = \{T(v) : v \in \mathscr{B}\}$$

is a basis for W.

6. Remember that we may consider the vector space $\mathbb{C}^2$ as a vector space over $\mathbb{R}$ or over $\mathbb{C}$. Let T be the function from $\mathbb{C}^2$ to $\mathbb{C}^2$ given by

$$T(z,\ w) = (z + iw,\ iz - w).$$

where $z,\ w \in \mathbb{C}$.

(a) Show that $(1,0)$, $(i,0)$, $(0,1)$ and $(0,i)$ form a basis for $\mathbb{C}^2$ over $\mathbb{R}$.

(b) Is T a linear transformation over $\mathbb{R}$? If yes, compute $\operatorname{rk} T$ and nullity T over $\mathbb{R}$.

(c) Is T a linear transformation over $\mathbb{C}$? If yes, compute $\operatorname{rk} T$ and nullity T over $\mathbb{C}$.

7. Let $T\colon V \to W$ be a linear transformation. Let $V' \subseteq V$ and $W' \subseteq W$ be subspaces.

(a) Show that
$$T(V') = \{T(v) \in W : v \in V'\}$$

is a subspace of W. This is called the **image** of V' under T. Note that $\operatorname{Range}(T) = T(V)$.

(b) Show that
$$T^{-1}(W') = \{v \in V : T(v) \in W'\}$$

is a subspace of V. This is called the **pre-image** of W' under T. Note that $\operatorname{Null}(T) = T^{-1}(\{O\})$.

3.4   Matrices and linear transformations

In this section we describe a method to better represent a linear transformation.

Matrices representing a linear transformation.   Let V and W be real vector spaces of dimension n and m respectively and let $T\colon V \to W$ be a linear transformation over $\mathbb{R}$. Let $\beta = (v_1, \ldots, v_n)$ be an ordered basis of V. Let $\gamma = (w_1, \ldots, w_m)$ be an ordered basis of W. Remember that T is uniquely determined by the image of β. Suppose

$$(3.5) \qquad\qquad T(v_j) = \sum_{i=1}^{m} a_{ij} w_i, \qquad j = 1, \ldots, n.$$

Let $v = \sum_{j=1}^{n} c_j v_j \in V$. Then

$$(3.6) \qquad\qquad T(v) = \sum_{j=1}^{n} c_j T(v_j) = \sum_{j=1}^{n} \left(c_j \sum_{i=1}^{m} a_{ij} w_i \right)$$

$$= \sum_{i,j=1}^{m,n} a_{ij} c_j w_i = \sum_{i=1}^{m} \left(\sum_{j=1}^{n} a_{ij} c_j \right) w_i.$$

Observe that

$$(3.7) \qquad
\begin{pmatrix}
a_{11} & a_{12} & \cdots & a_{1n} \\
a_{21} & a_{22} & \cdots & a_{2n} \\
\vdots & \vdots & \ddots & \vdots \\
a_{m1} & a_{m2} & \cdots & a_{mn}
\end{pmatrix}
\begin{pmatrix}
c_1 \\ c_2 \\ \vdots \\ c_n
\end{pmatrix}
=
\begin{pmatrix}
\sum_{j=1}^{n} a_{1j} c_j \\
\sum_{j=1}^{n} a_{2j} c_j \\
\vdots \\
\sum_{j=1}^{n} a_{mj} c_j
\end{pmatrix}.
$$

Remember that if $v = \sum_{j=1}^{n} c_j v_j \in V$, we call $(c_1, c_2, \ldots, c_n)$ the coordinate of v with respect to the ordered basis β. As a convention we will use $[v]_\beta$ for the coordinate of v with respect to β as a column vector:

$$[v]_\beta = \begin{pmatrix} c_1 \\ c_2 \\ \vdots \\ c_n \end{pmatrix}.$$

Similarly, $[T(v)]_\gamma$ stands for the coordinate of $T(v)$ with respect to the ordered basis γ as a column vector. By (3.6), we have that

$$[T(v)]_\gamma = \begin{pmatrix}
\sum_{j=1}^{n} a_{1j} c_j \\
\sum_{j=1}^{n} a_{2j} c_j \\
\vdots \\
\sum_{j=1}^{n} a_{mj} c_j
\end{pmatrix}.$$

Hence, (3.7) may be rewritten as

$$\left(a_{ij} \right)_{m \times n} [v]_\beta = [T(v)]_\gamma.$$

We will call $\left(a_{ij}\right)_{m \times n}$ the **matrix associated with** T (or **representing** T) **with respect to (relative to) the ordered bases** β **and** γ and denote it by $[T]_\beta^\gamma$. Note that by (3.5), we have that

$$[T(v_j)]_\gamma = \begin{pmatrix} a_{1j} \\ a_{2j} \\ \vdots \\ a_{mj} \end{pmatrix}$$

is the j-th column of $[T]_\beta^\gamma$. Hence

$$(3.8) \qquad [T]_\beta^\gamma = \Big([T(v_1)]_\gamma, \quad [T(v_2)]_\gamma, \quad \ldots, \quad [T(v_n)]_\gamma \Big).$$

If in particular $V = W$ and $\gamma = \beta$, we will simply write $[T]_\beta$ for $[T]_\beta^\beta$.

Theorem 3.4.1. *Let* $T\colon V \to W$ *be a linear linear transformation of finite dimensional vector spaces. Let* β *and* γ *be ordered bases of* V *and* W *respectively. Then*

$$[T(v)]_\gamma = [T]_\beta^\gamma [v]_\beta.$$

Example 3.4.2. Let $T\colon \mathbb{R}^2 \to \mathbb{R}^2$ be the linear transformation defined by

$$T(a, b) = (2b - a, 3a)$$

as in Example 3.3.11.

(a) Find the matrix associated with T with respect to the standard basis of $\mathbb{R}^2$.

(b) We leave it to the reader to verify that $\beta = ((1,0), (1,1))$ is an ordered basis of $\mathbb{R}^2$. Find the matrix associated with T with respect to the standard basis and β.

Solution. (a) Since $T(1,0) = (-1, 3)$ and $T(0,1) = (2,0)$, by (3.8) we have

$$[T]_\alpha = \begin{pmatrix} -1 & 2 \\ 3 & 0 \end{pmatrix}$$

where α is the standard basis of $\mathbb{R}^2$.

(b) We calculate that

$$T(1,0) = (-1,3) = -4(1,0) + 3(1,1),$$
$$T(2,0) = 2(1,0).$$

Thus

$$[T]_\alpha^\beta = \begin{pmatrix} -4 & 2 \\ 3 & 0 \end{pmatrix}$$

by (3.8).   ◇

Example 3.4.3. Let $T: \mathscr{P}_2 \to \mathscr{P}_3$ be the linear transformation defined by

$$T(f(x)) = 2f'(x) + \int_0^x 6f(t)\,dt$$

as in Example 3.3.12.

(a) Find the matrix of T with respect to the standard bases of $\mathscr{P}_2$ and of $\mathscr{P}_3$.

(b) Find $T(\sqrt{2} + 100x - 37.8x^2)$.

Solution. (a) Let $\alpha = (1, x, x^2)$ and $\alpha' = (1, x, x^2, x^3)$. These are the standard bases of $\mathscr{P}_2$ and $\mathscr{P}_3$ respectively. In the solution of Example 3.3.12, we have already seen that

$$T(1) = 6x,$$
$$T(x) = 2 + 3x^2,$$
$$T(x^2) = 4x + 2x^3.$$

It follows that

$$[T]_\alpha^{\alpha'} = \begin{pmatrix} 0 & 2 & 0 \\ 6 & 0 & 4 \\ 0 & 3 & 0 \\ 0 & 0 & 2 \end{pmatrix}.$$

(b) We use Theorem 3.4.1 to find

$$[T]_\alpha^{\alpha'} \begin{pmatrix} \sqrt{2} \\ 100 \\ -37.8 \end{pmatrix} = \begin{pmatrix} 0 & 2 & 0 \\ 6 & 0 & 4 \\ 0 & 3 & 0 \\ 0 & 0 & 2 \end{pmatrix} \begin{pmatrix} \sqrt{2} \\ 100 \\ -37.8 \end{pmatrix} = \begin{pmatrix} 200 \\ 6\sqrt{2} - 151.2 \\ 300 \\ -75.6 \end{pmatrix}.$$

We conclude that

$$T(\sqrt{2} + 100x - 37.8x^2) = 200 + (6\sqrt{2} - 151.2)x + 300x^2 - 75.6x^3$$

using the standard basis of $\mathscr{P}_3$.   ◇

We summarize the following easy but important facts.

Proposition 3.4.4.   (a) *The matrix associated with the zero transformation with respect to whatever bases is the zero matrix.*

(b) *If $T: V \rightarrow W$ is a linear transformation such that $[T]_\beta^\gamma$ is the zero matrix for any ordered bases β and γ of V and W, then T is the zero transformation.*

(c) *Let β be a base of the n-dimensional $\mathbb{R}$-vector space V. Then $[\mathbf{1}_V]_\beta = I_n$, the identity matrix of size n.*

(d) *If $T: V \rightarrow V$ is a linear transformation such that $[T]_\beta$ is the identity matrix for any ordered bases β of V, then $T = \mathbf{1}_V$.*

Proof. Parts (a) and (c) are trivial.

(b) Suppose that T maps all base elements to the zero vector. The zero transformation does the same thing. Now the result follows from Theorem 3.1.11.

(d) Each vector in the basis β is mapped to itself. The identity transformation $\mathbf{1}_S$ does the same thing. The result follows from Theorem 3.1.11 again.   □

Note that the identity transformation is not necessarily represented by the identity matrix.

Example 3.4.5. Let α be the standard basis of $\mathbb{R}^3$ and let

$$\beta = ((1,0,0),\ (1,1,0),\ (3,2,1)).$$

Find $[\mathbf{1}_{\mathbb{R}^3}]_\beta^\alpha$ and $[\mathbf{1}_{\mathbb{R}^3}]_\alpha^\beta$.

Solution. The matrix $[\mathbf{1}_{\mathbb{R}^3}]_\beta^\alpha$ is easy to find:

$$[\mathbf{1}_{\mathbb{R}^3}]_\beta^\alpha = \begin{pmatrix} 1 & 1 & 3 \\ 0 & 1 & 2 \\ 0 & 0 & 1 \end{pmatrix}.$$

The other matrix is not so transparent. We first find

$$1_{\mathbb{R}^3}(e_1) = e_1 = (1, 0, 0),$$
$$1_{\mathbb{R}^3}(e_2) = e_2 = -(1, 0, 0) + (1, 1, 0),$$
$$1_{\mathbb{R}^3}(e_3) = e_3 = -(1, 0, 0) - 2(1, 1, 0) + (3, 2, 1).$$

This implies that

$$[1_{\mathbb{R}^3}]_\alpha^\beta = \begin{pmatrix} 1 & -1 & -1 \\ 0 & 1 & -2 \\ 0 & 0 & 1 \end{pmatrix}$$

by (3.8).   ◇

The rank of a linear transformation.   The properties of a linear transformation T is often reflected by those of a matrix representing it. We will demonstrate this by the following discussion. First, we establish a lemma.

Lemma 3.4.6. *Let $F: V \to U$ be an isomorphism and let*

$$W = \mathrm{Sp}\,(w_1, w_2, \ldots, w_r)$$

be a subspace of V. Then

$$W \simeq \mathrm{Sp}\,(F(w_1), F(w_2), \ldots, F(w_r)).$$

Proof. The isomorphism between W and $F(W)$ is established by restricting the domain of F to W and the codomain to $F(W)$. The restriction map remains injective and onto by Proposition 3.3.4.   □

Proposition 3.4.7. *Let $T: V \to W$ be a linear transformation of finite dimensional vector spaces. Then*

$$\mathrm{rk}\,T = \mathrm{col}\,\mathrm{rk}\,[T]_\beta^\gamma$$

for any choice of ordered bases β and γ of V and W respectively.

Proof. Let $T: V \to W$ be a linear transformation between two finite dimensional vector spaces. For any choice of ordered bases $\beta = (v_j)_{j=1}^n$ and

$\gamma = (w_i)_{i=1}^m$ for V and W respectively, we can construct the matrix $[T]_\beta^\gamma$. Remember that

$$\text{Range}\,(T) = \text{Sp}\,(T(v_1), T(v_2), \dots, T(v_n))\,.$$

The isomorphism F between W and $\mathbb{R}^m$ sending w_i to e_i identifies $T(v_j)$ with the j-th column of $[T]_\beta^\gamma$. Thus, by Lemma 3.4.6 the subspace Range (T) is isomorphic to the column space of $[T]_\beta^\gamma$. The result now follows. $\square$

Linear transformations associated with a matrix. If we are given an $m \times n$ matrix $A = \left(a_{ij}\right)$ with entries in $\mathbb{R}$, it is naturally to look upon it as *the* linear transformation from $\mathbb{R}^n$ to $\mathbb{R}^m$ sending

$$T(e_j^{(n)}) = a_{1j}e_1^{(m)} + a_{2j}e_2^{(m)} + \cdots + a_{mj}e_m^{(m)}.$$

Or more generally, let V and W be of dimension n and m respectively. With your choice of ordered bases $\beta = (v_j)_j$ and $\gamma = (w_i)_i$ for V and W respectively, T is the linear transformation associated with A for which

$$T(v_j) = a_{1j}w_1 + a_{2j}w_2 + \cdots + a_{mj}w_m$$

with respect to the bases β and γ. Clearly, the construction of T is so that $[T]_\beta^\gamma = A$.

Example 3.4.8. Let $T \colon \mathbb{R}^3 \to \mathscr{P}_2$ be the linear transformation associated with the identity matrix I_3 with respect to

(i) the standard bases;

(ii) $((1,1,1),\ (1,0,1),\ (1,1,0))$ and $(1,\ 1+x,\ 1+x^2)$.

Determine for each case

(a) the value of $T(3,2,2)$,

(b) $\text{rk}\,T$ and nullity T, and

(c) whether T is an isomorphism.

Solution. For the case (i), the linear transformation T is clearly the identity map. Thus (a) $T(3,2,2) = 3 + 2x + 2x^2$, (b) $\operatorname{rk} T = 3$ and nullity $T = 0$, and (c) T is an isomorphism.

For the case (ii), the linear transformation T is such that

$$
\begin{array}{rcl}
T: & \mathbb{R}^3 & \longrightarrow \quad \mathcal{P}_2 \\
& (1,1,1) & \longmapsto \quad 1 \\
& (1,0,1) & \longmapsto \quad 1+x \\
& (1,1,0) & \longmapsto \quad 1+x^2.
\end{array}
$$

(a) The vector $(3,2,2) = (1,1,1) + (1,0,1) + (1,1,0)$. It follows that

$$T(3,2,2) = 1 + 1 + x + 1 + x^2 = 3 + x + x^2.$$

(b) By Proposition 3.4.7, we have that $\operatorname{rk} T = \operatorname{colrk} I_3 = 3$. By Theorem 3.3.8 we have nullity $T = 3 - \operatorname{rk} T = 0$.

(c) The rank of T is 3. This means T is onto. The nullity of T is 0. This says T is one-one. Hence T is an isomorphism.   ◇

Observe that the identity matrix does not always represent the identity transformation. However, it is nice to have a "simple" matrix representing the linear transformation you are working on.

The vector spaces $\mathcal{L}(V, W)$ and $M_{m \times n}(\mathbb{R})$.   Let V and W be real vector spaces of dimension n and m respectively. Fix a choice of ordered bases $\beta = (v_j)_{j=1}^n$ and $\gamma = (w_i)_{i=1}^m$ for V and W respectively. There is a one-to-one correspondence between the linear transformations from V to W and the $m \times n$ matrices with entries in $\mathbb{R}$:

$$
\begin{array}{ccc}
\mathcal{L}(V, W) & \longleftrightarrow & M_{m \times n}(\mathbb{R}) \\
T & \rightsquigarrow & [T]_\beta^\gamma.
\end{array}
$$

For the definition of $\mathcal{L}(V, W)$ see Exercise 6 in §3.1.

Let's take this discussion further. Suppose we are given two linear transformations S and T from V to W such that

$$T(v_j) = \sum_{i=1}^m a_{ij} w_i \quad \text{and} \quad S(v_j) = \sum_{i=1}^m b_{ij} w_i, \qquad j = 1, 2 \ldots, n.$$

Then

$$[T]_\beta^\gamma = \left(a_{ij} \right)_{m \times n} \quad \text{and} \quad [S]_\beta^\gamma = \left(b_{ij} \right)_{m \times n}.$$

Since

$$(T + S)(v_j) = \sum_{i=1}^{m} a_{ij} w_j + \sum_{i=1}^{m} b_{ij} w_j = \sum_{i=1}^{m} (a_{ij} + b_{ij}) w_i, \qquad j = 1, 2, \ldots, n,$$

we have that

$$[T + S]_\beta^\gamma = \left(a_{ij} + b_{ij} \right)_{m \times n} = [T]_\beta^\gamma + [S]_\beta^\gamma.$$

Similarly, for $c \in \mathbb{R}$ we have

$$(cT)(v_j) = c \sum_{i=1}^{m} a_{ij} w_i = \sum_{i=1}^{m} c a_{ij} w_i, \qquad i = 1, 2, \ldots, m.$$

Hence

$$[cT]_\beta^\gamma = \left(c a_{ij} \right)_{m \times n} = c[T]_\beta^\gamma.$$

We have seen that sending a linear transformation from V to W to $[T]_\beta^\gamma$ gives an isomorphism from $\mathcal{L}(V, W)$ to $M_{m \times n}(\mathbb{R})$. The sum and scalar multiplication of linear transformations are translated as the sum and scalar multiplication of their corresponding matrices. The result is summarized in the following theorem.

Theorem 3.4.9. *Let V and W be $\mathbb{R}$-vector spaces of dimension n and m respectively. Then*

$$\mathcal{L}(V, W) \simeq M_{m \times n}(\mathbb{R}).$$

Example 3.4.10. Let F be the linear transformation from $\mathcal{P}_2$ to $\mathcal{P}_3$ sending $f(x)$ to $f'(x)$. Let G be the linear transformation from $\mathcal{P}_2$ to $\mathcal{P}_3$ sending $f(x)$ to $\int_0^x f(t)\, dt$. Find the matrices representing F, G and $2F + 6G$ relative to the standard bases of $\mathcal{P}_2$ and of $\mathcal{P}_3$. Compare your answer with the answer of Example 3.4.3.

Solution. Let $\alpha = (1, x, x^2)$ and $\alpha' = (1, x, x^2, x^3)$. Since

$$F(1) = 0, \quad F(x) = 1, \quad F(x^2) = 2x,$$

the matrix associated with F is

$$[F]_\alpha^{\alpha'} = \begin{pmatrix} 0 & 1 & 0 \\ 0 & 0 & 2 \\ 0 & 0 & 0 \\ 0 & 0 & 0 \end{pmatrix}.$$

We also find

$$G(1) = x, \quad G(x) = \frac{x^2}{2}, \quad G(x^2) = \frac{x^3}{3}.$$

The matrix associated with G is

$$[G]_\alpha^{\alpha'} = \begin{pmatrix} 0 & 0 & 0 \\ 1 & 0 & 0 \\ 0 & 1/2 & 0 \\ 0 & 0 & 1/3 \end{pmatrix}.$$

From Example 3.4.3, we have seen that

$$[2F + 6G]_\alpha^{\alpha'} = \begin{pmatrix} 0 & 2 & 0 \\ 6 & 0 & 4 \\ 0 & 3 & 0 \\ 0 & 0 & 2 \end{pmatrix}.$$

Note that $[2F + 6G]_\alpha^{\alpha'} = 2[F]_\alpha^{\alpha'} + 6[G]_\alpha^{\alpha'}$.   ◇

Exercises 3.4

1. Let $T \colon \mathbb{R}^3 \to \mathbb{R}^3$ be the linear transformation such that

$$T(e_1) = (-1, 2, 0),$$
$$T(e_2) = (1, 1, -1),$$
$$T(e_3) = (1, -3, 4).$$

Calculate the matrix of T with respect to the standard basis of $\mathbb{R}^3$.

2. Suppose the matrix of the linear transformation $T: \mathbb{R}^3 \to \mathbb{R}^3$ is

$$\begin{pmatrix} 1 & 2 & 3 \\ 1 & 0 & 0 \\ 1 & -1 & 2 \end{pmatrix}$$

with respect to

(i) the standard basis;

(ii) the ordered basis $((1,1,1),\ (0,1,1),\ (0,0,1))$.

Compute $T(1,2,3)$ for each case.

3. If $T: \mathbb{R}^3 \to \mathbb{R}^3$ is the linear transformation with matrix

$$\begin{pmatrix} 0 & 1 & 1 \\ 1 & 0 & 1 \\ 1 & 1 & 0 \end{pmatrix}$$

with respect to the standard basis, is T an isomorphism?

4. Suppose given the following linear transformations from $\mathbb{R}^3$ to $\mathbb{R}^3$:

$$T(x, y, z) = (x + y + z, 0, 0);$$
$$Q(x, y, z) = (x, x + y, x + y + z);$$
$$F(x, y, z) = (x + 2y + 3z, 2x - y, z - 3x);$$
$$G(x, y, z) = (y - z, x + y, z - 2x).$$

(a) Give the matrices of T, Q, F and G relative to the standard basis of $\mathbb{R}^3$.

(b) Give the matrix of $T - 2Q + 3F - 4G$ relative to the standard basis of $\mathbb{R}^3$.

(c) Is $T - 2Q + 3F - 4G$ an isomorphism?

5. In this problem we discuss how to compute the dimension of the solution space of the homogeneous system of linear equations

(3.9)
$$\begin{array}{ccccccccc} a_{11}x_1 & + & a_{12}x_2 & + & \cdots & + & a_{1n}x_n & = & 0; \\ a_{21}x_1 & + & a_{22}x_2 & + & \cdots & + & a_{2n}x_n & = & 0; \\ & & & & \cdots & & & & \\ a_{m1}x_1 & + & a_{m2}x_2 & + & \cdots & + & a_{mn}x_n & = & 0. \end{array}$$

This system is often written as

$$
\begin{pmatrix}
a_{11} & a_{12} & \cdots & a_{1n} \\
a_{21} & a_{22} & \cdots & a_{2n} \\
\vdots & \vdots & \ddots & \vdots \\
a_{m1} & a_{m2} & \cdots & a_{mn}
\end{pmatrix}
\begin{pmatrix}
x_1 \\
x_2 \\
\vdots \\
x_n
\end{pmatrix}
=
\begin{pmatrix}
0 \\
0 \\
\vdots \\
0
\end{pmatrix}
$$

where $A = \left(a_{ij} \right)_{m \times n}$ is called the **coefficient matrix** of the system (3.9).

(a) Let $T \colon \mathbb{R}^n \to \mathbb{R}^m$ be the linear transformation associated with the A relative to the standard bases. Show that $\mathrm{Null}\,(T)$ is exactly the solution space of the system (3.9).

(b) Show that the dimension of the solution space of (3.9) is equal to $n - \mathrm{col\,rk}\ A$.

(c) Show that $\mathrm{col\,rk}\ A \le m$ and $\mathrm{col\,rk}\ A \le n$. Conclude that the dimension of the solution space (3.9) $\ge n - m$ (*Cf.* Problem (5) in §2.5).

3.5   Composites and inverses

In this section we want to discuss the matrices associated with composites and inverses of linear transformations. Throughout this section U, V and W denote real vector spaces and S, T denote linear transformations over $\mathbb{R}$.

Compositions of linear transformations.   Let $\dim U = n$, $\dim V = m$ and $\dim W = \ell$. Find ordered bases α, β and γ for U, V and W respectively. Let $S \colon U \to V$ and $T \colon V \to W$ be such that

$$
[S]_\alpha^\beta = \left(a_{ij} \right)_{m \times n} \qquad \text{and} \qquad [T]_\beta^\gamma = \left(b_{ij} \right)_{\ell \times m}.
$$

What is $[T \circ S]_\alpha^\gamma$?

Let $\alpha = (u_1, u_2, \ldots, u_n)$, $\beta = (v_1, v_2, \ldots, v_m)$ and $\gamma = (w_1, w_2, \ldots, w_\ell)$. To express $[T \circ S]_\alpha^\gamma$ we need to find $(S \circ T)(u_j)$ in terms of $w_1, w_2, \ldots, w_\ell$.

For $j = 1, 2, \ldots, n$,

$$(T \circ S)(u_j) = T\left(S(u_j)\right) = T\left(\sum_{k=1}^{m} a_{kj} v_k\right) = \sum_{k=1}^{m} a_{kj} T(v_k)$$

$$= \sum_{k=1}^{m} \left(a_{kj} \sum_{i=1}^{\ell} b_{ik} w_i\right) = \sum_{i,k} a_{kj} b_{ik} w_i = \sum_{i=1}^{\ell} \left(\sum_{k=1}^{m} b_{ik} a_{kj}\right) w_i.$$

Thus

$$[T \circ S]_\alpha^\gamma = \left(\sum_{k=1}^{m} b_{ik} a_{kj}\right)_{\ell \times n} = \left(b_{ij}\right)_{\ell \times m} \left(a_{ij}\right)_{m \times n} = [T]_\beta^\gamma [S]_\alpha^\beta.$$

Hence the composite of linear transformations is translated as the product of their corresponding matrices.

Theorem 3.5.1. *Let U, V and W be finite dimensional real vector spaces. Let α, β and γ be bases for U, V and W respectively. Let $S\colon U \to V$ and $T\colon V \to W$ be linear transformations. Then*

$$[T \circ S]_\alpha^\gamma = [T]_\beta^\gamma [S]_\alpha^\beta.$$

Example 3.5.2. Let $S\colon \mathbb{R}^3 \to \mathbb{R}^2$ be the linear transformation such that

$$S(e_1^{(3)}) = (1,1), \qquad S(e_2^{(3)}) = (2,-3) \qquad \text{and} \qquad S(e_3^{(3)}) = (-1,1).$$

Let $T\colon \mathbb{R}^2 \to \mathbb{R}^3$ be the linear transformation such that

$$T(e_1^{(2)}) = (1,-2,1) \qquad \text{and} \qquad T(e_2^{(2)}) = (0,1,0).$$

Find the matrices of $S \circ T$ and $T \circ S$ with respect to the standard bases.

Solution. Let $\alpha = \left(e_1^{(3)}, e_2^{(3)}, e_3^{(3)}\right)$ and $\alpha' = \left(e_1^{(2)}, e_2^{(2)}\right)$. Then

$$[S]_\alpha^{\alpha'} = \begin{pmatrix} 1 & 2 & -1 \\ 1 & -3 & 1 \end{pmatrix} \qquad \text{and} \qquad [T]_{\alpha'}^\alpha = \begin{pmatrix} 1 & 0 \\ -2 & 1 \\ 1 & 0 \end{pmatrix}.$$

We have

$$[T \circ S]_\alpha^\alpha = [T]_{\alpha'}^\alpha [S]_\alpha^{\alpha'} = \begin{pmatrix} 1 & 0 \\ -2 & 1 \\ 1 & 0 \end{pmatrix} \begin{pmatrix} 1 & 2 & -1 \\ 1 & -3 & 1 \end{pmatrix} = \begin{pmatrix} 1 & 2 & -1 \\ -1 & -7 & 3 \\ 1 & 2 & -1 \end{pmatrix}$$

and

$$[S \circ T]_{\alpha'}^{\alpha'} = [S]_{\alpha}^{\alpha'}[T]_{\alpha'}^{\alpha} = \begin{pmatrix} 1 & 2 & -1 \\ 1 & -3 & 1 \end{pmatrix} \begin{pmatrix} 1 & 0 \\ -2 & 1 \\ 1 & 0 \end{pmatrix} = \begin{pmatrix} -4 & 2 \\ 8 & -3 \end{pmatrix}$$

by Theorem 3.5.1. ◇

A few words on products of matrices. Remember that using Definition 2.2.14 and Proposition 2.2.15 we may write

$$A = \left(a_{ij} \right)_{m \times n} = \sum_{\substack{i=1,2,\dots,m \\ j=1,2,\dots,n}} a_{ij} e_{ij}^{(m,n)}.$$

The coefficient of $e_{ij}^{(m,n)}$ in A gives the (i,j)-entry of A. Note that

$$e_{ij}^{(m,n)} e_{kl}^{(n,p)} = \begin{cases} \mathbf{0}_{m \times p}, & \text{if } j \neq k, \\ e_{il}^{(m,p)}, & \text{if } j = k. \end{cases}$$

If we write a matrix as a linear combination of the e_{ij}'s, the multiplication of matrices is distributive with respect to addition (see Exercise 4).

Example 3.5.3. We now attempt to compute in $M_{9 \times 9}(\mathbb{R})$ the matrix

$$(e_{23} - 3.5 e_{37} + 9 e_{39} - e_{89})(\sqrt{2} e_{12} - e_{35} + 2 e_{91}).$$

The product may look complicated, but most terms can be omitted. For example, $e_{23} e_{12} = \mathbf{0}$. There is no need to list this term. Clearly, what remains is

$$-e_{23} e_{35} + 18 e_{39} e_{91} - 2 e_{89} e_{91} = -e_{25} + 18 e_{31} - 2 e_{81}.$$

When the entries are most 0 in a matrix, it is easier to carry out the computation in this notation.

Inverses of linear transformations. Suppose $T \colon V \to W$ is an isomorphism. This means that $\dim V = n = \dim W$. It follows that any matrix associated with T is a square matrix. The matrix T has an inverse transformation $T^{-1} \colon W \to V$ such that

$$T^{-1} \circ T = \mathbf{1}_V \qquad \text{and} \qquad T \circ T^{-1} = \mathbf{1}_W.$$

Hence we have that

$$[T^{-1}]^{\beta}_{\gamma}[T]^{\gamma}_{\beta} = [T^{-1} \circ T]^{\beta}_{\beta} = [\mathbf{1}_V]_{\beta} = I_n$$
$$[T]^{\gamma}_{\beta}[T^{-1}]^{\beta}_{\gamma} = [T \circ T^{-1}]^{\gamma}_{\gamma} = [\mathbf{1}_W]_{\gamma} = I_n$$

for the ordered bases β and γ for V and W respectively. We now pause to make a definition.

Definition 3.5.4. Let $A \in M_n(\mathbb{R})$. We say A is an **invertible** matrix if there is a $B \in M_n(\mathbb{R})$ such that $AB = BA = I_n$. The matrix B is unique (see Exercise 5). The matrix B is called the **inverse** matrix of A, denoted A^{-1}.

If a matrix is not a square matrix then it is never invertible. In fact you can find a 2×3 matrix A and a 3×2 matrix B such that $AB = I_2$ but I guarantee you that BA will never be I_3. For example, we have that

$$\begin{pmatrix} 1 & 0 & 0 \\ 0 & 1 & 0 \end{pmatrix} \begin{pmatrix} 1 & 0 \\ 0 & 1 \\ 0 & 0 \end{pmatrix} = \begin{pmatrix} 1 & 0 \\ 0 & 1 \end{pmatrix}$$

while

$$\begin{pmatrix} 1 & 0 \\ 0 & 1 \\ 0 & 0 \end{pmatrix} \begin{pmatrix} 1 & 0 & 0 \\ 0 & 1 & 0 \end{pmatrix} = \begin{pmatrix} 1 & 0 & 0 \\ 0 & 1 & 0 \\ 0 & 0 & 0 \end{pmatrix}.$$

However we will see later in this course that when A is a *square* matrix of size n, in order to show that A is invertible we only need to find a square matrix of the same size such that $AB = I_n$ or $BA = I_n$. We don't need to compute both AB and BA.

We now go back to our discussion regarding isomorphisms. We have just seen that $[T]^{\gamma}_{\beta}$ is an invertible matrix when T is an isomorphism. We have also seen that $[T^{-1}]^{\beta}_{\gamma} = \left([T]^{\gamma}_{\beta}\right)^{-1}$.

Conversely, suppose that $A = [T]_\beta^\gamma$ is invertible. Find the unique linear transformation $T' \colon W \to V$ such that $[T']_\gamma^\beta = A^{-1}$. Then

$$[T' \circ T]_\beta^\beta = [T']_\gamma^\beta [T]_\beta^\gamma = A^{-1} A = I_n.$$

By Proposition 3.4.4(d) we have that $T' \circ T = \mathbf{1}_V$. Similarly,

$$[T \circ T']_\gamma^\gamma = [T]_\beta^\gamma [T']_\gamma^\beta = A A^{-1} = I_n.$$

and we have $T \circ T' = \mathbf{1}_W$. Therefore, T' is the inverse transformation of T and T is an isomorphism.

We see now that an invertible linear transformation translates as an invertible matrix and that the inverse linear transformation translates as the inverse matrix. We now summarize all the results above in the following theorem.

Theorem 3.5.5. *Let V and W be finite dimensional vector spaces of equal dimension and let β and γ be bases for V and W respectively. Then $T \colon V \to W$ is an isomorphism if and only if $[T]_\beta^\gamma$ is invertible. In this case*

$$[T^{-1}]_\gamma^\beta = \left([T]_\beta^\gamma \right)^{-1}.$$

Now we can give a criterion for determining whether a matrix A is invertible.

Proposition 3.5.6. *Let $A \in M_n(\mathbb{R})$. The matrix A is invertible if and only if the columns of A form a basis for $\mathbb{R}^n$.*

Proof. Let $T \colon \mathbb{R}^n \to \mathbb{R}^n$ be the associated linear transformation of A relative to the standard basis. Then $\text{Range}(T)$ is the column space of A.

Suppose the columns of A form a basis for $\mathbb{R}^n$. Then the column space of A is $\mathbb{R}^n$. In other words, $\text{Range}(T) = \mathbb{R}^n$. It follows that nullity $T = 0$. Thus T is an isomorphism. By Theorem 3.5.5, A is invertible.

Conversely, when A is invertible, T is an isomorphism. Thus $\text{Range}(T) = \mathbb{R}^n$. The columns of A are n vectors that span $\mathbb{R}^n$. By Corollary 2.4.17, the columns of A form a basis for $\mathbb{R}^n$. □

Example 3.5.7. Let $T \colon M_{2 \times 3}(\mathbb{R}) \to M_{3 \times 2}(\mathbb{R})$ be the linear transformation sending A to A^t. Let α be the ordered basis $(e_{11}, e_{12}, e_{13}, e_{21}, e_{22}, e_{23})$ for $M_{2 \times 3}(\mathbb{R})$ and let β be the ordered basis $(e_{11}, e_{12}, e_{21}, e_{22}, e_{31}, e_{32})$ for $M_{3 \times 2}(\mathbb{R})$.

(a) Show that T is an isomorphism by giving the inverse transformation S of T.

(b) Find $[T]_\alpha^\beta$ and $[S]_\beta^\alpha$.

(c) Verify that $[T]_\alpha^\beta$ and $[S]_\beta^\alpha$ are inverse to each other.

Solution. (a) The inverse of T is clearly given by $S: M_{3\times 2}(\mathbb{R}) \to M_{2\times 3}(\mathbb{R})$ sending B to B^t.

(b) We have

$$[T]_\alpha^\beta = \begin{pmatrix} 1 & 0 & 0 & 0 & 0 & 0 \\ 0 & 0 & 0 & 1 & 0 & 0 \\ 0 & 1 & 0 & 0 & 0 & 0 \\ 0 & 0 & 0 & 0 & 1 & 0 \\ 0 & 0 & 1 & 0 & 0 & 0 \\ 0 & 0 & 0 & 0 & 0 & 1 \end{pmatrix} = e_{11} + e_{32} + e_{53} + e_{24} + e_{45} + e_{66}$$

and

$$[S]_\beta^\alpha = \begin{pmatrix} 1 & 0 & 0 & 0 & 0 & 0 \\ 0 & 0 & 1 & 0 & 0 & 0 \\ 0 & 0 & 0 & 0 & 1 & 0 \\ 0 & 1 & 0 & 0 & 0 & 0 \\ 0 & 0 & 0 & 1 & 0 & 0 \\ 0 & 0 & 0 & 0 & 0 & 1 \end{pmatrix} = e_{11} + e_{42} + e_{23} + e_{54} + e_{35} + e_{66}.$$

(c) We find that

$$\begin{aligned} [T]_\alpha^\beta [S]_\beta^\alpha &= (e_{11} + e_{42} + e_{23} + e_{54} + e_{35} + e_{66}) \\ &\quad \times (e_{11} + e_{32} + e_{53} + e_{24} + e_{45} + e_{66}) \\ &= e_{11} + e_{44} + e_{22} + e_{55} + e_{33} + e_{66} \\ &= I_6 \end{aligned}$$

and

$$\begin{aligned} [S]_\beta^\alpha [T]_\alpha^\beta &= (e_{11} + e_{32} + e_{53} + e_{24} + e_{45} + e_{66}) \\ &\quad \times (e_{11} + e_{42} + e_{23} + e_{54} + e_{35} + e_{66}) \\ &= e_{11} + e_{33} + e_{55} + e_{22} + e_{44} + e_{66} \\ &= I_6. \end{aligned}$$

We can see that $[T]_\alpha^\beta$ and $[S]_\beta^\alpha$ are indeed inverse to each other.   ◇

Example 3.5.8. Find the inverse matrix of

$$
\begin{pmatrix}
0 & 0 & 0 & 0 & 1 \\
1 & 0 & 0 & 0 & 0 \\
0 & 1 & 0 & 0 & 0 \\
0 & 0 & 1 & 0 & 0 \\
0 & 0 & 0 & 1 & 0
\end{pmatrix}.
$$

Solution. Relative to the standard basis $\alpha = (e_1, e_2, e_3, e_4, e_5)$, the associated linear transformation of the given matrix is $T: \mathbb{R}^5 \to \mathbb{R}^5$ sending

$$e_1 \mapsto e_2, \quad e_2 \mapsto e_3, \quad e_3 \mapsto e_4, \quad e_4 \mapsto e_5 \quad \text{and} \quad e_5 \mapsto e_1.$$

The inverse of T is clearly the linear transformation $S: \mathbb{R}^5 \to \mathbb{R}^5$ sending

$$e_1 \mapsto e_5, \quad e_2 \mapsto e_1, \quad e_3 \mapsto e_2, \quad e_4 \mapsto e_3 \quad \text{and} \quad e_5 \mapsto e_4.$$

Hence the inverse of A is

$$
[S]_\alpha^\alpha =
\begin{pmatrix}
0 & 1 & 0 & 0 & 0 \\
0 & 0 & 1 & 0 & 0 \\
0 & 0 & 0 & 1 & 0 \\
0 & 0 & 0 & 0 & 1 \\
1 & 0 & 0 & 0 & 0
\end{pmatrix}
$$

by Theorem 3.5.5.   ◇

Exercises 3.5

1. Let A be an invertible matrix. Show that $(A^{-1})^{-1} = A$ and $(A^t)^{-1} = (A^{-1})^t$.

2. Let

$$
A = \begin{pmatrix}
1 & 2 & -3 \\
0 & -1 & 2 \\
1 & 1 & 0
\end{pmatrix}
$$

be the matrix of $S\colon \mathbb{R}^3 \to \mathbb{R}^3$ relative to the standard bases. Let A also be the matrix of $T\colon \mathbb{R}^3 \to \mathbb{R}^3$ relative to the standard basis and the basis $((1,1,1),\ (1,1,0),\ (1,0,0))$. Compute $T \circ S(1,2,1)$.

3. Show that the matrix

$$
\begin{pmatrix}
0 & 1 & 0 & 0 & 0 \\
0 & 0 & 0 & 0 & 2 \\
0 & 0 & 3 & 0 & 0 \\
0 & 0 & 0 & 4 & 0 \\
5 & 0 & 0 & 0 & 0
\end{pmatrix}
$$

is invertible and find its inverse matrix.

4. Show that

$$
\left(\sum_{\substack{i=1,2,\ldots,l \\ j=1,2,\ldots,m}} a_{ij} e_{ij}^{(\ell,m)} \right) \left(\sum_{\substack{k=1,2,\ldots,m \\ l=1,2,\ldots,n}} b_{kl} e_{kl}^{(m,n)} \right)
$$

$$
= \sum_{\substack{i=1,2,\ldots,l \\ j=1,2,\ldots,n}} \left(\sum_{k=1}^{m} a_{ik} b_{kj} \right) e_{ij}^{(\ell,n)}
$$

if the multiplication is distributive with respect to the addition. This coincides with our usual definition of the multiplication of matrices.

5. Let $A,\ B,\ C \in M_n(\mathbb{R})$. Prove the following statements.

   (a) If $AB = AC = I_n = CA = BA$ then $B = C$.
   (b) If $AB = I_n = CA$ then $B = C$.

6. We say that a square matrix A is **idempotent** if $A^2 = A$. For the following problems assume that A is idempotent.

   (a) Show that $I - A$ is idempotent.
   (b) Show that $A = I$ if A is invertible.
   (c) Show that $1 - 2A$ is invertible.

7. We say a square matrix A is **nilpotent** if $A^n = 0$ for some positive integer n. Show that a nilpotent matrix is never invertible.

Review Exercises for Chapter 3

1. Does there exist a linear transformation $T\colon \mathbb{R}^3 \to \mathbb{R}^3$ such that $\operatorname{Range}(T) = \operatorname{Null}(T)$?

2. In $M_n(\mathbb{R})$, given the necessary and sufficient condition for i and j such that e_{ij} is

   (a) idempotent;

   (b) nilpotent.

3. In $M_n(\mathbb{R})$, show that

$$A = e_{12} + e_{23} + \cdots + e_{n-1,n}$$

   is nilpotent.

4. Let A be a square matrix of size n. Show that A is invertible if and only if $\operatorname{col}\operatorname{rk} A = n$ if and only if nullity $A = 0$.

5. Let $A \in M_n(\mathbb{R})$ be idempotent. Show that $I_n + A$ is invertible.

6. Let A, B and $A + B$ be invertible matrices of size n. Show that $A^{-1} + B^{-1}$ is invertible. (Hint: If you cannot think of anything else, try showing that $A(A+B)^{-1}B$ is the inverse matrix of $A^{-1} + B^{-1}$.)

CHAPTER 4

Elementary Matrix Operations

We have seen that linear transformations and matrices are the two faces of the same thing. We may solve a problem on linear transformations by working on matrices. We may also solve a problem on matrices by working on linear transformations. In this chapter we will discuss how to handle matrices. It will let us answer more difficult questions.

Throughout this chapter all vector spaces are over $\mathbb{R}$ and all matrices have entries in $\mathbb{R}$ unless otherwise noted.

4.1   Elementary matrix operations

In this section we will discuss some of the most important invertible matrices.

Remember that in §1.4 we used the *elementary row operations* to change the augmented matrix of a system of linear equations into its reduced echelon form. This will help us solve the given system of linear equations. We will carry on the discussion of elementary row (or column) operations in more detail in this section.

Row and column operations.   Before we start let's first make an observation. Let $A \in M_{m \times n}(\mathbb{R})$. For our purpose we will view A as a column m-vector with row n-vectors as its entries. In other words, we may view A as

$$A = \begin{pmatrix} \mathbf{r}_1 \\ \mathbf{r}_2 \\ \vdots \\ \mathbf{r}_m \end{pmatrix}$$

where $\mathbf{r}_i$ is a row n-vector for each i. Notice that

$$\begin{pmatrix} a_1, & a_2, & \ldots, & a_m \end{pmatrix} \begin{pmatrix} \mathbf{r}_1 \\ \mathbf{r}_2 \\ \vdots \\ \mathbf{r}_m \end{pmatrix} = \begin{pmatrix} a_1\mathbf{r}_1 + a_2\mathbf{r}_2 + \cdots + a_m\mathbf{r}_m \end{pmatrix}.$$

If we multiply a row m-vector to the left of A we obtain a linear combination of the rows of A. More generally,

$$\begin{pmatrix} a_{11} & a_{12} & \cdots & a_{1m} \\ a_{21} & a_{22} & \cdots & a_{2m} \\ \vdots & \vdots & \ddots & \vdots \\ a_{\ell 1} & a_{\ell 2} & \cdots & a_{\ell m} \end{pmatrix} \begin{pmatrix} \mathbf{r}_1 \\ \mathbf{r}_2 \\ \vdots \\ \mathbf{r}_m \end{pmatrix} = \begin{pmatrix} a_{11}\mathbf{r}_1 + a_{12}\mathbf{r}_2 + \cdots + a_{1m}\mathbf{r}_m \\ a_{21}\mathbf{r}_1 + a_{22}\mathbf{r}_2 + \cdots + a_{2m}\mathbf{r}_m \\ \vdots \\ a_{\ell 1}\mathbf{r}_1 + a_{\ell 2}\mathbf{r}_2 + \cdots + a_{\ell m}\mathbf{r}_m \end{pmatrix}.$$

If we multiply an $\ell \times m$ matrix to the left of A, we obtain an $\ell \times n$ matrix whose rows are linear combinations of the rows of A. Thus we can see that to multiply a matrix to the left of A is equivalent to performing a **row operation** on A.

Similarly, if we view A as a row n-vector with column m-vector as its entries,

$$A = \begin{pmatrix} \mathbf{c}_1, & \mathbf{c}_2, & \ldots, & \mathbf{c}_n \end{pmatrix},$$

we have that

$$\begin{pmatrix} \mathbf{c}_1 & \mathbf{c}_2 & \ldots & \mathbf{c}_n \end{pmatrix} \begin{pmatrix} b_{11} & b_{12} & \ldots & b_{1p} \\ b_{21} & b_{22} & \ldots & b_{2p} \\ \vdots & \vdots & \ddots & \vdots \\ b_{n1} & b_{n2} & \ldots & b_{np} \end{pmatrix}$$

$$= (b_{11}\mathbf{c}_1 + b_{21}\mathbf{c}_2 + \cdots + b_{n1}\mathbf{c}_n, \ b_{12}\mathbf{c}_1 + b_{22}\mathbf{c}_2 + \cdots + b_{n2}\mathbf{c}_n,$$
$$\ldots, \ b_{1p}\mathbf{c}_1 + b_{2p}\mathbf{c}_2 + \cdots + b_{np}\mathbf{c}_n).$$

If we multiply an $n \times p$ matrix to the right of A, we obtain an $m \times p$ matrix whose columns are linear combinations of the columns of A. Thus we can see that to multiply a matrix to the right of A is equivalent to performing a **column operation** on A.

Elementary matrix operations and elementary matrices. Among all row and column operations, some are of particular importance.

Let $A \in M_{m \times n}(\mathbb{R})$. Any one of the following three operations on the rows [columns] of A is called an **elementary row [column] operation**:

(Type I) exchanging any two rows [columns] of A;

(Type II) multiplying any row [column] of A by a nonzero scalar;

(Type III) adding a scalar multiple of a row [column] of A to another row [column].

Actually we have already encountered these elementary operations in §1.4.

Elementary row [column] operations can be obtained by multiplying a square matrix (of the appropriate size) to the left [right] of the matrix which is to be performed the operations on. The corresponding matrix are called **elementary matrices**. The elementary matrix is said to be of **type I, type II** or **type III** according to whether the elementary operation performed is of type I, type II or type III.

What are the elementary matrices? They are quite easy to figure out. To perform an elementary row operation to A is equivalent to multiplying an elementary matrix E to the left of A, that is, the new matrix is EA. How do we find E? First note that if A is of size $m \times n$, then we expect EA is also of size $m \times n$. This forces E to be of size $m \times m$. We only need to perform that elementary operation to I_m and we obtain $EI = E$. Similarly, if we are to obtain the matrix of an elementary column operation we simply perform the same elementary column operation to I_n.

Elementary Matrices of type I. If we are to exchange the i-th and the j-th rows of A we need to multiply the matrix

$$P_{ij} = I - e_{ii} - e_{jj} + e_{ij} + e_{ji}$$

to the left of A. If, instead, we multiply P_{ij} (of the appropriate size) to the right of A, we exchange the i-th and the j-th column of A.

Elementary Matrices of type II. If we are to multiply the i-th row of A by a scalar $u \neq 0$, we need to multiply the matrix

$$D_i(u) = I + (u - 1)e_{ii}$$

to the left of A. If instead we multiply $D_i(u)$ to the right of A then we multiply the i-th column of A by the nonzero scalar u.

Elementary Matrices of type III. If we want to add to the i-th row the row obtained by b times the j-th row, we need to multiply the matrix

$$T_{ij}(b) = I + be_{ij}$$

to the left of A. If instead we multiply $T_{ij}(b)$ to the right of A we obtain a matrix obtained by replacing the j-th column of A by b times the i-th column of A plus the j-th column of A.

Example 4.1.1. Compute the following matrices:

$$(1) \ A = \begin{pmatrix} 0 & 0 & 1 \\ 0 & 1 & 0 \\ 1 & 0 & 0 \end{pmatrix} \begin{pmatrix} 1 & 2 \\ -7 & 2 \\ -4 & 9 \end{pmatrix}$$

$$(2) \ B = \begin{pmatrix} 1 & 2 \\ -7 & 2 \\ -4 & 9 \end{pmatrix} \begin{pmatrix} 1 & 0 \\ 0 & -0.9 \end{pmatrix}$$

$$(3) \ C = \begin{pmatrix} 1 & 0 & 0 \\ 0 & 1 & 0 \\ 7 & 0 & 1 \end{pmatrix} \begin{pmatrix} 1 & 2 \\ -7 & 2 \\ -4 & 9 \end{pmatrix}$$

$$(4) \ D = \begin{pmatrix} 0 & 0 & 1 \\ 0 & 1 & 0 \\ 1 & 0 & 0 \end{pmatrix} \begin{pmatrix} 1 & 0 & 0 \\ 0 & -0.9 & 0 \\ 0 & 0 & 1 \end{pmatrix} \begin{pmatrix} 1 & 0 & 0 \\ 0 & 1 & 0 \\ 7 & 0 & 1 \end{pmatrix} \begin{pmatrix} 1 & 2 \\ -7 & 2 \\ -4 & 9 \end{pmatrix}$$

$$(5) \ E = \begin{pmatrix} 1 & 2 & 1 \\ -7 & 2 & 3 \\ -4 & 9 & 7 \end{pmatrix} \begin{pmatrix} 1 & 0 & 0 \\ 0 & 3 & 0 \\ 7 & 0 & 1 \end{pmatrix}$$

Solution. To compute these matrices efficiently, it is essential that one recognizes the elementary matrices.

(1) The matrix on the left is an elementary matrix of type I, and it performs an elementary row operation to the matrix on the right. Observe that the matrix on the left is obtained by exchanging the first row and the third row of the I_3. Thus A is obtained by exchanging the first row and the third row of the matrix on the right. We conclude that

$$A = \begin{pmatrix} -4 & 9 \\ -7 & 2 \\ 1 & 2 \end{pmatrix}.$$

(2) The matrix on the right is an elementary matrix of type II. It is obtained by multiplying the second column by -0.9. Hence

$$B = \begin{pmatrix} 1 & -1.8 \\ -7 & -1.8 \\ -4 & -8.1 \end{pmatrix}.$$

(3) The matrix on the left performs an elementary row operation of type III on the matrix on the right. It adds 7 times the first row to the third row. Hence

$$C = \begin{pmatrix} 1 & 2 \\ -7 & 2 \\ 3 & 23 \end{pmatrix}.$$

(4) The three matrices on the left are all elementary matrices. They perform three consecutive elementary row operations on the last matrix. Thus,

$$
\begin{pmatrix} 1 & 2 \\ -7 & 2 \\ -4 & 9 \end{pmatrix} \rightsquigarrow \begin{pmatrix} 1 & 2 \\ -7 & 2 \\ 3 & 23 \end{pmatrix} \rightsquigarrow \begin{pmatrix} 1 & 2 \\ 6.3 & -1.8 \\ 3 & 23 \end{pmatrix}
$$

$$
\rightsquigarrow D = \begin{pmatrix} 3 & 23 \\ 6.3 & -1.8 \\ 1 & 2 \end{pmatrix}.
$$

(5) Although that matrix on the right is not an elementary matrix, one can observe that it is obtained by performing two elementary column operations on I_3: adding 7 times the third column to the first column, and multiplying the second column by 3. We may perform these two operations to obtain

$$
E = \begin{pmatrix} 8 & 6 & 1 \\ 14 & 6 & 3 \\ 45 & 27 & 7 \end{pmatrix}.
$$

As we can see from these examples, there is no need to carry out the multiplication of the matrices when some of the matrices are elementary matrices.   ◇

The inverse of an elementary matrix is an elementary matrix. You have probably noticed that all the elementary row [column] operations can be reversed by an elementary row [column] operation. This means that the corresponding elementary matrices must be invertible.

If we want to reverse the elementary row operation of exchanging the i-th and the j-th row of A, we simply exchange it back. This tells us that

$$
(P_{ij})^{-1} = P_{ij}.
$$

If we want to reverse the elementary row operation of multiplying the i-th row by $u \neq 0$, we multiply the i-th row by u^{-1}. Hence we have

$$
(D_i(u))^{-1} = D_i(u^{-1}).
$$

If we want to reverse the elementary row operation of adding b times the j-th row to the i-th row of A, we add again to the i-th row of A by $(-b)$ times the j-th row of A. Hence we have

$$(T_{ij}(b))^{-1} = T_{ij}(-b).$$

We leave it as an exercise (see Exercise (1)) to verify the assertions by actually multiplying the matrices. We summarize these results in the theorem below.

Theorem 4.1.2. *Elementary matrices are invertible, and the inverse of an elementary matrix is an elementary matrix of the same type.*

We also rephrase an old result in the following theorem.

Theorem 4.1.3. *Let $A \in M_{m \times n}(\mathbb{R})$. We can find a series of elementary matrices $E_1, E_2, \ldots, E_N$ of size m such that*

$$E_N \cdots E_2 E_1 A$$

is the reduced row echelon form of A. We can also find a series of elementary matrices $E_1', E_2', \ldots, E_M'$ of size n such that

$$A E_1' E_2' \cdots E_M'$$

is the reduced column echelon form of A.

Exercises 4.1

1. Verify the following identities by actually multiplying the matrices:

   (a) $(P_{ij})^{-1} = P_{ij}$;

   (b) $(D_i(u))^{-1} = D_i(u^{-1})$;

   (c) $(T_{ij}(b))^{-1} = T_{ij}(-b)$.

2. Suppose that $A = B_1 B_2 \cdots B_{n-1} B_n$ where B_i is an invertible matrix for each i. Show that

$$A^{-1} = B_n^{-1} B_{n-1}^{-1} \cdots B_2^{-1} B_1^{-1}.$$

Hence A is invertible. Conclude that a product of invertible matrices is invertible. Give a different proof for this fact using Proposition 3.2.3(c).

3. Let

$$A = \begin{pmatrix} 1 & 0 & -2 \\ 0 & 1 & 0 \\ 0 & 0 & 1 \end{pmatrix} \begin{pmatrix} 1 & 0 & 0 \\ 0 & -2 & 0 \\ 0 & 0 & 1 \end{pmatrix} \begin{pmatrix} 0 & 0 & 1 \\ 0 & 1 & 0 \\ 1 & 0 & 0 \end{pmatrix}.$$

Compute A and A^{-1}.

4. Factorize

$$A = \begin{pmatrix} 1 & 0 & -3 & 1 \\ 0 & 0 & 1 & 0 \\ 0 & 1 & 0 & 0 \\ 1 & 0 & 0 & 5 \end{pmatrix}$$

as a product of elementary matrices. Compute A^{-1}.

5. Show that the transpose of an elementary matrix is still an elementary matrix.

4.2   The rank and the inverse of a matrix

The rank of a matrix.   Remember that the *rank* of the linear transformation $T \colon V \to W$ is defined to be the dimension of Range (T). To compute $\operatorname{rk} T$, it is sufficient to compute the column rank of any matrix representing T. Now we elaborate on how to do this.

As a previous exercise, we have the following result.

Lemma 4.2.1. *Let $S \colon U \to V$ and $T \colon V \to W$ be linear transformations. Then*

$$\operatorname{rk} T \circ S \leq \operatorname{rk} S \quad and \quad \operatorname{rk} T \circ S \leq \operatorname{rk} T.$$

See Exercise(4)(c) in §3.3.

We may further improve Lemma 4.2.1.

Lemma 4.2.2. *Let* $S\colon U \to V$ *and* $T\colon V \to W$ *be linear transformations. The following are true.*

(a) *If* S *is onto then* $\operatorname{rk} T \circ S = \operatorname{rk} T$.

(b) *If* T *is one-one then* $\operatorname{rk} T \circ S = \operatorname{rk} S$.

Proof. (a) By definition, $\operatorname{rk} T \circ S = \dim T(S(U))$. Since S is onto, we have $\operatorname{rk} T \circ S = \dim T(V) = \operatorname{rk} T$.

(b) We consider the linear transformation

$$T|_{S(U)}\colon S(U) \to W,$$

which is obtained by restricting the domain of T to $S(U)$. Since T is one-one, so is $T|_{S(U)}$. Hence nullity $T|_{S(U)} = 0$. Note that

$$\begin{aligned} \operatorname{rk} T \circ S = \dim T(S(U)) &= \operatorname{rk} T|_{S(U)} \\ &= \dim S(U) - \text{nullity } T|_{S(U)} \\ &= \dim S(U) = \operatorname{rk} S \end{aligned}$$

by applying Proposition 3.3.8 to $T|_{S(U)}$. $\qquad\square$

Proposition 4.2.3. *Let* $S\colon V' \to V$, $T\colon V \to W$ *and* $S'\colon W \to W'$ *be linear transformations. Suppose* S *and* S' *are isomorphisms. Then*

$$\operatorname{rk}(S' \circ T \circ S) = \operatorname{rk} T.$$

Proof. The result follows from Lemma 4.2.2(a) and Lemma 4.2.2(b) applied to S and S' respectively. $\qquad\square$

Definition 4.2.4. Let $A \in M_{m \times n}(\mathbb{R})$. We define $L_A\colon \mathbb{R}^n \to \mathbb{R}^m$ to be the unique linear transformation whose associated matrix with respect to the standard bases of $\mathbb{R}^n$ and $\mathbb{R}^m$ is A.

Remark. We may also define Range (A) and Null (A) to be the range and null space of L_A respectively. However, Range (A) is exactly the column space of A, and so the notation is redundant. Similarly, we define nullity A to be nullity L_A. The rank of A is more tricky as we will see shortly.

Let E be an elementary matrix of size m. Since E is invertible, L_E is an isomorphism. This implies that

$$\operatorname{col rk} \ A = \operatorname{rk} L_A = \operatorname{rk} L_E \circ L_A = \operatorname{col rk} \ EA.$$

Similarly, we have that

$$\operatorname{col rk} \ A = \operatorname{rk} L_A = \operatorname{rk} L_A \circ L_{E'} = \operatorname{col rk} \ AE'$$

for an elementary matrix E' of size n. Thus, we have the following result.

Lemma 4.2.5. *The column rank of a matrix does not change when elementary row or column operations are performed.*

We may perform a series of elementary row operations on A to obtain its reduced row echelon form B. Assume $r = \operatorname{row rk} \ A$ is the number of the nonzero rows in B. If we perform more elementary column operations on B, we can further transform B into the matrix

$$(4.1) \qquad C = \begin{pmatrix} 1 & & & & & \\ & \ddots & & & & \\ & & 1 & & & \\ & & & 0 & & \\ & & & & \ddots & \\ & & & & & 0 \end{pmatrix} = \begin{pmatrix} I_r & \\ & \mathbf{0}_{(m-r) \times (n-r)} \end{pmatrix}.$$

(It is customary to sometimes leave a blank space where the entry is 0 in a matrix.) Thus, we may conclude that $\operatorname{row rk} \ A = r = \operatorname{col rk} \ C = \operatorname{col rk} \ A$. We now summarize the result below.

Theorem 4.2.6. *Let $A \in M_{m \times n}(\mathbb{R})$. Then there are elementary matrices $E_1, \ldots, E_k$ and $E'_1, \ldots, E'_l$ such that*

$$E_1 \cdots E_k A E'_1 \cdots E'_l = \begin{pmatrix} I_r & \\ & \mathbf{0}_{(m-r) \times (n-r)} \end{pmatrix},$$

where $r = \operatorname{row rk} \ A = \operatorname{col rk} \ A$.

Definition 4.2.7. We define the **rank** of a matrix A, denoted $\operatorname{rk} A$, to be the row (or equivalently, column) rank of A.

Corollary 4.2.8. *The rank of a matrix does not change when elementary operations are performed.*

Example 4.2.9. Find the rank of

$$A = \begin{pmatrix} 1 & & & -4 \\ & 1 & & -3 \\ & & 1 & -2 \\ 2 & & -4 & \\ & 2 & -3 & \\ 3 & -4 & & \end{pmatrix}.$$

Solution. We perform appropriate elementary row- and column operations:

$$A \rightsquigarrow \begin{pmatrix} 1 & & & -4 \\ & 1 & & -3 \\ & & 1 & -2 \\ 0 & & -4 & 8 \\ & 2 & -3 & \\ 0 & -4 & & 12 \end{pmatrix} \quad \begin{array}{l} (-2) \times \text{row(i)} + \text{row(iv)} \\[2em] (-3) \times \text{row(i)} + \text{row(vi)} \end{array}$$

$$\rightsquigarrow \begin{pmatrix} 1 & & & -4 \\ & 1 & & -3 \\ & & 1 & -2 \\ & & -4 & 8 \\ & 0 & -3 & 6 \\ & 0 & & 0 \end{pmatrix} \quad \begin{array}{l} (-2) \times \text{row(ii)} + \text{row(v)} \\ 4 \times \text{row(ii)} + \text{row(vi)} \end{array}$$

$$\rightsquigarrow \begin{pmatrix} 1 & & & -4 \\ & 1 & & -3 \\ & & 1 & -2 \\ & & 0 & 0 \\ & & 0 & 0 \\ 0 & 0 & 0 & 0 \end{pmatrix} \quad \begin{array}{l} 4 \times \text{row(iii)} + \text{row(iv)} \\ 3 \times \text{row(iii)} + \text{row(v)} \end{array}$$

we may further perform elementary column operations so that the matrix becomes

$$\begin{pmatrix} I_3 & \mathbf{0}_{3\times 1} \\ \mathbf{0}_{3\times 3} & \mathbf{0}_{1\times 3} \end{pmatrix}.$$

However, this is not necessary. Once we reach the reduced row echelon form, we already know that $\operatorname{rk} A = 3$.   ◇

The inverse of a matrix. In this subsection we provide a method to compute the inverse of a matrix.

Proposition 4.2.10. *Let $A \in M_n(\mathbb{R})$. Then A is invertible if and only if $\operatorname{rk} A = n$ if and only if the reduced echelon form of A is I_n.*

Proof. By Theorem 3.5.5, A is invertible if and only if L_A is an isomorphism. By Corollary 3.3.10, L_A is an isomorphism if and only if L_A is onto. In other words, this is true if and only if $\operatorname{rk} A = \operatorname{rk} L_A = n$. Remember that $\operatorname{rk} A$ is the number of non-zero rows in the reduced row echelon form of A. Thus A is invertible if and only if the reduced echelon form of A has n nonzero rows. In this case, the reduced row echelon form of A is I_n since A has only n columns.   □

The proposition above tells us we only need elementary row operations to transform an invertible matrix into the identity matrix. Of course, we can also perform elementary column operations exclusively to an invertible matrix to obtain the identity matrix.

As an immediate corollary to Proposition 4.2.10 and Theorem 4.1.3 we have the following results.

Corollary 4.2.11. *Let $A \in M_n(\mathbb{R})$. Then A is invertible if and only if there are elementary matrices $E_1, \ldots, E_N$ such that*

$$E_N \cdots E_1 A = I_n.$$

In this case

$$A^{-1} = E_N \cdots E_1.$$

Corollary 4.2.12. *Any invertible matrix over $\mathbb{R}$ is a product of elementary matrices.*

Proof. Use the notation in Corollary 4.2.11, we have $A = E_1^{-1} \cdots E_n^{-1}$. □

Example 4.2.13. Express $A = \begin{pmatrix} 1 & 2 \\ 3 & 4 \end{pmatrix}$ as a product of elementary matrices.

Solution. Adding $(-3) \times$ row(i) to row(ii), we have

$$\begin{pmatrix} 1 & 0 \\ -3 & 1 \end{pmatrix} \begin{pmatrix} 1 & 2 \\ 3 & 4 \end{pmatrix} = \begin{pmatrix} 1 & 2 \\ 0 & -2 \end{pmatrix}.$$

Next we multiply row(ii) by $(-1/2)$,

$$\begin{pmatrix} 1 & 0 \\ 0 & -1/2 \end{pmatrix} \begin{pmatrix} 1 & 2 \\ 0 & -2 \end{pmatrix} = \begin{pmatrix} 1 & 2 \\ 0 & 1 \end{pmatrix}.$$

Finally, we add $(-2) \times$ row(ii) to row(i),

$$\begin{pmatrix} 1 & -2 \\ 0 & 1 \end{pmatrix} \begin{pmatrix} 1 & 2 \\ 0 & 1 \end{pmatrix} = \begin{pmatrix} 1 & 0 \\ 0 & 1 \end{pmatrix}.$$

Combining the process together, we have

$$\begin{pmatrix} 1 & -2 \\ 0 & 1 \end{pmatrix} \begin{pmatrix} 1 & 0 \\ 0 & -1/2 \end{pmatrix} \begin{pmatrix} 1 & 0 \\ -3 & 1 \end{pmatrix} \begin{pmatrix} 1 & 2 \\ 3 & 4 \end{pmatrix} = \begin{pmatrix} 1 & 0 \\ 0 & 1 \end{pmatrix}.$$

Hence

$$\begin{pmatrix} 1 & 2 \\ 3 & 4 \end{pmatrix} = \begin{pmatrix} 1 & 0 \\ -3 & 1 \end{pmatrix}^{-1} \begin{pmatrix} 1 & 0 \\ 0 & -1/2 \end{pmatrix}^{-1} \begin{pmatrix} 1 & -2 \\ 0 & 1 \end{pmatrix}^{-1}$$

$$= \begin{pmatrix} 1 & 0 \\ 3 & 1 \end{pmatrix} \begin{pmatrix} 1 & 0 \\ 0 & -2 \end{pmatrix} \begin{pmatrix} 1 & 2 \\ 0 & 1 \end{pmatrix}$$

using what we know about inverses of elementary matrices.   ◇

We may use Corollary 4.2.11 to find the inverse of an invertible matrix. First we form the $n \times (2n)$ matrix

$$\left(A \mid I_n \right).$$

We then perform a series of elementary row operations to this matrix to transform A on the left into its reduced echelon form. If A is not invertible then the rank of A would be less than n. The number of nonzero rows will be less than n, and there will be zero rows in the reduced row echelon form of A. On the other hand, if A is invertible, the reduced row echelon form of A is I_n. Suppose that

$$E_N \cdots E_1 A = I_n$$

where the E_i's are elementary matrices. Then

$$E_N \cdots E_1 \left(A \,\middle|\, I_n \right)$$
$$= \left(E_N \cdots E_1 A \,\middle|\, E_N \cdots E_1 I_n \right)$$
$$= \left(I_n \,\middle|\, E_N \cdots E_1 \right)$$
$$= \left(I_n \,\middle|\, A^{-1} \right).$$

When we obtain I_n on the left, the I_n on the right becomes the inverse matrix of A. This is often the simplest way to verify whether a matrix is invertible, and to compute the inverse of a matrix on the side, especially for large size matrices.

Example 4.2.14. Let

$$A = \begin{pmatrix} 0 & 1 & 1 \\ 1 & 0 & 1 \\ 1 & 1 & 0 \end{pmatrix} \in M_3(\mathbb{Q}).$$

We perform elementary row operations on

$$\begin{pmatrix} 0 & 1 & 1 & \big| & 1 & 0 & 0 \\ 1 & 0 & 1 & \big| & 0 & 1 & 0 \\ 1 & 1 & 0 & \big| & 0 & 0 & 1 \end{pmatrix}$$

$$\rightsquigarrow \begin{pmatrix} 1 & 0 & 1 & \big| & 0 & 1 & 0 \\ 0 & 1 & 1 & \big| & 1 & 0 & 0 \\ 1 & 1 & 0 & \big| & 0 & 0 & 1 \end{pmatrix} \begin{array}{l} \text{row(ii)} \\ \text{row(i)} \\ \end{array}$$

$$\rightsquigarrow \begin{pmatrix} 1 & 0 & 1 & \big| & 0 & 1 & 0 \\ 0 & 1 & 1 & \big| & 1 & 0 & 0 \\ 0 & 1 & -1 & \big| & 0 & -1 & 1 \end{pmatrix} \begin{array}{l} \\ \\ \text{row(iii)} - \text{row(i)} \end{array}$$

$$\rightsquigarrow \begin{pmatrix} 1 & 0 & 1 & \big| & 0 & 1 & 0 \\ 0 & 1 & 1 & \big| & 1 & 0 & 0 \\ 0 & 0 & -2 & \big| & -1 & -1 & 1 \end{pmatrix} \begin{array}{l} \\ \\ \text{row(iii)} - \text{row(ii)} \end{array}$$

$$\rightsquigarrow \begin{pmatrix} 1 & 0 & 1 & \big| & 0 & 1 & 0 \\ 0 & 1 & 1 & \big| & 1 & 0 & 0 \\ 0 & 0 & 1 & \big| & 1/2 & 1/2 & -1/2 \end{pmatrix} \begin{array}{l} \\ \\ (-1/2) \times \text{row(iii)} \end{array}$$

$$\leadsto \left(\begin{array}{ccc|ccc} 1 & 0 & 0 & -1/2 & 1/2 & 1/2 \\ 0 & 1 & 0 & 1/2 & -1/2 & 1/2 \\ 0 & 0 & 1 & 1/2 & 1/2 & -1/2 \end{array}\right) \begin{array}{l} \text{row(i)} - \text{row(iii)} \\ \text{row(ii)} - \text{row(iii)} \end{array} .$$

Hence A is invertible and

$$A^{-1} = \left(\begin{array}{ccc} -1/2 & 1/2 & 1/2 \\ 1/2 & -1/2 & 1/2 \\ 1/2 & 1/2 & -1/2 \end{array}\right).$$

To make it more fun, let's try the same matrix over a different field.

Example 4.2.15. Consider

$$A = \begin{pmatrix} 0 & 1 & 1 \\ 1 & 0 & 1 \\ 1 & 1 & 0 \end{pmatrix}$$

as a matrix in $M_3(\mathbb{Z}_2)$. We want to determine whether this matrix is invertible.

We perform elementary row operations on

$$A \leadsto \begin{pmatrix} 1 & 0 & 1 \\ 0 & 1 & 1 \\ 1 & 1 & 0 \end{pmatrix} \begin{array}{l} \text{row(ii)} \\ \text{row(i)} \end{array}$$

$$\leadsto \begin{pmatrix} 1 & 0 & 1 \\ 0 & 1 & 1 \\ 0 & 1 & 1 \end{pmatrix} \begin{array}{l} \\ \\ \text{row(iii)} + \text{row(i)} \end{array}$$

$$\leadsto \begin{pmatrix} 1 & 0 & 1 \\ 0 & 1 & 1 \\ 0 & 0 & 0 \end{pmatrix} \begin{array}{l} \\ \\ \text{row(iii)} + \text{row(ii)} \end{array} .$$

Contrary to the case over $\mathbb{Q}$, the matrix A is not invertible. We can see that over $\mathbb{Z}_2$, the computing process is effortless.

Example 4.2.16. Let

$$A = \begin{pmatrix} 0 & 2 & 4 \\ 2 & 4 & 2 \\ 3 & 3 & 1 \end{pmatrix}.$$

To find its inverse, we perform elementary row operations on

$$\begin{pmatrix} 0 & 2 & 4 & | & 1 & 0 & 0 \\ 2 & 4 & 2 & | & 0 & 1 & 0 \\ 3 & 3 & 1 & | & 0 & 0 & 1 \end{pmatrix}$$

$$\rightsquigarrow \begin{pmatrix} 1 & 2 & 1 & | & 0 & 1/2 & 0 \\ 0 & 1 & 2 & | & 1/2 & 0 & 0 \\ 3 & 3 & 1 & | & 0 & 0 & 1 \end{pmatrix} \begin{array}{l} (1/2) \times \text{row(ii)} \\ (1/2) \times \text{row(i)} \end{array}$$

$$\rightsquigarrow \begin{pmatrix} 1 & 2 & 1 & | & 0 & 1/2 & 0 \\ 0 & 1 & 2 & | & 1/2 & 0 & 0 \\ 0 & -3 & -2 & | & 0 & -3/2 & 1 \end{pmatrix} \begin{array}{l} \\ \\ \text{row(iii)} - 3 \times \text{row(i)} \end{array}$$

$$\rightsquigarrow \begin{pmatrix} 1 & 0 & -3 & | & -1 & 1/2 & 0 \\ 0 & 1 & 2 & | & 1/2 & 0 & 0 \\ 0 & 0 & 4 & | & 3/2 & -3/2 & 1 \end{pmatrix} \begin{array}{l} \text{row(i)} - 2 \times \text{row(ii)} \\ \\ \text{row(iii)} + 3 \times \text{row(ii)} \end{array}$$

$$\rightsquigarrow \begin{pmatrix} 1 & 0 & -3 & | & -1 & 1/2 & 0 \\ 0 & 1 & 2 & | & 1/2 & 0 & 0 \\ 0 & 0 & 1 & | & 3/8 & -3/8 & 1/4 \end{pmatrix} \begin{array}{l} \\ \\ (1/4) \times \text{row(iii)} \end{array}$$

$$\rightsquigarrow \begin{pmatrix} 1 & 0 & 0 & | & 1/8 & -5/8 & 3/4 \\ 0 & 1 & 0 & | & -1/4 & 3/4 & -1/2 \\ 0 & 0 & 1 & | & 3/8 & -3/8 & 1/4 \end{pmatrix} \begin{array}{l} \text{row(i)} + 3 \times \text{row(iii)} \\ \text{row(ii)} - 2 \times \text{row(iii)} \end{array} .$$

Hence A is invertible and

$$A^{-1} = \begin{pmatrix} 1/8 & -5/8 & 3/4 \\ -1/4 & 3/4 & -1/2 \\ 3/8 & -3/8 & 1/4 \end{pmatrix} .$$

Example 4.2.17. Let $T \colon \mathscr{P}_2 \to \mathscr{P}_2$ be given by

$$T(f(x)) = f(x) + f'(x) + f''(x).$$

We want to determine whether T is an isomorphism.

Let $\alpha = (1, x, x^2)$ be the standard basis for $\mathscr{P}_2$. Then

$$T(1) = 1,$$
$$T(x) = x + 1,$$
$$T(x^2) = x^2 + 2x + 2$$

and

$$[T]_\alpha^\alpha = \begin{pmatrix} 1 & 1 & 2 \\ 0 & 1 & 2 \\ 0 & 0 & 1 \end{pmatrix}.$$

At this point, it is already clear that $[T]_\alpha^\alpha$ is invertible since it is upper triangular with nonzero entries on the diagonal. Hence T is indeed an isomorphism.

Now we proceed to find the inverse of T. To do so we need to find $([T]_\alpha^\alpha)^{-1}$. We perform elementary row operations on

$$\begin{pmatrix} 1 & 1 & 2 & \bigm| & 1 & 0 & 0 \\ 0 & 1 & 2 & \bigm| & 0 & 1 & 0 \\ 0 & 0 & 1 & \bigm| & 0 & 0 & 1 \end{pmatrix}$$

$$\rightsquigarrow \begin{pmatrix} 1 & 1 & 0 & \bigm| & 1 & 0 & -2 \\ 0 & 1 & 0 & \bigm| & 0 & 1 & -2 \\ 0 & 0 & 1 & \bigm| & 0 & 0 & 1 \end{pmatrix} \quad \begin{matrix} \text{row(i)} - 2 \times \text{row(iii)} \\ \text{row(ii)} - 2 \times \text{row(iii)} \end{matrix}$$

$$\rightsquigarrow \begin{pmatrix} 1 & 0 & 0 & \bigm| & 1 & -1 & 0 \\ 0 & 1 & 0 & \bigm| & 0 & 1 & -2 \\ 0 & 0 & 1 & \bigm| & 0 & 0 & 1 \end{pmatrix} \quad \text{row(i)} - \text{row(ii)}$$

.

Hence

$$([T]_\alpha^\alpha)^{-1} = \begin{pmatrix} 1 & -1 & 0 \\ 0 & 1 & -2 \\ 0 & 0 & 1 \end{pmatrix}.$$

Thus T^{-1} is the linear transformation such that

$$T^{-1}(1) = 1,$$
$$T^{-1}(x) = -1 + x,$$
$$T^{-1}(x^2) = -2x + x^2.$$

Example 4.2.18. Is there a 3×5 matrix A whose first, second and fourth columns of A are

$$\begin{pmatrix} 1 \\ -1 \\ 3 \end{pmatrix}, \quad \begin{pmatrix} 0 \\ -1 \\ 1 \end{pmatrix} \quad \text{and} \quad \begin{pmatrix} 1 \\ -2 \\ 0 \end{pmatrix}$$

respectively, and whose reduced echelon form is

$$\begin{pmatrix} 1 & 0 & 2 & 0 & -2 \\ 0 & 1 & -5 & 0 & -3 \\ 0 & 0 & 0 & 1 & 6 \end{pmatrix}?$$

Solution. By Theorem 4.1.3 we may find a series of elementary matrices $E_1, \ldots, E_n$ such that

$$E_n \cdots E_1 A = \begin{pmatrix} 1 & 0 & 2 & 0 & -2 \\ 0 & 1 & -5 & 0 & -3 \\ 0 & 0 & 0 & 1 & 6 \end{pmatrix}.$$

If we ignore the third and the fifth column, we have that

$$E_n \cdots E_1 \begin{pmatrix} 1 & 0 & 1 \\ -1 & -1 & -2 \\ 3 & 1 & 0 \end{pmatrix} = \begin{pmatrix} 1 & 0 & 0 \\ 0 & 1 & 0 \\ 0 & 0 & 1 \end{pmatrix}.$$

This says that

$$E_n \cdots E_1 = \begin{pmatrix} 1 & 0 & 1 \\ -1 & -1 & -2 \\ 3 & 1 & 0 \end{pmatrix}^{-1}.$$

However, this is true only if

$$\begin{pmatrix} 1 & 0 & 1 \\ -1 & -1 & -2 \\ 3 & 1 & 0 \end{pmatrix}$$

is invertible. To verify this, we do not need to find the inverse. It suffices to find its rank. We perform elementary row operations on

$$B = \begin{pmatrix} 1 & 0 & 1 \\ -1 & -1 & -2 \\ 3 & 1 & 0 \end{pmatrix} \rightsquigarrow \begin{pmatrix} 1 & 0 & 1 \\ 0 & -1 & -1 \\ 0 & 1 & -3 \end{pmatrix} \rightsquigarrow \begin{pmatrix} 1 & 0 & 1 \\ 0 & -1 & -1 \\ 0 & 0 & -4 \end{pmatrix}.$$

The rank of B is 3. The matrix B is indeed invertible. Hence we have

$$A = \begin{pmatrix} 1 & 0 & 1 \\ -1 & -1 & -2 \\ 3 & 1 & 0 \end{pmatrix} \begin{pmatrix} 1 & 0 & 2 & 0 & -2 \\ 0 & 1 & -5 & 0 & -3 \\ 0 & 0 & 0 & 1 & 6 \end{pmatrix}$$

$$= \begin{pmatrix} 1 & 0 & 2 & 1 & 4 \\ -1 & -1 & 3 & -2 & -7 \\ 3 & 1 & 1 & 0 & -9 \end{pmatrix}.$$

We leave it to the reader to figure out why A is the only matrix satisfying the requirement.   ◇

Systems of linear equations revisited.   Suppose given the system of linear equations

$$A\mathbf{x} = \mathbf{b},$$

where $A = \left(a_{ij}\right)_{m \times n}$, $\mathbf{x} = \begin{pmatrix} x_1 \\ \vdots \\ x_n \end{pmatrix}$ and $\mathbf{b} = \begin{pmatrix} b_1 \\ \vdots \\ b_m \end{pmatrix}$. If A is invertible, the

solution is

$$\mathbf{x} = I_n\mathbf{x} = (A^{-1}A)\mathbf{x} = A^{-1}(A\mathbf{x}) = A^{-1}\mathbf{b},$$

and it is the unique solution of the given system.

One may use the method of this section to compute A^{-1} to obtain the final answer. This is a preferred method to the well-known *Cramer's Rule* (if you know what it is). Cramer's rule is cumbersome to use in comparison. We leave the construction of Cramer's rule in an exercise. See Exercise 9 in S4.3.

Example 4.2.19. Solve the system

$$2y + 4z = 1$$
$$2x + 4y + 2z = 1$$
$$3x + 3y + z = 0.$$

Solution. This system may be written as

$$A \begin{pmatrix} x \\ y \\ z \end{pmatrix} = \begin{pmatrix} 1 \\ 1 \\ 0 \end{pmatrix} \quad \text{where} \quad A = \begin{pmatrix} 0 & 2 & 4 \\ 2 & 4 & 2 \\ 3 & 3 & 1 \end{pmatrix}.$$

From Example 4.2.16, we have that A is invertible and

$$
\begin{pmatrix} x \\ y \\ z \end{pmatrix} = A^{-1} \begin{pmatrix} 1 \\ 1 \\ 0 \end{pmatrix} = \begin{pmatrix} 1/8 & -5/8 & 3/4 \\ -1/4 & 3/4 & -1/2 \\ 3/8 & -3/8 & 1/4 \end{pmatrix} \begin{pmatrix} 1 \\ 1 \\ 0 \end{pmatrix}
$$

$$
= \begin{pmatrix} -1/2 \\ 1/2 \\ 0 \end{pmatrix}.
$$

This is the unique solution for the system.   ◇

Exercises 4.2

1. Let $S: V \to W$ be a linear transformation and let $T: W \to V$ be an isomorphism. Prove or disprove the following conjectures:

   (a) nullity $T \circ S =$ nullity S;

   (b) nullity $T \circ S =$ nullity T;

   (c) nullity $T \circ S =$ nullity $S \circ T$.

2. Let A be an $k \times m$ matrix and B be an $m \times n$ matrix. Show that $\mathrm{rk}\, AB \leq \min\{\mathrm{rk}\, A,\ \mathrm{rk}\, B\}$.

3. Consider the matrix

$$
A = \begin{pmatrix} i & 1 \\ 1 & i \end{pmatrix}
$$

   over $\mathbb{C}$.

   (a) Show that A is invertible by finding its inverse.

   (b) Express A as a product of elementary matrices.

4. Suppose given the system of linear equations

$$
\begin{aligned}
3x + \quad\ \ z &= b_1 \\
-6x + \ y - 2z &= b_2 \\
x + 3y \quad\ \ &= b_3.
\end{aligned}
$$

(a) Let A be the coefficient matrix of the system above. Verify that A is an invertible matrix and find A^{-1}.

(b) Solve the system when $\mathbf{b} = \begin{pmatrix} b_1 \\ b_2 \\ b_3 \end{pmatrix} = \begin{pmatrix} 1 \\ 2 \\ 3 \end{pmatrix}$?

(c) Solve the system when $\mathbf{b} = \begin{pmatrix} b_1 \\ b_2 \\ b_3 \end{pmatrix} = \begin{pmatrix} 1 \\ 0 \\ -1 \end{pmatrix}$?

5. Suppose $T\colon \mathbb{R}^3 \to \mathbb{R}^3$ is the linear transformation sending (x, y, z) to $(3x+z, -6x+y-2z, x+3y)$. Determine whether T is an isomorphism. If yes, find $T^{-1}(1, 0, 1)$.

6. Suppose given a homogeneous system of m linear equations in n variables

(4.2) $a_{i1}x_1 + a_{i2}x_2 + \cdots + a_{in}x_n = b_i, \qquad i = 1, 2, \ldots, n.$

Let A be the coefficient matrix of the system and let r be the number of nonzero rows in the reduced echelon form of A.

(a) Show that the solution space of (4.2) is Null (L_A).

(b) Show that the dimension of the solution space of (4.2) is $n - r$. Since r is the number of pivot variables, $n - r$ is exactly the number of free parameters in the reduced row echelon form of A.

7. Let
$$A = \begin{pmatrix} 1 & -1 & 7 \\ 0 & 1 & -4 \\ 1 & 2 & -5 \\ 0 & -1 & 4 \end{pmatrix}.$$

Find bases β and γ for $\mathbb{R}^3$ and $\mathbb{R}^4$ respectively such that the matrix representing $L_A\colon \mathbb{R}^3 \to \mathbb{R}^4$ relative to β and γ is of the form (4.1).

8. Let $\dim V = n$ and $\dim W = m$. Let $T\colon V \to W$ be a linear transformation such that $\operatorname{rk} T = r$. In this problem we intend to find the simplest matrix you can choose to represent T.

(a) Find a basis $\{w_1, \ldots, w_r\}$ for Range (T). Find $\{v_1, \ldots, v_r\}$ in V such that $T(v_i) = w_i$. Find a basis $\{v_{r+1}, \ldots, v_{r+s}\}$ for Null (T). Remember that $r + s = n$. Show that $\{v_1, \ldots, v_r, v_{r+1}, \ldots, v_n\}$ forms a basis for V.

(b) Expand $\{w_1, \ldots, w_r\}$ to a basis $\{w_1, \ldots, w_r, \ldots, w_m\}$ for W. Let $\beta = (v_1, \ldots, v_n)$ and $\gamma = (w_1, \ldots, w_m)$. Show that

$$[T]_\beta^\gamma = \begin{pmatrix} I_r & \\ & \mathbf{0}_{(m-r)\times(n-r)} \end{pmatrix}.$$

9. Is there a 3×5 matrix A whose first, second and fourth columns of A are

$$\begin{pmatrix} 1 \\ -1 \\ 3 \end{pmatrix}, \quad \begin{pmatrix} 0 \\ -1 \\ 1 \end{pmatrix} \quad \text{and} \quad \begin{pmatrix} 1 \\ -2 \\ 0 \end{pmatrix}$$

respectively, and whose reduced echelon form is

$$\begin{pmatrix} 1 & 0 & 2 & 0 & -2 \\ 0 & 1 & -5 & 0 & -3 \\ 0 & 0 & 0 & 0 & 0 \end{pmatrix}?$$

10. Is there a 3×5 matrix A whose first, third and fourth columns of A are

$$\begin{pmatrix} 1 \\ -1 \\ 3 \end{pmatrix}, \quad \begin{pmatrix} 0 \\ -1 \\ 1 \end{pmatrix} \quad \text{and} \quad \begin{pmatrix} 1 \\ -2 \\ 0 \end{pmatrix}$$

respectively, and whose reduced echelon form is

$$\begin{pmatrix} 1 & 2 & 0 & 0 & 2 \\ 0 & 0 & 1 & 0 & -3 \\ 0 & 0 & 0 & 1 & 1 \end{pmatrix}.$$

11. Is there a 3×6 matrix A whose first, third and fourth columns of A are

$$\begin{pmatrix} 1 \\ -1 \\ 3 \end{pmatrix}, \quad \begin{pmatrix} 0 \\ -1 \\ 1 \end{pmatrix} \quad \text{and} \quad \begin{pmatrix} 3 \\ 2 \\ 4 \end{pmatrix}$$

respectively, and whose reduced echelon form is

$$\begin{pmatrix} 1 & 2 & 0 & 0 & 2 & 1 \\ 0 & 0 & 1 & 0 & -3 & -1 \\ 0 & 0 & 0 & 1 & 1 & 3 \end{pmatrix}.$$

4.3 A description of determinants

In this section we attempt to "describe" the behavior of determinants of square matrices. For this we mean we will only present the rules without rigorous proof. A formal treatment of determinants requires some knowledge of symmetric groups which will be covered in an introductory course of Algebra.

Minors, cofactors and determinants. Let

$$A = \left(a_{ij} \right)_{n \times n} \in M_n(\mathbb{R}).$$

Let $i, j = 1, \ldots n$. We define the (i, j)-**minor** of A, denoted M_{ij}, to be the $(n-1) \times (n-1)$ matrix obtained by deleting the ith row and the jth column of A.

Remark. In some books *minors* refer to the determinant of the $(n-1) \times (n-1)$ matrix we just obtained. Please mind the convention when you start a new book on Linear Algebra.

When $n = 1$ we define the **determinant** of A, denoted by $\det A$, to be a_{11}. We will now describe the determinant of matrices by induction on its size n. Suppose we know how to find the determinant of any square matrix of size $n - 1$ for $n \geq 2$. The (i, j)-**cofactor** of A, denoted A_{ij}, is $(-1)^{i+j}$ times the determinant of M_{ij}. We define the determinant of A to be one of the following expressions (which are all equal)

(4.3)
$$a_{i1}A_{i1} + a_{i2}A_{i2} + \cdots + a_{in}A_{in},$$
$$a_{1j}A_{1j} + a_{2j}A_{2j} + \cdots + a_{nj}A_{nj}.$$

This formula is called the **cofactor expansion** or **Laplace expansion**, named after Pierre-Simon Laplace.

The following is a list of basic properties regarding determinant which we will assume without proof. Although some of them can be explained easily, others will require more mathematical maturity.

PROPERTIES OF DETERMINANT:

(1) $\det \begin{pmatrix} a & b \\ c & d \end{pmatrix} = ad - bc.$

(2) $\det \begin{pmatrix} a & b & c \\ d & e & f \\ g & h & i \end{pmatrix} = aei + cdh + bfg - ceg - afh - bdi.$

(3) If one of the rows or of the columns of a square matrix is zero then its determinant is 0.

(4) If a square matrix has two identical rows (or columns), its determinant is 0.

(5) If the ith row of the square matrix A has only one nonzero entry at the jth column then
$$\det A = a_{ij} A_{ij}.$$

Similarly, If the jth column of the square matrix A has only one nonzero entry at the ith row then

$$\det A = a_{ij} A_{ij}.$$

(6) The determinant of any square zero matrices is 0 and the determinant of any identity matrices is 1.

(7) If A is a triangular square matrix then $\det A$ is the product of the entries on the diagonal.

(8) For any square matrix A, $\det A = \det A^t$.

(9) If the square matrix B is obtained by applying an elementary row (or column) operations of type I on the matrix A (switching two rows or columns), then $\det B = -\det A$.

(10) If the square matrix B is obtained by applying an elementary row (or column) operation of type II on the matrix A (multiplying the i-th row or column by u), then $\det B = u \det A$.

(11) If the square matrix B is obtained by applying an elementary row (or column) operation of type III (adding b times the j-th row or column to its i-th row, or column respectively), then $\det B = \det A$.

Example 4.3.1. Find the $(2, 1)$- and $(3, 3)$-minors and cofactors of

$$A = \begin{pmatrix} 1 & 2 & 3 \\ 6 & 5 & 4 \\ 1 & 1 & 1 \end{pmatrix}.$$

Solution. The $(2, 1)$-minor of A is

$$\begin{pmatrix} 2 & 3 \\ 1 & 1 \end{pmatrix}$$

and the $(2, 1)$-cofactor of A is $(-1)^{2+1}(2 - 3) = 1$.

The $(3, 3)$-minor of A is

$$\begin{pmatrix} 1 & 2 \\ 6 & 5 \end{pmatrix}$$

and the $(3, 3)$-cofactor of A is $(-1)^{3+3}(5 - 12) = -7$.   ◇

Example 4.3.2. Find the following determinants of the following matrices.

(1) $A = \begin{pmatrix} 1 & 3 \\ -3 & 2 \end{pmatrix}$

(2) $B = \begin{pmatrix} 0 & 1 & 3 \\ -2 & -3 & -5 \\ 4 & -4 & 4 \end{pmatrix}$

(3) $C = \begin{pmatrix} 2 & 0 & 0 & 1 \\ 0 & 1 & 3 & -3 \\ -2 & -3 & -5 & 2 \\ 4 & -4 & 4 & -6 \end{pmatrix}$

Solution. (1) The determinant of A is $1 \cdot 2 - 3(-3) = 11$.

(2) Using Property (2) we have

$$\det B = (-2)(-4)3 + 4(-5) - 4(-3)3 - 4(-2)$$
$$= 48.$$

(3) Using Property (10) we have

$$\det C = 2 \det \begin{pmatrix} 1 & 0 & 0 & 1 \\ 0 & 1 & 3 & -3 \\ -1 & -3 & -5 & 2 \\ 2 & -4 & 4 & -6 \end{pmatrix}$$

$$= 4 \det \begin{pmatrix} 1 & 0 & 0 & 1 \\ 0 & 1 & 3 & -3 \\ -1 & -3 & -5 & 2 \\ 1 & -2 & 2 & -3 \end{pmatrix}.$$

Since there are two zero entries in the first row, it is convenient to compute $\det C$ using the first row:

$$\det \begin{pmatrix} 1 & 0 & 0 & 1 \\ 0 & 1 & 3 & -3 \\ -1 & -3 & -5 & 2 \\ 1 & -2 & 2 & -3 \end{pmatrix}$$

$$= \det \begin{pmatrix} 1 & 3 & -3 \\ -3 & -5 & 2 \\ -2 & 2 & -3 \end{pmatrix} - \det \begin{pmatrix} 0 & 1 & 3 \\ -1 & -3 & -5 \\ 1 & -2 & 2 \end{pmatrix}$$

$$= 20 - 12 = 8.$$

Hence $\det C = 32$.   ◇

Lemma 4.3.3. *For elementary matrices of size n, we have the following results:*

(a) $\det P_{ij} = -1$;

(b) $\det D_i(u) = u$ *for all $u \in \mathbb{R}$ (even when $u = 0$);*

(c) $\det T_{ij}(b) = 1$.

In particular, the determinant of any elementary matrix is nonzero!

Proof. These are immediate results from facts (9)–(11). □

Determinants of products of matrices. Properties (9)–(11) can be rephrased as the lemma below.

Lemma 4.3.4. *Let A be a square matrix and all the elementary matrices appearing in this corollary are of the same size. The following are true:*

(a) $\det(P_{ij}A) = \det(AP_{ij}) = -\det A = \det P_{ij} \det A;$

(b) $\det(D_i(u)A) = \det(AD_i(u)) = u \det A = \det D_i(u) \det A;$

(c) $\det(T_{ij}(b)A) = \det(AT_{ij}(b)) = \det A = \det T_{ij}(b) \det A.$

This lemma is not only a special case of but a crucial instrument in proving Theorem 4.3.6. However, to prove it we also need the following result.

Proposition 4.3.5. *A square matrix A is invertible if and only if $\det A \neq 0$.*

Proof. Let A be a square matrix of size n. By Theorem 4.1.3, there exist elementary matrices $E_1, \ldots, E_N$ such that

$$E_N \cdots E_1 A$$

is the reduced row echelon form of A. If A is invertible, then the reduced row echelon form of A is the identity matrix. By Lemma 4.3.4, we have

$$\det E_N \cdots \det E_1 \det A = 1.$$

Thus $\det A \neq 0$. On the other hand, if A is not invertible, then $\operatorname{rk} A < n$ by Proposition 4.2.10. The reduced row echelon form of A has zero rows. From Property (3), we have

$$\det E_N \cdots \det E_1 \det A = 0.$$

We conclude that $\det A = 0$ since $\det E_i \neq 0$ for all i by Lemma 4.3.3. □

Theorem 4.3.6. *For* $A, B \in M_n(\mathbb{R})$,

$$\det(AB) = (\det A)(\det B).$$

Proof. First, we deal with the case where $\det A = 0$. In this case, A is not invertible by Proposition 4.3.5. This says that $\operatorname{rk} A < n$ by Proposition 4.2.10. Use the result from Exercise 2 in §4.2, $\operatorname{rk} AB < n$. Thus $\det AB = 0$.

Next, we assume A is invertible. By Corollary 4.2.12, we have

$$A = E_1 E_2 \cdots E_n$$

where E_i are elementary matrices. Hence

$$
\begin{aligned}
\det(AB) &= \det(E_1 E_2 \cdots E_n B) \\
&= \det E_1 \det E_2 \cdots \det E_n \det B \\
&= \det(E_1 \cdots E_n) \det B \\
&= \det A \det B
\end{aligned}
$$

by Lemma 4.3.4. $\qquad\qquad\square$

Corollary 4.3.7. *Let* A *be an invertible matrix. Then* $\det A^{-1} = (\det A)^{-1}$.

Proof. Since $AA^{-1} = I$, we have $\det A \det A^{-1} = \det AA^{-1} = \det I = 1$ by Theorem 4.3.6. $\qquad\qquad\square$

The adjoint of a matrix. Remember we may use (4.3) to compute the determinant of a matrix. As a variation we have the following lemma.

Lemma 4.3.8. *Let* $A = (a_{ij}) \in M_n(\mathbb{R})$ *where* $n \geq 2$. *For* $i \neq j$, *we have that*

$$(4.4) \qquad a_{i1}A_{j1} + a_{i2}A_{j2} + \cdots + a_{in}A_{jn} = 0,$$
$$a_{1i}A_{1j} + a_{2i}A_{2j} + \cdots + a_{ni}A_{nj} = 0.$$

Proof. If we replace the j-th row in $\left(a_{ij}\right)_{n \times n}$ by its i-th row, the we obtain a matrix with two identical rows. The determinant of this matrix is 0. It is also the first expression in (4.4). We can say the same thing if we replace the j-th column by the ith column. $\qquad\qquad\square$

Definition 4.3.9. Let $A \in M_n(\mathbb{R})$ where $n \geq 2$. We define the **adjoint** of the matrix A, denoted adj A, to be $\left(A_{ij} \right)^t_{n \times n} = \left(A_{ji} \right)$.

Example 4.3.10. Compute adj $\begin{pmatrix} 1 & 1 & 1 \\ 1 & 2 & 2 \\ 1 & 2 & 3 \end{pmatrix}$.

Solution. We find

$$A_{11} = \det \begin{pmatrix} 2 & 2 \\ 2 & 3 \end{pmatrix} = 2, \qquad A_{12} = -\det \begin{pmatrix} 1 & 2 \\ 1 & 3 \end{pmatrix} = -1,$$

$$A_{13} = \det \begin{pmatrix} 1 & 2 \\ 1 & 2 \end{pmatrix} = 0, \qquad A_{21} = -\det \begin{pmatrix} 1 & 1 \\ 2 & 3 \end{pmatrix} = -1,$$

$$A_{22} = \det \begin{pmatrix} 1 & 1 \\ 1 & 3 \end{pmatrix} = 2, \qquad A_{23} = -\det \begin{pmatrix} 1 & 1 \\ 1 & 2 \end{pmatrix} = -1,$$

$$A_{31} = \det \begin{pmatrix} 1 & 1 \\ 2 & 2 \end{pmatrix} = 0, \qquad A_{32} = -\det \begin{pmatrix} 1 & 1 \\ 1 & 2 \end{pmatrix} = -1,$$

$$A_{33} = \det \begin{pmatrix} 1 & 1 \\ 1 & 2 \end{pmatrix} = 1.$$

We have

$$\mathrm{adj}\, A = \begin{pmatrix} 2 & -1 & 0 \\ -1 & 2 & -1 \\ 0 & -1 & 1 \end{pmatrix}.$$

◇

Proposition 4.3.11. *Let $A \in M_n(\mathbb{R})$ where $n \geq 2$. Then*

$$A(\mathrm{adj}\, A) = (\mathrm{adj}\, A)A = (\det A)I_n = \begin{pmatrix} \det A & & & \\ & \det A & & \mathbf{0} \\ & & \ddots & \\ \mathbf{0} & & & \det A \end{pmatrix}.$$

Proof. This follows from (4.3) and (4.4). □

Corollary 4.3.12. *Let $A \in M_n(\mathbb{R})$ where $n \geq 2$. If $\det A \neq 0$, then*

$$(4.5) \qquad A^{-1} = \frac{1}{\det A}(\mathrm{adj}\, A).$$

Corollary 4.3.13. *Let A and B be square matrices. Then $AB = I_n$ if and only if $BA = I_n$. In either case, we have $B = \dfrac{1}{\det A}(\operatorname{adj} A)$.*

Proof. Assume $AB = I_n$. Then

$$B = I_n B = (\det A)^{-1}(\operatorname{adj} A)AB$$
$$= (\det A)^{-1}(\operatorname{adj} A)I_n = (\det A)^{-1}(\operatorname{adj} A).$$

This implies that $BA = I_n$.

The converse is similar. □

Example 4.3.14. Does the set

$$S = \left\{ \begin{pmatrix} 0 \\ -2 \\ 4 \end{pmatrix}, \begin{pmatrix} 1 \\ -3 \\ -4 \end{pmatrix}, \begin{pmatrix} 3 \\ -5 \\ 4 \end{pmatrix} \right\}$$

form a basis for $\mathbb{R}^3$?

Solution. By Proposition 3.5.6, the set S forms a basis if the matrix

$$A = \begin{pmatrix} 0 & 1 & 3 \\ -2 & -3 & -5 \\ 4 & -4 & 4 \end{pmatrix}$$

is invertible. To check if this is so, we only need to find out if $\det A \neq 0$. We have

$$\det A = 2 \det \begin{pmatrix} 0 & 1 & 3 \\ -1 & -3 & -5 \\ 2 & -4 & 4 \end{pmatrix}$$

$$= 48.$$

Thus S forms a basis for $\mathbb{R}^3$. ◇

Example 4.3.15. For $n \geq 2$, let A_n be the $n \times n$ matrix

$$\begin{pmatrix} 0 & 1 & 1 & \cdots & 1 \\ 1 & 0 & 1 & \cdots & 1 \\ 1 & 1 & 0 & \cdots & 1 \\ \vdots & \vdots & \vdots & \ddots & \vdots \\ 1 & 1 & 1 & \cdots & 0 \end{pmatrix}$$

which has 0 on the diagonal and 1 elsewhere.

(a) Show that the inverse of A_n is

$$\frac{1}{n-1} \begin{pmatrix} -(n-2) & 1 & 1 & \cdots & 1 \\ 1 & -(n-2) & 1 & \cdots & 1 \\ 1 & 1 & -(n-2) & \cdots & 1 \\ \vdots & \vdots & \vdots & \ddots & \vdots \\ 1 & 1 & 1 & \cdots & -(n-2) \end{pmatrix}.$$

(b) Show that $\det A_n = (-1)^{n-1}(n-1)$.

(c) Find $\operatorname{adj} A_n$.

Solution. Let $B_n = \sum_{i,j=1}^n e_{ij}^{(n)}$. Note that $A_n = B_n - I_n$.

(a) It is not hard to see that

$$B_n^2 = nB_n.$$

The problem asks us to check whether

$$\frac{1}{n-1}\left(B_n - (n-1)I_n\right)$$

is the inverse of $A_n = B_n - I_n$. We find

$$\begin{aligned}
&\left(B_n - I_n\right)\left(B_n - (n-1)I_n\right) \\
&= B_n^2 - (n-1)B_n - B_n + (n-1)I_n \\
&= nB_n - nB_n + (n-1)I_n \\
&= (n-1)I_n.
\end{aligned}$$

Thus

$$A_n^{-1} = (B_n - I_n)^{-1} = \frac{1}{n-1}\left(B_n - (n-1)I_n\right)$$

as claimed.

(b) We may add all the columns to the first column of A_n without changing its determinant. Thus

$$
\det A_n = \det \begin{pmatrix}
n-1 & 1 & 1 & \cdots & 1 \\
n-1 & 0 & 1 & \cdots & 1 \\
n-1 & 1 & 0 & \cdots & 1 \\
\vdots & \vdots & \vdots & \ddots & \vdots \\
n-1 & 1 & 1 & \cdots & 0
\end{pmatrix}
$$

$$
= (n-1) \det \begin{pmatrix}
1 & 1 & 1 & \cdots & 1 \\
1 & 0 & 1 & \cdots & 1 \\
1 & 1 & 0 & \cdots & 1 \\
\vdots & \vdots & \vdots & \ddots & \vdots \\
1 & 1 & 1 & \cdots & 0
\end{pmatrix}
$$

$$
= (n-1)\left((A_n)_{11} + (A_n)_{21} + \cdots + (A_n)_{n1} \right)
$$

where $(A_n)_{ij}$ is the (i,j)-cofactor of A_n. We have

$$
\det A_n = (n-1)\left(\det A_{n-1} + \det A_n \right).
$$

This implies that

$$
(n-2)\det A_n = -(n-1)\det A_{n-1}.
$$

It is easy to check that $\det A_2 = -1$. It now follows by induction on n,

$$
\det A_n = -\frac{n-1}{n-2} \cdot (-1)^{n-2}(n-2) = (-1)^{n-1}(n-1).
$$

(c) We have

$$
\operatorname{adj} A = - \begin{pmatrix}
-(n-2) & 1 & 1 & \cdots & 1 \\
1 & -(n-2) & 1 & \cdots & 1 \\
1 & 1 & -(n-2) & \cdots & 1 \\
\vdots & \vdots & \vdots & \ddots & \vdots \\
1 & 1 & 1 & \cdots & -(n-2)
\end{pmatrix}
$$

from (a), (b) and Corollary 4.3.12.   ◇

Although (4.5) gives a formula for finding the inverse matrix, it is not a practical method. It is really more of theoretic significance. To compute the inverse of a size n invertible matrix using (4.5), one needs to compute the determinant of a square matrix of size n and the determinants of n^2 square matrices of size $n - 1$. This is too much work. We still recommend the method discussed in §4.2.

Exercises 4.3

1. Find the adjoint of
$$\begin{pmatrix} 1 & 3 & 0 \\ 0 & 1 & 2 \\ 1 & 2 & 1 \end{pmatrix}.$$

2. Find the determinant, the $(1, 4)$-cofactor and the $(2, 3)$-cofactor of
$$\begin{pmatrix} 1 & 0 & 3 & 4 & 0 \\ 0 & 2 & 0 & 4 & 0 \\ 1 & 0 & 0 & 0 & 5 \\ 2 & 0 & 1 & 4 & 0 \\ 1 & 0 & 1 & 2 & 0 \end{pmatrix}.$$

3. Use (4.5) to verify that
$$\begin{pmatrix} a & b \\ c & d \end{pmatrix}^{-1} = \frac{1}{ad - bc} \begin{pmatrix} d & -b \\ -c & a \end{pmatrix}$$

If $ad - bc \neq 0$. This is a formula worth memorizing.

4. Show that
$$\det \begin{pmatrix} 1 & a_1 & a_1^2 & \cdots & a_1^{n-2} & a_1^n \\ 1 & a_2 & a_2^2 & \cdots & a_2^{n-2} & a_2^n \\ & \multicolumn{4}{c}{\dots\dots\dots\dots\dots} & \\ & \multicolumn{4}{c}{\dots\dots\dots\dots} & \\ 1 & a_n & a_n^2 & \cdots & a_n^{n-2} & a_n^n \end{pmatrix} = \prod_{i<j}(a_j - a_i).$$

This is the so-called Vandermonde determinant.

5. Use the the Vandermonde determinant to give a criterion for

$$(1, a_1, a_1^2, \ldots, a_1^{n-1}), \ (1, a_2, a_2^2, \ldots, a_2^{n-1}), \ \ldots, \ (1, a_n, a_n^2, \ldots, a_n^{n-1})$$

to be linear independent in $\mathbb{R}^n$ over $\mathbb{R}$. (*Cf.* Problem (3)(c), §2.5.)

6. Let $n \geq 2$ and let A be the $n \times n$ matrix

$$\begin{pmatrix} i & 1 & 1 & \cdots & 1 \\ 1 & i & 1 & \cdots & 1 \\ 1 & 1 & i & \cdots & 1 \\ \vdots & \vdots & \vdots & \ddots & \vdots \\ 1 & 1 & 1 & \cdots & i \end{pmatrix}$$

with complex entries. Find A^{-1}, $\det A$ and $\operatorname{adj} A$.

7. Let $A \in M_n(\mathbb{R})$ where $n \geq 2$. Show that $\det \operatorname{adj} A = (\det A)^{n-1}$.

8. Show that

$$\det \begin{pmatrix} A_{m \times m} & B_{m \times n} \\ \mathbf{0}_{n \times m} & C_{n \times n} \end{pmatrix}_{(m+n) \times (m+n)} = \det A \det C.$$

9. (Cramer's rule) Suppose given the system of linear equations

$$a_{i1} x_1 + a_{i2} x_2 + \cdots + a_{in} x_n = b_i, \qquad i = 1, 2, \ldots, n,$$

and suppose $A = (a_{ij})_{n \times n}$ is invertible. Show that

$$x_i = \frac{\det A_i}{\det A}, \qquad i = 1, \ldots, n,$$

where A_i is the matrix obtained by replacing the i-th column of A by

$$\begin{pmatrix} b_1 \\ \vdots \\ b_n \end{pmatrix}.$$

4.4   Applications in linear programming

In this section we give a brief introduction on low linear algebra is used in solving linear optimization problems.

Linear programming problems.   **Linear programming** or **linear optimization** is a method to achieve the best outcome (often a maximum or a minimum) of a linear function under linear constraints. That is, the problem can often be described as:

$$\text{Maximize} \quad f = c_1 x_1 + c_2 x_2 + \cdots + c_n x_n,$$
$$\text{subject to} \quad a_{11} x_1 + a_{12} x_2 + \cdots + a_{1n} x_n \leq b_1,$$
$$a_{21} x_1 + a_{22} x_2 + \cdots + a_{2n} x_n \leq b_2,$$
$$\dots\dots\dots\dots\dots\dots\dots\dots\dots\dots\dots$$
$$a_{m1} x_1 + a_{m2} x_2 + \cdots + a_{mn} x_n \leq b_m,$$
$$x_1 \geq 0, \ x_2 \geq 0, \ \dots, \ x_n \geq 0,$$

or by its dual problem (which will be described later in this section).

We use two typical linear programming problems as examples.

Example 4.4.1 (A production problem). A phone company produces two types of smart phones, type I phone and type II phone. The type I phone is sold at 525 US dollars apiece while the the type II phone is sold at 875 US dollars apiece. The main cost of producing smart phones comes from design and manufacture as shown below

	cost of design	cost of manufacture
type I phone	50	75
type II phone	150	100

The total funds available for the production is $750,000$ dollars, $350,000$ dollars for cost of design and $400,000$ for cost of manufacture.

Suppose that the company is going to produce x pieces of type I phone and y pieces of type II phone. Then

$$\text{total cost of design} = 50x + 150y;$$
$$\text{total cost of manufacture} = 75x + 100y.$$

This leads to the following constraints

$$50x + 150y \leq 350,000;$$
$$75x + 100y \leq 400,000;$$
$$x \geq 0, \ y \geq 0.$$

The profit of type I phone is $525 - 50 - 75 = 400$ apiece while the profit of type II phone is $875 - 150 - 100 = 625$ apiece. The total profit of producing x pieces of type I phone and y pieces of type II phone is

$$400x + 625y.$$

The corresponding linear programming problem is thus:

$$\begin{aligned}
\textbf{Maximize} \quad & f = 400x + 625y, \\
\textbf{subject to} \quad & 50x + 150y \le 350,000, \\
& 75x + 100y \le 400,000, \\
& x \ge 0, \ y \ge 0.
\end{aligned}$$

Example 4.4.2 (A transportation problem). A company owns two plants A and B to produce canned dog food for the two stores I and II. Each month, plant A produces $150,000$ cans and plant B produces $200,000$ cans. However, store I can sell at least $180,000$ cans and the store II can sell at least $160,000$ cans. The cost of shipments is shown below

	store I	store II
plant A	50	60
plant B	70	40

(dollars/per 1000 cans)

Suppose that x thousand cans and y thousand cans are distributed from plant A to stores I and II, respectively, while z thousand cans and w thousand cans are distributed from plant B to stores I and II, respectively. The problem can be formulated as

$$\begin{aligned}
\textbf{Minimize} \quad & f = 50x + 60y + 70z + 40w, \\
\textbf{subject to} \quad & x + y \le 150, \\
& z + w \le 200, \\
& x + z \ge 180, \\
& y + w \ge 160, \\
& x \ge 0, \ y \ge 0, \ z \ge 0, \ w \ge 0.
\end{aligned}$$

The graphical method. The constraints on the production problem in Example 4.4.1 are described by the following system of linear inequalities in two variables:

$$\begin{cases} 50x + 150y \leq 350,000, \\ 75x + 100y \leq 400,000, \\ x \geq 0,\ y \geq 0. \end{cases}$$

In the Cartesian plane, each linear inequality

$$ax + by + c > 0$$

represents a half plane. The inequality $x \geq 0$ represents the half plane at the right side of the y-axis. The inequality $y \geq 0$ represents the upper half plane above the x-axis. The inequality

$$50x + 150y \leq 350,000 \quad \Leftrightarrow \quad x + 3y \leq 7,000$$

represents the half plane below the line $x + 3y = 7,000$. Finally, the inequality

$$75x + 100y \leq 400,000 \quad \Leftrightarrow \quad 3x + 4y \leq 16,000$$

represents the half plane below the line $3x + 4y = 16,000$.

Therefore, the four constraints of the production problem can be represented in the Cartesian plane as an intersection of four half planes which is a region with four sides and four vertices $(0,0)$, $(0,7000/3)$, $(16000/3,0)$ and $(4000,1000)$ (see Figure 4.1).

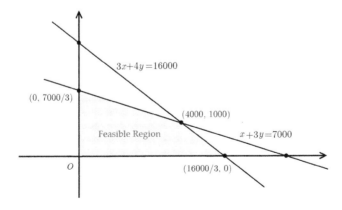

Figure 4.1: The feasible region imposed by the constraints in Example 4.4.1

In general, the graphical representation of the constraints of a linear programming problem is called the **feasible region** of the problem and points in the region are called **feasible solutions**. Among the points in the feasible region, there is at least one point that gives a maximum (or minimum) value to the objective function. Such a point is called an **optimal solution** to the linear programming problem.

A general theorem in linear programming asserted that if the feasible region is nonempty, the optimal solution occurs at one vertex (or corner point) of the feasible region.

The feasible region in Example 4.4.1 has four vertices. They are $O(0,0)$, $A(0, 7000/3)$, $B(4000, 1000)$ and $C(16000/3, 0)$. The values of the objective function $f = 400x + 625y$ at the vertexes are as follows:

vertex	$f = 400x + 625y$
$O(0,0)$	0
$A(0, 7000/3)$	$1,458,333$
$B(4000, 1000)$	$2,025,000$
$C(16000/3, 0)$	$2,133,333$

The linear function f achieves maximal profit when $x = 16000/3$ and $y = 0$.

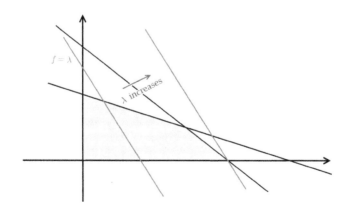

Figure 4.2: The optimal solution occurs at the farthest vertex from the origin.

This optimal solution can also be detected visually. Construct a family of parallel lines

$$f = 400x + 625y = \lambda$$

by assigning different λ to the objective function. As one can see, the value of f increases as the line $f = \lambda$ moves to the right. The line $f = \lambda$ that is farthest from the origin and still intersects the feasible set gives the optimal solution. See Figure 4.2.

The simplex method. Solving linear programming problems with the graphical method has its limitations. It is alright with problems involving only two variables, or even three variables. To deal with problems with more than three variables, the geometric approach becomes impractical since the feasible region cannot be graphed effectively.

The **simplex method** was first developed in 1947 by George B. Dantzig, an American Mathematical scientist. Dr. Dantzig made important contributions in many areas such as operations research, computer science, economics and statistics, but he is best known for his development of the simplex method for solving linear programming. It is an algebraic method that can be used for any number of variables and has the advantage of being readily programmable. Here we outline the simplex method with a linear programming problem in three variables.

Example 4.4.3. Maximize the function $f = x + 2y + 2z$ under the constraints

$$2x + y + 2z \leq 28,$$
$$x + 4y + 2z \leq 52,$$
$$3x + 2y + z \leq 56,$$
$$x \geq 0, \ y \geq 0, \ z \geq 0.$$

STEP I. The first step of the simplex method is to introduces non-negative **slack variables** so that the constraints described by a system of inequalities becomes a system of linear equations. For example,

$$x + y \leq 10 \quad \rightsquigarrow \quad x + y + u = 10;$$
$$x + y \geq 10 \quad \rightsquigarrow \quad x + y - u = 10,$$

where u is the slack variable used to compensate the inequality.

For Example 4.4.3 we may introduce the slack variables u, v, w so that the constraints are redescribed as the following system of linear equations

$$
\begin{aligned}
2x + \ y + 2z + u \qquad\qquad\qquad &= 28, \\
x + 4y + 2z \quad\ \ + v \qquad\qquad &= 52, \\
3x + 2y + \ z \qquad\quad + w \qquad &= 56, \\
-x - 2y - 2z \qquad\qquad\quad + f &= 0.
\end{aligned}
$$

Build the initial simplex tableau

x	y	z	u	v	w	f	constant
2	1	2	1	0	0	0	28
1	4	2	0	1	0	0	52
3	2	1	0	0	1	0	56
-1	-2	-2	0	0	0	1	0

from the system above.

STEP II. We now describe how to choose the **pivot element** in the simplex tableau. Among the negative entries at the bottom row, look for the least number. In our example, it occurs both at the y-column and the z-column. Choose either one of the columns to be the **pivot column**. We chose the z-column for now. In the z-column, we compare the ratios of the entries in the constant column to the entries in the pivot column, except for the bottom row. Here they are

$$
\frac{28}{2} = 14, \quad \frac{52}{2} = 26 \quad \text{and} \quad \frac{56}{1} = 56.
$$

We pick the row where the least ratio lies as the **pivot row**. In our example, the pivot row is the first row.

	x	y	z	u	v	w	f	constant
pivot →	2	1	$\boxed{2}$	1	0	0	0	28
row	1	4	2	0	1	0	0	52
	3	2	1	0	0	1	0	56
	-1	-2	-2	0	0	0	1	0

$$\uparrow$$
$$\text{pivot col}$$

The entry where the pivot column and the pivot row meet is the **pivot element**. Here in this example, the pivot element lies at the first row of the z-column.

STEP III. Next we perform a series of elementary row operations once we have chosen the pivot element. The goal is so that the pivot element becomes 1 and the other entries at the pivot column become 0. In this example, we may achieve this by (a) dividing the first row by 2, (b) subtracting 2 times of row (i) from row (ii), (c) subtracting row (i) from row (iii) and (d) adding 2 times of row (i) to row (iv). The end result is

x	y	z	u	v	w	f	constant
1	1/2	1	1/2	0	0	0	14
-1	3	0	-1	1	0	0	24
2	3/2	0	$-1/2$	0	1	0	42
1	-1	0	1	0	0	1	28

STEP IV. Repeat STEP I–STEP III if there still are negative entries at the bottom row. In our case, there is still a negative entry at the bottom row and the y-column. Hence, the y-column is the new pivot column. Compare the ratios as described in STEP II:

$$\frac{14}{1/2} = 28, \quad \frac{24}{3} = 8 \quad \text{and} \quad \frac{42}{3/2} = 28.$$

The new pivot row is the second row. We choose the element in the y-column and the second row as a new pivot element.

	x	y	z	u	v	w	f	constant
	1	1/2	1	1/2	0	0	0	14
pivot →	-1	$\boxed{3}$	0	-1	1	0	0	24
row	2	3/2	0	$-1/2$	0	1	0	42
	1	-1	0	1	0	0	1	28

$$\uparrow$$
$$\text{pivot col}$$

Perform more elementary row operations so that the new pivot element becomes 1 and the other entries at the y-column become 0. We then have

x	y	z	u	v	w	f	constant
7/6	0	1	2/3	−1/6	0	0	10
−1/3	1	0	−1/3	1/3	0	0	8
5/2	0	0	0	−1/2	1	0	30
2/3	0	0	2/3	1/3	0	1	36

All entries at the bottom row are nonnegative now. This means that we have reached the optimal solution

$$x = 0, \; y = 8, \; z = 10, \; u = 0, \; v = 0, \; w = 30 \text{ and } f = 36.$$

This is similar to how we interpret the augmented matrix associated with a system of linear equations. For a special solution, we set the non-pivotal variables to be 0 and obtain values for the pivotal variables.

Duality for the optimization problems. Another special class of linear programming problems in practical applications is characterized by the following conditions.

- The objective function is to be minimized, however, it is still linear.

- Each linear constraint can be expressed as a linear combination of variables is greater than or equal to a constant.

- All the variables are nonnegative.

A convenient method for solving this type of problem is to consider its dual problem. Here is an example to illustrate the general process.

The primal problem is

$$
\begin{aligned}
\textbf{Minimize} \quad & f = 7x + 8y, \\
\textbf{subject to} \quad & 30x + 10y \geq 24, \\
& 10x + 16y \geq 21, \\
& 9x + 7y \geq 12, \\
& x \geq 0, \; y \geq 0.
\end{aligned}
$$

First we build the following tableau for the given primal problem.

x	y	constant
30	10	24
10	16	21
9	7	12
7	8	

Take the transpose of the matrix appearing in the tableau and introduce new variables u, v and w.

u	v	w	constant
30	10	9	7
10	16	7	8
24	21	12	

The dual problem is now

$$\textbf{Maximize} \quad g = 24u + 21v + 12w,$$
$$\textbf{subject to} \quad 30u + 10v + 9w \leq 7,$$
$$10u + 16v + 7w \leq 8,$$
$$u \geq 0, \ v \geq 0, \ w \geq 0.$$

The connection between the solution of the primal problem and that of its dual problem is given by the following theorem which is attributed to John von Neumann (1903–1957).

Theorem 4.4.4 (Fundamental Theorem of Duality). *A primal problem has an optimal solution if and only if its dual problem has an optimal solution. Furthermore, if an optimal solution exists, the following statements are true.*

(1) *The objective functions of both problems attain the same optimal value.*

(2) *The optimal solution to the primal problem appears under the slack variables in the last row of the final simplex tableau of the dual problem.*

Exercises 4.4

1. Solve the following linear programming problems.

    (a) Maximize $f = 36x + 30y$ subject to the constraints

$$2x + y \le 70, \quad x + y \le 40,$$
$$y + 3x \le 90, \quad x \ge 0, \quad y \ge 0.$$

    (b) Maximize $f = 4x + 3y$ subject to the constraints

$$x + 2y \ge 18, \quad 2x + y \ge 12,$$
$$x + y \le 15, \quad 2x + y \le 24,$$
$$x \ge 0, \quad y \ge 0.$$

    (c) Minimize $f = 2x - 3y$ subject to the constraints

$$x + y \le 5, \quad x + 3y \ge 9,$$
$$-2x + y \le 2, \quad x \ge 0, \quad y \ge 0.$$

2. Solve the following linear programming problems.

    (a) Maximize $p = 4x + 5y + 6z$ subject to the constraints

$$x + y + 2z \le 360, \quad 2x + 3y + z \le 300,$$
$$2x + y + 2z \le 240, \quad x \ge 0, \quad y \ge 0, \quad z \ge 0.$$

    (b) Minimize $f = -30x - 20y$ subject to the constraints

$$4x + 5y \le 320, \quad 2x + y \le 100, \quad x \ge 0, \quad y \ge 0.$$

    (c) Minimize $f = 20x_1 + 8x_2 + 10x_3 + 12x_4 + 22x_5 + 18x_6$ subject to the constraints

$$x_1 + x_2 + x_3 \le 40, \quad x_4 + x_5 + x_6 \le 60,$$
$$x_1 + x_4 \ge 20, \quad x_2 + x_5 \ge 30, \quad x_3 + x_6 \ge 40,$$
$$x_1 \ge 0, \; x_2 \ge 0, \; \ldots, \; x_6 \ge 0.$$

(d) Minimize $f = x - 2y + z$ subject to the constraints

$$x - 2y + 3z \leq 10, \quad 2x + y - 2z \leq 15,$$
$$2x + y + 3z \leq 20, \quad x \geq 0, \quad y \geq 0, \quad z \geq 0.$$

(e) Maximize $p = x + 2y + 3z$ subject to the constraints

$$x + 2y + z \leq 20, \quad 2x + y + z = 10,$$
$$3x + y \leq 30, \quad x \geq 0, \quad y \geq 0, \quad z \geq 0.$$

3. A manufacture produces two types of fertilizers A and B from chemicals C and D. Fertilizer A is made up of 70% chemical C and 30% chemical D. Fertilizer B is made up of 55% chemical C and 45% chemical D. The manufacturer requires at least 40 tons of fertilizer A and 60 tons of fertilizer B. He has 100 tons of chemical C and 60 tons of chemical D available. What quantities of A and B should be produced if he wants to produce as much fertilizer as possible? Write down the mathematical model to describe the whole problem.

4. A man is suffering from iron and vitamin-B deficiency. A nutritionist advises him to take at least 2400 mg of iron, 2100 mg of vitamin B_1 and 2500 mg of vitamin B_2 over a period of time. Two brands of vitamin pills are avaiable: brand A and brand B. The cost of the two brands of pills and their contents are shown in the following table.

	Brand A	Brand B	Minimum Requirement
Iron	40 mg	15 mg	2400 mg
Vitamin B_1	10 mg	10 mg	2100 mg
Vitamin B_2	10 mg	15 mg	2500 mg
Cast/Pill	8¢	10¢	

What combination of pill should the man purchase in order to meet the minimum iron and vitamin requirements at the lowest cost?

Review Exercises for Chapter 4

1. In this problem we intend to show that the multiplication of two matrices can be done in block form. Show that

$$
\begin{array}{c}
\ell \\
m
\end{array}
\begin{pmatrix}
A_{\ell \times n} & B_{\ell \times p} \\
C_{m \times n} & D_{m \times p}
\end{pmatrix}
\begin{pmatrix}
E_{n \times r} & F_{n \times s} \\
G_{p \times r} & H_{p \times s}
\end{pmatrix}
\begin{array}{c}
n \\
p
\end{array}
$$

$$
= \begin{array}{c}
\ell \\
m
\end{array}
\begin{pmatrix}
AE + BG & AF + BH \\
CE + DG & CF + DH
\end{pmatrix} .
$$

2. Is it true that

$$
\det \begin{pmatrix}
A_{n \times n} & B_{n \times n} \\
C_{n \times n} & D_{n \times n}
\end{pmatrix}_{(2n) \times (2n)} = \det A \det D - \det B \det C?
$$

Prove your assertion or give a counterexample.

3. Let $a \in \mathbb{C}$. Find a necessary and sufficient condition for

$$
\begin{pmatrix}
a & 1 & 1 & \cdots & 1 \\
1 & a & 1 & \cdots & 1 \\
1 & 1 & a & \cdots & 1 \\
\vdots & \vdots & \vdots & \ddots & \vdots \\
1 & 1 & 1 & \cdots & a
\end{pmatrix}_{n \times n}
$$

to be invertible.

4. Let A and B be square matrices which are also upper triangular.

  (a) Show that AB is upper triangular.

  (b) If A is invertible, show that A^{-1} is also an upper triangular matrix.

  These results are valid for lower triangular matrices as well.

5. Observe that the elementary matrices are all triangular (upper or lower). Use this to show that any square matrix A may be rewritten as a product of two matrices LU where L is lower triangular and U is upper triangular. This is called an **LU decomposition** of A.

CHAPTER 5

Diagonalization

In this chapter we describe what the *diagonalization problem* is and the develop the technique that is required for solving the diagonalization problem.

Throughout this chapter, we assume all vector spaces are finite dimensional unless otherwise noted.

5.1   Base change

Throughout this section V and W denote finite dimensional real vector spaces unless otherwise noted.

What is the diagonalization problem.   So far we have answered two of the three problems regarding *classification*.

Classification of vector spaces. Vector spaces are classified by the dimension. Two vector spaces over the same field are isomorphic if and only if they are of the same dimension.

Classification of linear transformations. The properties of a linear transformation reflects on and are reflected by any of its representing matrix. Hence, any two linear transformations represented by the same matrix are considered to be in the same class. Linear transformations (and respectively matrices) from V to W are classified by the rank. Assume $\dim V = n$ and $\dim W = m$. If the rank of a linear transformation T is r, we can find appropriate ordered bases such that T is represented by the matrix

$$\begin{pmatrix} I_r & \\ & \mathbf{0}_{(m-r)\times(n-r)} \end{pmatrix},$$

which is the simplest matrix fulfilling this qualification (see Problem (8), §4.2).

For the rest of this course we deal with the last and the most difficult classification problem: the classification of *linear endomorphisms*.

A **linear endomorphism** is a linear transformation from a vector space V into itself, that is, of which the domain and the codomain are the same. We want to find the *simplest* matrix which can represent and classify any given linear endomorphism.

How is this problem different form the second classification problem? The catch is that since the domain and the codomain are the same, we are required to use only one fixed basis for V. Now the problem becomes this: For any linear endomorphism $T : V \to V$, how do we find a basis β for V such that $[T]_\beta^\beta$ is as simple as possible? Furthermore, we wish that one "simple" matrix representing a linear endomorphism should not be able to represent the linear endomorphisms represented by other "simple" matrices.

Before diving into this problem in full force, in this chapter we deal with a simpler problem (part of the bigger problem): When is it possible to find a basis for V so that the matrix representing $T\colon V \to V$ is *diagonal*? This is the so-called **diagonalization problem**.

Base change. Here we derive a formula for computing matrices representing the same linear transformation but with respect to different bases.

Definition 5.1.1. Let β_1 and β_2 be bases for V. We call $[1_V]_{\beta_2}^{\beta_1}$ the **base change** matrix from the base β_2 to β_1.

Let $\beta_1 = \{u_1, u_2, \ldots, u_n\}$ and $\beta_2 = \{v_1, v_2, \ldots, v_n\}$ be bases for V. If

$$v_1 = a_{11}u_1 + a_{21}u_2 + \cdots + a_{n1}u_n$$
$$v_2 = a_{12}u_1 + a_{22}u_2 + \cdots + a_{n2}u_n$$
$$\vdots \qquad\qquad \vdots$$
$$v_n = a_{1n}u_1 + a_{2n}u_2 + \cdots + a_{nn}u_n$$

then

$$[1_V]_{\beta_2}^{\beta_1} = \Big(\ a_{ij}\ \Big)_{n \times n}.$$

Furthermore, by Theorem 3.5.5 we have

$$[1_V]_{\beta_1}^{\beta_2} = \Big(\ a_{ij}\ \Big)_{n \times n}^{-1}.$$

Theorem 5.1.2. *Let $T\colon V \to W$ be a linear transformation. Let β_1 and β_2 be bases for V and let γ_1 and γ_2 be bases for W. Then*

$$[T]_{\beta_2}^{\gamma_2} = [1_W]_{\gamma_1}^{\gamma_2}[T]_{\beta_1}^{\gamma_1}[1_V]_{\beta_2}^{\beta_1}.$$

Hence if P is the base change matrix from β_2 to β_1 and Q is the base change matrix from γ_1 to γ_2. Then

$$[T]_{\beta_2}^{\gamma_2} = Q[T]_{\beta_1}^{\gamma_1}P.$$

Proof. Since

$$T = 1_W \circ T \circ 1_V,$$

this is an immediate result of Theorem 3.5.1.   □

Definition 5.1.3. Let $A, B \in M_{m \times n}(\mathbb{R})$. We say A is **equivalent to** B if and only if there are invertible matrices P and Q such that $B = QAP$.

Thus, any two matrices representing the same linear transformation are equivalent.

Example 5.1.4. Remember that polynomials of different degrees are linearly independent (see Review Exercise 2 for Chapter 2). Hence

$$\beta = (1, \; 2 + x, \; 3 + x + 5x^2)$$

and

$$\beta' = (1, \; 3 - 4x, \; 1 + x - 2x^2, \; 1 + x^3)$$

are ordered bases for $\mathscr{P}_2$ and $\mathscr{P}_3$ respectively .

(a) Find the base change matrix of $\mathscr{P}_2$ form β to the standard basis of $\mathscr{P}_2$.

(b) Find the base change matrix of $\mathscr{P}_3$ form β' to the standard basis of $\mathscr{P}_3$.

(c) Find the matrix of

$$\begin{array}{rcl} T: & \mathscr{P}_2 & \longrightarrow & \mathscr{P}_3 \\ & f(x) & \longmapsto & 2f'(x) + \int_0^x 3f(t)\,dt \end{array}$$

with respect to the bases $(1, \; 2 + x, \; 3 + x + 5x^2)$ and $(1, \; 3 - 4x, \; 1 + x - 2x^2, \; 1 + x^3)$ using the base change formula. (*Cf.* Example 3.4.3.)

Solution. Let α and α' be the standard bases for $\mathscr{P}_2$ and $\mathscr{P}_3$ respectively.

(a) The base change matrix from β to α is

$$[1_{\mathscr{P}_2}]_\beta^\alpha = \begin{pmatrix} 1 & 2 & 3 \\ 0 & 1 & 1 \\ 0 & 0 & 5 \end{pmatrix}.$$

(b) The base change matrix from β' to α' is

$$[\mathbf{1}_{\mathscr{P}_3}]_{\beta'}^{\alpha'} = \begin{pmatrix} 1 & 3 & 1 & 1 \\ 0 & -4 & 1 & 0 \\ 0 & 0 & -2 & 0 \\ 0 & 0 & 0 & 1 \end{pmatrix}.$$

(c) From Example 3.4.3, we have

$$[T]_{\alpha}^{\alpha'} = \begin{pmatrix} 0 & 2 & 0 \\ 3 & 0 & 4 \\ 0 & 3 & 0 \\ 0 & 0 & 2 \end{pmatrix}.$$

From Theorem 5.1.2, we have

$$[T]_{\beta}^{\beta'} = [\mathbf{1}_{\mathscr{P}_3}]_{\alpha'}^{\beta'}[T]_{\alpha}^{\alpha'}[\mathbf{1}_{\mathscr{P}_2}]_{\beta}^{\alpha}$$

$$= \left([\mathbf{1}_{\mathscr{P}_3}]_{\beta'}^{\alpha'}\right)^{-1}[T]_{\alpha}^{\alpha'}[\mathbf{1}_{\mathscr{P}_2}]_{\beta}^{\alpha}$$

$$= \begin{pmatrix} 1 & 3 & 1 & 1 \\ 0 & -4 & 1 & 0 \\ 0 & 0 & -2 & 0 \\ 0 & 0 & 0 & 1 \end{pmatrix}^{-1} \begin{pmatrix} 0 & 2 & 0 \\ 3 & 0 & 4 \\ 0 & 3 & 0 \\ 0 & 0 & 2 \end{pmatrix} \begin{pmatrix} 1 & 2 & 3 \\ 0 & 1 & 1 \\ 0 & 0 & 5 \end{pmatrix}$$

$$= \frac{1}{8}\begin{pmatrix} 8 & 6 & 7 & -8 \\ 0 & -2 & -1 & 0 \\ 0 & 0 & -4 & 0 \\ 0 & 0 & 0 & 8 \end{pmatrix}\begin{pmatrix} 0 & 2 & 0 \\ 3 & 0 & 4 \\ 0 & 3 & 0 \\ 0 & 0 & 2 \end{pmatrix}\begin{pmatrix} 1 & 2 & 3 \\ 0 & 1 & 1 \\ 0 & 0 & 5 \end{pmatrix}$$

$$= \frac{1}{8}\begin{pmatrix} 18 & 73 & 131 \\ -6 & -15 & -61 \\ 0 & -12 & -12 \\ 0 & 0 & 80 \end{pmatrix}.$$

Here the computation might take a long time, it is still better than trying to solve it from definition. ◇

When is a square matrix a base change? Let A be a square matrix of size n. If A is a base change matrix, we know it is invertible. The columns of A must form a basis in $\mathbb{R}^n$ by Theorem 3.5.6. Conversely, if the columns of A form a basis β of $\mathbb{R}^n$, it is invertible by Theorem 3.5.6 again. Obviously, in this case we may consider A as the base change matrix from β to the standard basis. We have the following theorem.

Theorem 5.1.5. *A matrix in $M_n(\mathbb{R})$ is a base change matrix if and only if its columns form a basis for $\mathbb{R}^n$ if and only if it is invertible.*

We knew that an invertible matrix represents an isomorphism. Here, Theorem 5.1.5 says an invertible matrix also represents a base change. It might sound confusing at first, but we will see that this flexibility in viewpoints is quite convenient.

Now Exercise 8 in §4.2 can be rewritten as the following corollary.

Corollary 5.1.6. *Let $A \in M_{m \times n}(\mathbb{R})$. Then $\operatorname{rk} A = r$ if and only if you can find invertible matrices P and Q such that*

$$QAP = \begin{pmatrix} I_r & \\ & \mathbf{0}_{(m-r) \times (n-r)} \end{pmatrix}.$$

Example 5.1.7. Suppose given two matrices

$$P = \begin{pmatrix} 0 & 0 & 1 \\ 0 & 1 & -1 \\ -1 & 2 & -1 \end{pmatrix} \quad \text{and} \quad Q = \begin{pmatrix} -1 & 2 & -1 \\ 0 & 1 & -1 \\ 0 & 0 & 1 \end{pmatrix}.$$

Since $\det P = 1$ and $\det Q = -1$, they are both invertible. From Theorem 5.1.5, we have that

$$\beta = \left(\begin{pmatrix} 0 \\ 0 \\ -1 \end{pmatrix}, \begin{pmatrix} 0 \\ 1 \\ 2 \end{pmatrix}, \begin{pmatrix} 1 \\ -1 \\ -1 \end{pmatrix} \right) \quad \text{and} \quad \gamma = \left(\begin{pmatrix} -1 \\ 0 \\ 0 \end{pmatrix}, \begin{pmatrix} 2 \\ 1 \\ 0 \end{pmatrix}, \begin{pmatrix} -1 \\ -1 \\ 1 \end{pmatrix} \right)$$

are two ordered bases of $\mathbb{R}^3$. Thus, P may be considered as the base change matrix from β to the standard basis α of $\mathbb{R}^3$, and Q may be considered as the base change matrix from γ to α as well.

Let $A = [T]_\beta^\beta$. Then

$$[T]_\alpha^\alpha = [\mathbf{1}_{\mathbb{R}^3}]_\beta^\alpha [T]_\beta^\beta [\mathbf{1}_{\mathbb{R}^3}]_\alpha^\beta = PAP^{-1}.$$

On the other hand,

$$[T]_\gamma^\gamma = [\mathbf{1}_{\mathbb{R}^3}]_\alpha^\gamma [T]_\alpha^\alpha [\mathbf{1}_{\mathbb{R}^3}]_\gamma^\alpha = Q^{-1} P A P^{-1} Q.$$

We may switch between bases freely without getting lost.

Similar matrices. For a linear endomorphism T, we prefer to use the same basis for the domain and the codomain. One obvious reason is that the domain and the codomain are considered as *one* space. Another equally important reason is that when dealing with a linear endomorphism T, we often also need to investigate other linear endomorphisms such as T^2, T^3, *etc.* Remember that we use $[T]_\beta$ for $[T]_\beta^\beta$, and we call it the matrix representing T relative to β twice. Using this notation we have

$$[T^n]_\beta = [T]_\beta^n$$

for $n \in \mathbb{N}$. As an immediate result of Theorem 5.1.2, we also have

Corollary 5.1.8. *Let $T: V \to V$ be a linear endomorphism and let β_1 and β_2 be bases of V. Let Q be the base change from β_2 to β_1. Then*

$$[T]_{\beta_2} = Q^{-1} [T]_{\beta_1} Q.$$

Definition 5.1.9. Let $A, B \in M_n(\mathbb{R})$. We say A is **similar to** B, denoted $A \sim B$, if and only if there exists an invertible matrix Q such that $B = Q^{-1} A Q$.

We now make a comparison between equivalent matrices and similar matrices.

- Two matrices are equivalent if and only if they represent the same linear transformation.

- Two square matrices A and B are similar if and only if they represent the same linear endomorphism relative to possibly different bases twice.

Example 5.1.10. (a) In $\mathbb{R}^2$ show that $\beta = ((2,4),\ (3,1))$ is a basis.

(b) Find the matrix of base change Q from β to the standard basis α.

(c) Let

$$A = \begin{pmatrix} 3 & 1 \\ -1 & 3 \end{pmatrix}$$

be the matrix of T with respect to the standard basis of $\mathbb{R}^2$. Find the matrix of T with respect to β.

Solution. (a) Since

$$\det \begin{pmatrix} 2 & 3 \\ 4 & 1 \end{pmatrix} = 2 - 12 = -10 \neq 0,$$

we know that β is a basis.

(b) The base change matrix Q is easily seen to be $\begin{pmatrix} 2 & 3 \\ 4 & 1 \end{pmatrix}$.

(c) Let α be the standard basis of $\mathbb{R}^2$. First we find

$$\begin{pmatrix} 2 & 3 \\ 4 & 1 \end{pmatrix}^{-1} = -\frac{1}{10} \begin{pmatrix} 1 & -3 \\ -4 & 2 \end{pmatrix}.$$

The matrix of T with respect to β is

$$Q^{-1}AQ = -\frac{1}{10} \begin{pmatrix} 1 & -3 \\ -4 & 2 \end{pmatrix} \begin{pmatrix} 3 & 1 \\ -1 & 3 \end{pmatrix} \begin{pmatrix} 2 & 3 \\ 4 & 1 \end{pmatrix} = \begin{pmatrix} 2 & -1 \\ 2 & 4 \end{pmatrix}.$$

As we can see from this example, there is no need to go through T at all. ◇

The many faces of invertible matrices. Before we leave this section, we would like to point out something. So far we have encountered several seemingly unrelated concepts which eventually all lead to the same thing: invertible matrices. Here we make a summary of these concepts.

The following statements about a square matrix A of size n are all equivalent:

- The matrix A is invertible.

- The rank of A is n.

- The nullity of A is 0.

- The linear endomorphism L_A is an isomorphism.

- The linear endomorphism L_A is onto.

- The linear endomorphism L_A is one-one.

- The rank of L_A is n.

- The nullity of L_A is 0.

- Any linear transformation $T\colon V \to W$ represented by A is an isomorphism.

- Any linear transformation $T\colon V \to W$ represented by A is onto.

- Any linear transformation $T\colon V \to W$ represented by A is one-one.

- The rank of any linear transformation $T\colon V \to W$ represented by A is n.

- The nullity of any linear transformation $T\colon V \to W$ represented by A is 0.

- The reduced row echelon form of A has n nonzero rows.

- The columns of A form a basis for $\mathbb{R}^n$.

- The rows of A form a basis for $\mathbb{R}^n$.

- The determinant of A is nonzero.

- The matrix A is a base change matrix.

Exercises 5.1

1. Let $T\colon \mathbb{R}^3 \to \mathbb{R}^2$ be a linear transformation such that the matrix representing T with respect to the standard bases is

$$\begin{pmatrix} 6 & -1 & 2 \\ 2 & 4 & 1 \end{pmatrix}.$$

Let $\beta_1 = ((1,2),\ (1,1))$ and $\beta_2 = ((1,1,1),\ (1,-1,0),\ (1,1,-2))$. Find $[T]_{\beta_2}^{\beta_1}$.

2. Let $T\colon \mathbb{R}^3 \to \mathbb{R}^3$ be the linear endomorphism defined by

$$T(x, y, z) = (y + z, x + z, y + x).$$

Let α be the standard basis and let $\beta = ((1, 1, 1),\ (1, -1, 0),\ (1, 1, -2))$.

(a) Calculate $[T]_\alpha$.

(b) Calculate $[T]_\beta$.

(c) Calculate $[T]_\beta^\alpha$.

(d) Calculate $[T]_\alpha^\beta$.

3. Show that a square matrix of size n is invertible if and only if its rows form a basis for $\mathbb{R}^n$.

4. Let β be a basis for the vector space V. If a linear endomorphism $T\colon V \to V$ is also an isomorphism, show that

$$[T^{-n}]_\beta = [T]_\beta^{-n}$$

for $n \in \mathbb{N}$.

5. Let A and B be square matrices.

(a) Show that if A and B are equivalent, then so are A^{-1} and B^{-1}.

(b) Show that if A and B are similar, then so are A^{-1} and B^{-1}.

(c) Suppose A and B are equivalent. Is it true that A^2 and B^2 are also equivalent?

(d) Suppose A and B are similar. Is it true that A^2 and B^2 are also similar?

6. We say two linear transformations from V to W are equivalent if they can be represented by the same matrix.

(a) Show that $S\colon V \to W$ and $T\colon V \to W$ are equivalent if and only if any of their matrices are equivalent.

(b) Show that two matrices of the same size are equivalent if and only if they are of the same rank.

(c) Show that two linear transformations from V to W are equivalent if and only if they are of the same rank.

7. We say two linear endomorphisms from V to V are similar if they can be represented by the same matrix relative to their respective basis twice. Show that $S\colon V \to V$ and $T\colon V \to V$ are similar if and only if their matrices relative to some basis twice are similar.

5.2 Eigenvalues and eigenvectors

Throughout this section V stands for a finite dimensional vector space.

Eigenvalues and eigenvectors. A linear endomorphism on a finite dimensional vector space is particularly simple if it can be represented by a diagonal matrix (relative to a basis twice)!

Definition 5.2.1. We say a linear endomorphism T on V is **diagonalizable** if there is an ordered basis β for V such that

$$
[T]_\beta = \begin{pmatrix} \lambda_1 & 0 & \cdots & 0 \\ 0 & \lambda_1 & \cdots & 0 \\ \vdots & \vdots & \ddots & \vdots \\ 0 & 0 & \cdots & \lambda_n \end{pmatrix}
$$

is a diagonal matrix. A square matrix A is called **diagonalizable** if L_A is diagonalizable.

A matrix A is diagonalizable if and only if there is an invertible matrix P such that $P^{-1}AP$ is diagonal, that is, if and only if A is similar to a diagonal matrix. If an linear endomorphism T is diagonalizable, it means that we can find n linearly independent vectors $v_1, \ldots, v_n$ such that

$$
T(v_i) = \lambda_i v_i, \qquad i = 1, 2, \ldots, n
$$

for some scalars $\lambda_1, \ldots, \lambda_n$.

Definition 5.2.2. Let T be a linear endomorphism on V (not necessarily finite dimensional). A *nonzero* vector $v \in V$ is called an **eigenvector** of T

if there exists a scalar λ such that $T(v) = \lambda v$. The scalar λ is called the **eigenvalue** of T corresponding to the eigenvector v.

Let A be in $M_n(\mathbb{R})$. A nonzero vector $v \in \mathbb{R}^n$ is called an **eigenvector** of A if v is an eigenvector of L_A, that is, if $Av = \lambda v$ for some scalar λ. The scalar λ is called an **eigenvalue** of A (or L_A) corresponding to the eigenvector v.

Proposition 5.2.3. *A linear endomorphism T on V is diagonalizable if and only if there is a basis for V consisting of eigenvectors. Furthermore, if $\beta = (v_1, v_2, \ldots, v_n)$ is an ordered basis consisting of eigenvectors, then*

$$[T]_\beta = \begin{pmatrix} \lambda_1 & 0 & \cdots & 0 \\ 0 & \lambda_1 & \cdots & 0 \\ \vdots & \vdots & \ddots & \vdots \\ 0 & 0 & \cdots & \lambda_n \end{pmatrix}$$

where λ_i is the eigenvalues of T corresponding to the eigenvector v_i.

Proof. This is just a reorganization of facts and definitions regarding eigenvalues, eigenvectors and diagonalizable linear endomorphisms. $\square$

Hence, to diagonalize a linear endomorphism or a square matrix is to find a basis consisting of eigenvectors.

Example 5.2.4. Any nonzero vector in the null space of a linear endomorphism is an eigenvector and its corresponding eigenvalue is 0.

Example 5.2.5. Let $T \colon V \to V$ be a linear endomorphism which is **nilpotent**, that is, $T^n = 0_V$ for some positive integer n. Suppose T has an eigenvector v and λ is its corresponding eigenvalue. Then $T^n(v) = \lambda^n v = O$. Since v is a nonzero vector, this means that $\lambda^n = 0$. It follows that $\lambda = 0$. We conclude that the only eigenvalue of a nilpotent linear endomorphism is 0.

Example 5.2.6. Let A be a square matrix. If $A - I$ is invertible, We claim that 1 is not an eigenvalue of A. Assuming the opposite, we may find a nonzero vector v such that $Av = v$. This means that $v \in \text{Null}\,(A - I)$. The matrix $A - I$ is not invertible, a contradiction.

Characteristic polynomials. The previous example certainly suggests a method to find eigenvectors and eigenvalues for a matrix.

Proposition 5.2.7. *Let $A \in M_n(\mathbb{R})$. A scalar λ is an eigenvalue of A if and only if*

$$\det(A - \lambda I_n) = 0.$$

Proof. We have

$$\lambda \text{ is an eigenvalue of } A$$
$$\Longleftrightarrow \quad Av = \lambda v \text{ for some } v \in \mathbb{R}^n \setminus \{0\}$$
$$\Longleftrightarrow \quad (A - \lambda I)v = O \text{ for some } v \in \mathbb{R}^n \setminus \{0\}$$
$$\Longleftrightarrow \quad v \in \text{Null}\,(A - \lambda I) \text{ for some } v \in \mathbb{R}^n \setminus \{0\}$$
$$\Longleftrightarrow \quad A - \lambda I \text{ is not one-one}$$
$$\Longleftrightarrow \quad A - \lambda I \text{ is not invertible}$$
$$\Longleftrightarrow \quad \det(A - \lambda I) = 0.$$

Now we have an easy method to find eigenvalues of A. $\square$

Definition 5.2.8. Let $A \in M_n(\mathbb{R})$. The polynomial

$$f(t) = \det(A - tI_n),$$

being a polynomial in t, is called the **characteristic polynomial** of A.

The matrix $A - tI_n$ does not have entries in $\mathbb{R}$. We consider it as a matrix in $M_n(\mathbb{R}(t))$. The notation $\mathbb{R}(t)$ denotes the set of all rational functions over $\mathbb{R}$. In other words,

$$\mathbb{R}(t) = \left\{ \frac{f(x)}{g(x)} : \begin{array}{l} f(x),\ g(x) \text{ are polynomials with} \\ \text{real coefficients,}\quad g(x) \neq 0 \end{array} \right\}$$

The set $\mathbb{R}(t)$ with its usual addition and multiplication is a field. The leading coefficient of $f(t)$ is $(-1)^n$ and the constant term of $f(t)$ is $\det A$. The eigenvalues of A are the zeros of the characteristic polynomial of A.

Example 5.2.9. Find the eigenvalues of

$$A = \begin{pmatrix} 1 & 1 \\ 4 & 1 \end{pmatrix} \in M_2(\mathbb{R}).$$

Find an eigenvector for each eigenvalue. Can you conclude that A is diagonalizable? If yes, find a diagonal matrix B and an invertible matrix P in $M_2(\mathbb{R})$ such that $B = P^{-1}AP$.

Solution. The characteristic polynomial of A is

$$f(t) = \det \begin{pmatrix} 1-t & 1 \\ 4 & 1-t \end{pmatrix}$$
$$= (1-t)^2 - 4$$
$$= t^2 - 2t - 3$$
$$= (t-3)(t+1).$$

The eigenvalues of A are 3 and -1.

To find an eigenvector for the eigenvalue 3, we need to look in Null $(A - 3I)$. We need to solve

$$\begin{pmatrix} -2 & 1 \\ 4 & -2 \end{pmatrix} \begin{pmatrix} a \\ b \end{pmatrix} = \begin{pmatrix} 0 \\ 0 \end{pmatrix}.$$

Thus $(1,2)$ is an eigenvector with eigenvalue 3. Note that this is not the only choice.

To find an eigenvector for the eigenvalue -1, we need to solve

$$\begin{pmatrix} 2 & 1 \\ 4 & 2 \end{pmatrix} \begin{pmatrix} a \\ b \end{pmatrix} = \begin{pmatrix} 0 \\ 0 \end{pmatrix}.$$

The vector $(1, -2)$ is an obvious choice. The vector $(1, -2)$ is an eigenvector with eigenvalue -1.

The set $\mathscr{B} = \{(1,2), (1,-2)\}$ is clearly linearly independent. It is a basis for $\mathbb{R}^2$. It consists of eigenvectors of A. Thus A is diagonalizable. Let α be the standard basis for $\mathbb{R}^2$, and let $\beta = ((1,2),(1,-2))$. Then

$$B = [L_A]_\beta^\beta = \begin{pmatrix} 3 & \\ & -1 \end{pmatrix}.$$

We know that

$$B = [1_{\mathbb{R}^2}]_\alpha^\beta [L_A]_\alpha^\alpha [1_{\mathbb{R}^2}]_\beta^\alpha.$$

Let P be the base change matrix from β to α, i.e.,

$$P = \begin{pmatrix} 1 & 1 \\ 2 & -2 \end{pmatrix}.$$

Thus $B = P^{-1}AP$.   ◇

Proposition 5.2.10. *Let A and B both be matrices representing the same linear endomorphism. In other words, A and B are similar to each other. Then the characteristic polynomials of both matrices are the same.*

Proof. By assumption, $B = P^{-1}AP$ for some invertible matrix P. Then

$$\begin{aligned}
\det(B - I) &= \det(P^{-1}AP - I) \\
&= \det(P^{-1}AP - P^{-1}IP) \\
&= \det P^{-1}(A - I)P \\
&= (\det P)^{-1}\det(A - I)\det P \\
&= \det(A - I).
\end{aligned}$$

Thus the characteristic polynomial of A is the same as the characteristic polynomial of B.   □

Let T be a linear endomorphism on a finite dimensional vector space and let A be any matrix representing T. By virtue of Proposition 5.2.10, we may define the characteristic polynomial of T to be the characteristic polynomial of A. Hence the eigenvalues of T are exactly the zeros of the characteristic polynomial of T.

Eigenspaces. For each eigenvalue of a square matrix A, the corresponding eigenvectors actually form a subspace if you also throw in the zero vector.

Proposition 5.2.11. *Let T be a linear endomorphism on V and let λ be an eigenvalue of T. Then*

$$E_\lambda = \{v \in V : T(v) = \lambda v\}$$

is a subspace of V. Similarly, if $A \in M_n(\mathbb{R})$ and λ is an eigenvalue of A, then

$$E_\lambda = \{v \in \mathbb{R}^n : A(v) = \lambda v\}$$

is a subspace of $\mathbb{R}^n$.

Proof. This is clear since $E_\lambda = \text{Null}\,(T - \lambda 1_V)$ in the first case, and $E_\lambda = \text{Null}\,(A - \lambda I_n)$ in the second case.   □

The E_λ in the proposition above is called the **eigenspace** of T (or of A) corresponding to the eigenvalue λ. Note that if λ is indeed an eigenvalue of T, then E_λ is of dimension at least 1!

Example 5.2.12. Let T be the linear endomorphism on $\mathscr{P}_2$ defined by

$$T(f(x)) = f(x) + (x+1)f'(x).$$

Find the eigenvalues of T. Compute the eigenspace for each eigenvalue of T.

Solution. Note that

$$T(1) = 1,$$
$$T(x) = 1 + 2x,$$
$$T(x^2) = 2x + 3x^2.$$

The matrix of T relative to the standard basis α is

$$A = \begin{pmatrix} 1 & 1 & 0 \\ 0 & 2 & 2 \\ 0 & 0 & 3 \end{pmatrix}.$$

The characteristic polynomial of T is

$$\det \begin{pmatrix} 1-t & 1 & \\ & 2-t & 2 \\ & & 3-t \end{pmatrix} = -(t-1)(t-2)(t-3).$$

The eigenvalues of T are 1, 2 and 3.

The eigenspace of A corresponding to the eigenvalue 1 is

$$\mathrm{Null}\left(\begin{pmatrix} 0 & 1 & 0 \\ 0 & 1 & 2 \\ 0 & 0 & 2 \end{pmatrix} \right).$$

For this, we need to solve

$$\begin{pmatrix} 0 & 1 & 0 \\ 0 & 1 & 2 \\ 0 & 0 & 2 \end{pmatrix} \begin{pmatrix} x \\ y \\ z \end{pmatrix} = \begin{pmatrix} 0 \\ 0 \\ 0 \end{pmatrix}.$$

The solution space is $\{(a,0,0) \in \mathbb{R}^3 : a \in \mathbb{R}\} = \mathrm{Sp}\,((1,0,0))$. Hence $E_1 = \mathrm{Sp}\,(1)$, the set of all constant polynomials.

Next, we find E_2. We first find

$$\mathrm{Null}\left(\begin{pmatrix} -1 & 1 & 0 \\ 0 & 0 & 2 \\ 0 & 0 & 1 \end{pmatrix}\right).$$

We need to solve

$$\begin{pmatrix} -1 & 1 & 0 \\ 0 & 0 & 2 \\ 0 & 0 & 1 \end{pmatrix}\begin{pmatrix} x \\ y \\ z \end{pmatrix} = \begin{pmatrix} 0 \\ 0 \\ 0 \end{pmatrix}.$$

The solution space is $\{(a,a,0) \in \mathbb{R}^3 : a \in \mathbb{R}\} = \mathrm{Sp}\,((1,1,0))$. Hence the eigenspace E_2 of T is $\mathrm{Sp}\,(1+x)$.

Finally, to find E_3, we solve

$$\begin{pmatrix} -2 & 1 & 0 \\ 0 & -1 & 2 \\ 0 & 0 & 0 \end{pmatrix}\begin{pmatrix} x \\ y \\ z \end{pmatrix} = \begin{pmatrix} 0 \\ 0 \\ 0 \end{pmatrix}.$$

The solution space is $\{(a,2a,a) \in \mathbb{R}^3 : a \in \mathbb{R}\} = \mathrm{Sp}\,((1,2,1))$. Hence the eigenspace E_3 of T is $\mathrm{Sp}\,(1+2x+x^2)$. $\diamond$

Example 5.2.13. Find the characteristic polynomial and all possible eigenvalues of $A = \begin{pmatrix} 0 & -1 \\ 1 & 0 \end{pmatrix}$. Find all corresponding eigenspaces. Can you determine if it is diagonalizable over $\mathbb{R}$? How about over $\mathbb{C}$?

Solution. The characteristic polynomial of A is

$$\det\begin{pmatrix} -t & -1 \\ 1 & -t \end{pmatrix} = t^2 + 1.$$

This polynomial has no zeros in $\mathbb{R}$. Thus A has no eigenvalue as a matrix over $\mathbb{R}$. However, if we consider the problem over $\mathbb{C}$, the matrix has two eigenvalues i and $-i$. To find E_i, we need to solve

$$\begin{pmatrix} -i & -1 \\ 1 & -i \end{pmatrix}\begin{pmatrix} x \\ y \end{pmatrix} = \begin{pmatrix} 0 \\ 0 \end{pmatrix}.$$

We use the augmented matrix for the system

$$\left(\begin{array}{cc|c} -i & -1 & 0 \\ 1 & -i & 0 \end{array}\right)$$

$$\rightsquigarrow \left(\begin{array}{cc|c} 1 & -i & 0 \\ -i & -1 & 0 \end{array}\right) \begin{array}{l} \text{row(ii)} \\ \text{row(i)} \end{array}$$

$$\rightsquigarrow \left(\begin{array}{cc|c} 1 & -i & 0 \\ 0 & 0 & 0 \end{array}\right) \begin{array}{l} \\ \text{row(ii)} + i \times \text{row(i)} \end{array} .$$

This says that $x - iy = 0$ or $x = iy$. The solution space is

$$E_i = \{(ai, a) \in \mathbb{C}^2 : a \in \mathbb{C}\} = \mathrm{Sp}\left((i, 1)\right).$$

Similarly, E_{-i} can be found by solving

$$\begin{pmatrix} i & -1 \\ 1 & i \end{pmatrix} \begin{pmatrix} x \\ y \end{pmatrix} = \begin{pmatrix} 0 \\ 0 \end{pmatrix}.$$

The solution space is

$$E_{-i} = \{(-ai, a) \in \mathbb{C}^2 : a \in \mathbb{C}\} = \mathrm{Sp}\left((-i, 1)\right).$$

Since $\beta = \{(i, 1), (-i, 1)\}$ is a basis for $\mathbb{C}^2$ over $\mathbb{C}$, it is a basis consisting of eigenvectors of A. Thus A is diagonalizable. In fact,

$$\begin{pmatrix} i & \\ & -i \end{pmatrix} = \begin{pmatrix} i & -i \\ 1 & 1 \end{pmatrix}^{-1} \begin{pmatrix} 0 & -1 \\ 1 & 0 \end{pmatrix} \begin{pmatrix} i & -i \\ 1 & 1 \end{pmatrix}$$

where

$$\begin{pmatrix} i & -i \\ 1 & 1 \end{pmatrix}$$

is the base change matrix from β to the standard basis of $\mathbb{C}^2$.   ◇

Example 5.2.14. Find the characteristic polynomial and all possible eigenvalues of $A = \begin{pmatrix} 0 & 1 \\ 0 & 0 \end{pmatrix}$. Find all corresponding eigenspaces. Can you determine if it is diagonalizable over $\mathbb{R}$? How about over $\mathbb{C}$?

Solution. The characteristic polynomial of A is

$$\det \begin{pmatrix} -t & 1 \\ 0 & -t \end{pmatrix} = t^2.$$

The matrix A has a unique eigenvalue 0. The eigenspace E_0 is Null $(A) =$ Sp $((1,0))$. The dimension of E_0 is 1. There is no way to find a basis consisting of eigenvectors of A. Thus A is *not* diagonalizable over $\mathbb{R}$. The discussion applies also to the case when we consider A as a matrix with complex entries.   ◇

Exercises 5.2

Throughout the following problems T is a linear endomorphism on a finite dimensional real vector space V.

1. Find the characteristic polynomial, eigenvalues with corresponding eigenspaces for each of the following linear endomorphisms.

    (a) $T \colon \mathbb{R}^2 \to \mathbb{R}^2$, $T(x,y) = (y,-x)$.

    (b) $T \colon \mathbb{R}^3 \to \mathbb{R}^3$, $T(x,y,z) = (y,-x,z)$.

    (c) $T \colon \mathbb{R}^4 \to \mathbb{R}^4$, $T(x,y,z,w) = (x+y,x,z+2w,2z+w)$.

2. Show that $T \colon \mathbb{R}^3 \to \mathbb{R}^3$ has at least one eigenvalue.

3. Determine whether each of the following statements is true.

    (a) Any multiple of an eigenvector is still an eigenvector.

    (b) The sum of two eigenvector is still an eigenvector.

    (c) If v is an eigenvector of T^2 with eigenvalue 4, then v is an eigenvector of T with eigenvalue 2 or -2.

4. Let v be an eigenvector of $T \colon V \to V$ with eigenvalue λ. Show that v is an eigenvector of T^n with eigenvalue λ^n for $n \in \mathbb{Z}_+$.

5. Suppose A is an invertible matrix.

(a) What relation is there between the eigenvalues of A and those of A^{-1}.

(b) If A is diagonalizable, show that so is A^{-1}.

6. Let v and v' be eigenvectors of T with eigenvalues λ and λ'. If $\lambda \neq \lambda'$, is $v + v'$ still an eigenvector? If yes, what is the eigenvalue of $v + v'$?

7. Suppose T^2 has a positive eigenvalue. Show that T has at least one real eigenvalue. (Hint: $A^2 - a^2 I = (A - aI)(A + aI)$.)

8. Suppose W is an eigenspace of T^k for some $k \in \mathbb{Z}_+$. Show that $T(W) \subseteq W$.

5.3   Diagonalizability

In this section we give sufficient conditions for diagonalizable matrices. We will be mostly using real vector spaces and complex vector spaces as examples. However, the reader will see that the discussion is valid for any field.

The splitting of the characteristic polynomial.   To diagonalize a square matrix, we need to find enough eigenvectors to form a basis. In this section we analyze when this is possible.

Definition 5.3.1. A polynomial $f(t)$ in the variable t with coefficients in a field F is said to **split** over F if there are scalars $c, a_1, \ldots, a_n$ (not necessarily distinct) in F such that

$$f(t) = c(t - a_1)(t - a_2) \cdots (t - a_n).$$

We call $a_1, a_2, \ldots, a_n$ the **zeros** of $f(t)$.

Example 5.3.2. The polynomial $f(t) = t^2 - 1 = (t+1)(t-1)$ splits over $\mathbb{R}$ while $g(t) = t^2 + 1$ does not split over $\mathbb{R}$. However, $g(t) = t^2 + 1 = (t+i)(t-i)$ does split over $\mathbb{C}$. The splitting of polynomials depends on the underlying field.

Theorem 5.3.3 (Fundamental Theorem of Algebra). *Every polynomial of one variable with coefficient in $\mathbb{C}$ splits over $\mathbb{C}$.*

We will simply assume this theorem. In spite of its name, this theorem will be proved in a course of Complex Analysis.

Proposition 5.3.4. *If the characteristic polynomial of a linear endomorphism (or of a square matrix) does not split over a field F then the given linear endomorphism (or matrix) is not diagonalizable over F.*

Proof. Suppose A is diagonalizable. Then A is similar to a diagonal matrix

$$B = \begin{pmatrix} \lambda_1 & & & \\ & \lambda_2 & & \\ & & \ddots & \\ & & & \lambda_n \end{pmatrix}.$$

By Proposition 5.2.10, the characteristic polynomial of A is also the characteristic polynomial of B. Now

$$\det \begin{pmatrix} \lambda_1 - t & & & \\ & \lambda_2 - t & & \\ & & \ddots & \\ & & & \lambda_n - t \end{pmatrix} = (-1)^n \prod_{i=1}^{n} (t - \lambda_i).$$

Hence the characteristic polynomial of A splits over the underlying field. $\square$

Example 5.3.5. The matrix $\begin{pmatrix} 0 & -1 \\ 1 & 0 \end{pmatrix} \in M_2(\mathbb{R})$ is not diagonalizable since its characteristic polynomial $t^2 + 1$ does not split over $\mathbb{R}$. Compare this result with Example 5.2.13. However, Proposition 5.3.4 does not tell us whether this matrix is diagonalizable over $\mathbb{C}$.

Independence of Eigenspaces. We continue to discuss when a matrix is diagonalizable when its characteristic polynomial splits.

Proposition 5.3.6. *Let T be a linear endomorphism on a vector space V. For $i = 1, \ldots, k$, let the λ_i's be distinct eigenvalues of T and the E_{λ_i}'s be the corresponding eigenspaces. For each i, let $v_i \in E_{\lambda_i}$. If $v_1 + \cdots + v_k = O$ then $v_1 = \cdots = v_k = O$.*

Proof. We will prove a more general version of this proposition. In fact, we claim that for $t = 1, 2, \ldots, k$,

$$v_1 + \cdots + v_t = O \quad \Longrightarrow \quad v_1 = \cdots = v_t = O.$$

We will prove this claim by induction on t.

The case $t = 1$ is trivial. We now assume the induction hypothesis for $t - 1$.

Let $v_1 + \cdots + v_t = O$. Then

$$T(v_1 + \cdots + v_t) = \lambda_1 v_1 + \cdots + \lambda_{t-1} v_{t-1} + \lambda_t v_t = O.$$

We also have

$$\lambda_t v_1 + \cdots + \lambda_t v_{t-1} + \lambda_t v_t = O.$$

Taking the difference of the two identities, we have

$$(\lambda_t - \lambda_1) v_1 + \cdots + (\lambda_t - \lambda_{t-1}) v_{t-1} = O.$$

By induction on t, we know that

$$(\lambda_t - \lambda_1) v_1 = \cdots = (\lambda_t - \lambda_{t-1}) v_{t-1} = O.$$

The eigenvalues are assumed to be distinct. Thus $\lambda_t - \lambda_1, \ldots, \lambda_t - \lambda_{t-1}$ are all nonzero. This implies that $v_1 = \cdots = v_{t-1} = O$. It follows that $v_t = O$ as well. $\qquad\square$

Corollary 5.3.7. *Let T be a linear endomorphism on a vector space V and let $\lambda_1, \ldots, \lambda_k$ be distinct eigenvalues of T. Let v_i be an eigenvector corresponding to the eigenvalue λ_i. Then $v_1, \ldots, v_k$ are linearly independent.*

Proposition 5.3.8. *Let V be an n-dimensional vector space. If the characteristic polynomial of a linear endomorphism $T \colon V \to V$ (or of a square matrix) splits with n distinct zeros, then T (or the square matrix) is diagonalizable.*

Proof. Let $\lambda_1, \ldots, \lambda_n$ be the distinct zeros of the characteristic polynomial of T. These are the eigenvalues of T. For each i, find a corresponding eigenvector v_i. By Corollary 5.3.7, $\mathscr{B} = \{v_1, \ldots, v_n\}$ is linearly independent. Since $\dim V = n$, the set $\mathscr{B}$ is a basis consisting of eigenvectors of T. We conclude that T is diagonalizable. $\qquad\square$

Example 5.3.9. The characteristic polynomial of

$$A = \begin{pmatrix} 0 & -1 \\ 1 & 0 \end{pmatrix}$$

is $t^2 + 1$. Over $\mathbb{C}$ it splits into $(t - i)(t + i)$. This polynomial has two distinct zeros. Thus by Proposition 5.3.8, A is diagonalizable. However, in order to find P such that $P^{-1}AP$ is diagonal, we still need to find the eigenspaces of A. See Example 5.2.13.

When is a matrix diagonalizable? To completely answer the *diagonalization problem*, it remains to deal with the case when the characteristic polynomial (of degree n) splits but does not have n distinct zeros.

Definition 5.3.10. The **multiplicity** of λ in a polynomial $f(t)$ is the highest nonnegative integer k for which $(t - \lambda)^k$ is a factor of $f(t)$.

Example 5.3.11. The polynomial $f(t) = (t - 1)(t - 2)^2(t - 3)^4$ has 3 distinct zeros. The zero 1 is of multiplicity 1, the zero 2 is of multiplicity 2 and the zero 3 is of multiplicity 4.

Definition 5.3.12. Let $f(t)$ be the characteristic polynomial of the linear endomorphism T (or of the matrix A). Let λ be an eigenvalue. The **algebraic multiplicity** of λ is defined to be the multiplicity of λ in $f(t)$. The **geometric multiplicity** of λ is defined to be the dimension of E_λ.

Note that when a polynomial of degree n splits, the sum of the algebraic multiplicities of all its zeros equals n.

Example 5.3.13. Find the algebraic multiplicity and the geometric multiplicity of the eigenvalues in $A = \begin{pmatrix} 0 & 1 \\ 0 & 0 \end{pmatrix}$.

Solution. The characteristic polynomial of A is

$$\det \begin{pmatrix} -t & 1 \\ & -t \end{pmatrix} = t^2.$$

There is only one eigenvalue 0. The algebraic multiplicity of 0 is 2. On the other hand,

$$E_0 = \text{Null} \left(\begin{pmatrix} 0 & 1 \\ 0 & 0 \end{pmatrix} \right) = \text{Sp}\,(e_1).$$

The dimension of E_0 is 1. Hence the geometric multiplicity of 0 is 1. The algebraic multiplicity and the geometric multiplicity of the eigenvalue 0 are not the same!   ◇

There is a relation between the algebraic multiplicity and the geometric multiplicity.

Lemma 5.3.14. *Let λ be a eigenvalue of a linear endomorphism. The geometric multiplicity of λ never exceeds the algebraic multiplicity.*

Proof. Let $T\colon V \to V$ be a linear endomorphism, where λ is an eigenvalue. Let $\dim E_\lambda = d$. Find a basis $\{v_1, \ldots, v_d\}$ for E_λ. Expand to an ordered basis $\beta = (v_1, \ldots, v_d, v_{d+1}, \ldots, v_n)$ for V. Relative to β, the associated matrix is of the form

$$
\left(
\begin{array}{ccc|c}
\lambda & & & \\
 & \ddots & & B_{d\times(n-d)} \\
 & & \lambda & \\
\hline
 & \mathbf{0}_{(n-d)\times d} & & C_{(n-d)\times(n-d)}
\end{array}
\right)
$$

Thus the characteristic polynomial $f(t)$ of T is

$$
\left(
\begin{array}{ccc|c}
\lambda - t & & & \\
 & \ddots & & B_{d\times(n-d)} \\
 & & \lambda - t & \\
\hline
 & \mathbf{0}_{(n-d)\times d} & & C - I_{n-d}
\end{array}
\right)
$$

From identity (4.4) we have that

$$
f(t) = (\lambda - t)^d g(t).
$$

It follows that the algebraic multiplicity of λ is at least d.   □

We are now ready to give a complete solution to the diagonalization problem.

Lemma 5.3.15. *Let $\lambda_1, \ldots, \lambda_r$ be distinct eigenvalues of the linear endomorphism T (or of the square matrix A). For each eigenspace E_{λ_i} find a basis $\{v_{i1}, v_{i2}, \ldots, v_{ik_i}\}$. Then*

$$
v_{11}, v_{12}, \ldots, v_{1k_1}, v_{21}, \ldots, v_{2k_2}, \ldots, v_{ij}, \ldots, v_{r1}, \ldots, v_{rk_r}
$$

are linearly independent.

Proof. Find scalars a_{ij} such that

$$a_{11}v_{11} + a_{12}v_{12} + \cdots + a_{1k_1}v_{1k_1} + a_{21}v_{21} + \cdots + a_{2k_2}v_{2k_2}$$
$$+ \cdots + a_{ij}v_{ij} + \cdots + a_{r1}v_{r1} + \cdots + a_{rk_r}v_{rk_r} = O.$$

Note that $a_{i1}v_{i1} + \cdots + a_{ik_k}v_{ik_i} \in E_i$ for each i. Using Proposition 5.3.6, we have

$$a_{i1}v_{i1} + \cdots + a_{ik_k}v_{ik_i} = O$$

for all i. But this is a relation for basis elements of E_{λ_i}. It follows that

$$a_{i1} = \cdots = a_{ik_i} = 0$$

for all i. This completes the proof. $\square$

Theorem 5.3.16. *A linear endomorphism or matrix is diagonalizable if and only if the following two conditions are both satisfied.*

- *The characteristic polynomial splits.*

- *The algebraic multiplicity and geometric multiplicity of each eigen-value are the same.*

Proof. The "only if" part: If the linear endomorphism T is diagonalizable, there exists an ordered basis β such that the matrix of T with respect to β is

$$B = \begin{pmatrix} \lambda_1 & & & & & & & & & \\ & \ddots & & & & & & & & \\ & & \lambda_1 & & & & & & & \\ & & & \lambda_2 & & & & & & \\ & & & & \ddots & & & & & \\ & & & & & \lambda_2 & & & & \\ & & & & & & \ddots & & & \\ & & & & & & & \lambda_s & & \\ & & & & & & & & \ddots & \\ & & & & & & & & & \lambda_s \end{pmatrix}$$

where $\lambda_1, \ldots, \lambda_s$ are distinct eigenvalues of T. The characteristic polynomial of T is

$$\prod_{i=1}^{s} (\lambda_i - t_i)^{d_i}$$

where d_i is the number of λ_i appearing in the diagonal matrix. Clearly, the algebraic multiplicity of λ_i is d_i. Note also that E_{λ_i} is spanned by the basis elements in β corresponding to the columns where λ_i lies. Hence $\dim E_{\lambda_i} = d_i$ since λ_i appears in d_i columns of B. The geometric multiplicity of λ_i is also d_i. Thus the algebraic multiplicity equals the geometric multiplicity for λ_i. This is true for all i.

The "if" part: Let $T \colon V \to V$ be a linear endomorphism satisfying the requirement. Let $\lambda_1, \ldots, \lambda_s$ be the distinct eigenvalues of T. Assume the algebraic (geometric) multiplicity of λ_i is d_i for each i. Then the degree of the characteristic polynomial of T is

$$d_1 + d_2 + \cdots + d_s.$$

Hence $\dim V = d_1 + d_2 + \cdots + d_s$. For each i, $\dim E_{\lambda_i} = d_i$. Find a basis

$$B_i = \{v_{11}, v_{12}, \ldots, v_{1d_i}\}$$

for E_{λ_i}. By Lemma 5.3.15, the set $B = \bigcup_{i=1}^{s} B_i$ is a linearly independent set. This set contains $d_1 + d_2 + \cdots + d_s$ elements. It is thus a basis for V. It is also a basis consisting of eigenvectors for T. We conclude that T is diagonalizable. $\qquad\square$

Example 5.3.17. Determine if the matrix

$$A = \begin{pmatrix} 3 & 1 & 0 \\ 0 & 3 & 0 \\ 0 & 0 & 4 \end{pmatrix} \in M_3(\mathbb{R})$$

is diagonalizable.

Solution. The characteristic polynomial of A is

$$\det \begin{pmatrix} 3-t & 1 & 0 \\ 0 & 3-t & 0 \\ 0 & 0 & 4-t \end{pmatrix} = -(t-3)^2(t-4).$$

The algebraic multiplicity of 3 is 2 and the algebraic multiplicity of 4 is 1.
The eigenspaces are

$$E_3 = \text{Null} \left(\begin{pmatrix} 0 & 1 & 0 \\ 0 & 0 & 0 \\ 0 & 0 & 1 \end{pmatrix} \right) = \text{Sp}\,(e_1),$$

$$E_4 = \text{Null} \left(\begin{pmatrix} -1 & 1 & 0 \\ 0 & -1 & 0 \\ 0 & 0 & 0 \end{pmatrix} \right) = \text{Sp}\,(e_3).$$

The geometric multiplicity of 3 is 1 and the geometric multiplicity of 4 is 1.

We conclude from Theorem 5.3.16 that A is not diagonalizable.   ◇

Example 5.3.18. Determine if the matrix

$$A = \begin{pmatrix} 2 & 1 & 1 \\ 0 & 1 & -1 \\ 0 & 2 & 4 \end{pmatrix} \in M_3(\mathbb{R})$$

is diagonalizable.

Solution. The characteristic polynomial of A is

$$\det \begin{pmatrix} 2-t & 1 & 1 \\ 0 & 1-t & -1 \\ 0 & 2 & 4-t \end{pmatrix}$$

$$= -t^3 + 7t^2 - 16t + 12$$

$$= -(t-2)^2(t-3).$$

The algebraic multiplicity of 2 is 2 and the algebraic multiplicity of 3 is 1.
The geometric multiplicity of 2 is

$$\text{nullity} \begin{pmatrix} 0 & 1 & 1 \\ 0 & -1 & -1 \\ 0 & 2 & 2 \end{pmatrix} = 3 - \text{rk} \begin{pmatrix} 0 & 1 & 1 \\ 0 & -1 & -1 \\ 0 & 2 & 2 \end{pmatrix}$$

$$= 3 - 1 = 2.$$

The algebraic multiplicity of 2 equals the geometric multiplicity of 2. There is no need to compute the geometric multiplicity of 3. It must be 1 by Lemma 5.3.14. Now we conclude that A is diagonalizable by Theorem 5.3.16.   ◇

Example 5.3.19. Let T be a function on $\mathscr{P}_2$ defined by

$$T(f(x)) = f(1) + f'(0)x + (f'(0) + f''(0))x^2.$$

We leave it to the reader to verify that T is a linear endomorphism.

(a) Let α be the standard ordered basis of $\mathscr{P}_2$. Compute $[T]_\alpha$.

(b) Show that T is diagonalizable.

(c) Find an ordered basis β such that $[T]_\beta$ is diagonal.

(d) Find an invertible matrix P such that $P^{-1}[T]_\alpha P$ is diagonal.

Solution. (a) Since

$$T(1) = 1,$$
$$T(x) = 1 + x + x^2,$$
$$T(x^2) = 1 + 2x^2,$$

we have

$$A = [T]_\alpha = \begin{pmatrix} 1 & 1 & 1 \\ 0 & 1 & 0 \\ 0 & 1 & 2 \end{pmatrix}.$$

(b) The characteristic polynomial of T is

$$\det \begin{pmatrix} 1-t & 1 & 1 \\ 0 & 1-t & 0 \\ 0 & 1 & 2-t \end{pmatrix} = -(t-1)^2(t-2).$$

We only need to check whether the geometric multiplicity of 1 is 2. The geometric multiplicity of 1 is

$$\text{nullity} \begin{pmatrix} 0 & 1 & 1 \\ 0 & 0 & 0 \\ 0 & 1 & 1 \end{pmatrix} = 3 - \text{rk} \begin{pmatrix} 0 & 1 & 1 \\ 0 & 0 & 0 \\ 0 & 1 & 1 \end{pmatrix} = 3 - 1 = 2.$$

Hence T is diagonalizable.

(c) We need to actually find E_1 and E_2 for L_A:

$$E_1 = \text{Null}\left(\begin{pmatrix} 0 & 1 & 1 \\ 0 & 0 & 0 \\ 0 & 1 & 1 \end{pmatrix}\right) = \text{Sp}\left(\begin{pmatrix} 1 \\ 0 \\ 0 \end{pmatrix}, \begin{pmatrix} 0 \\ 1 \\ -1 \end{pmatrix}\right),$$

$$E_2 = \text{Null}\left(\begin{pmatrix} -1 & 1 & 1 \\ 0 & -1 & 0 \\ 0 & 1 & 0 \end{pmatrix}\right) = \text{Sp}\left(\begin{pmatrix} 1 \\ 0 \\ 1 \end{pmatrix}\right).$$

Thus $\gamma = \left(\begin{pmatrix} 1 \\ 0 \\ 0 \end{pmatrix}, \begin{pmatrix} 0 \\ 1 \\ -1 \end{pmatrix}, \begin{pmatrix} 1 \\ 0 \\ 1 \end{pmatrix}\right)$ is a basis for $\mathbb{R}^3$, and

$$[L_A]_\beta = \begin{pmatrix} 1 & & \\ & 1 & \\ & & 2 \end{pmatrix}.$$

However, our original problem is about T. Let $\beta = (1, x - x^2, 1 + x^2)$. Then

$$[T]_\beta = \begin{pmatrix} 1 & & \\ & 1 & \\ & & 2 \end{pmatrix}.$$

(d) The matrix P is the base change matrix from β to α. Thus

$$P = \begin{pmatrix} 1 & 0 & 1 \\ 0 & 1 & 0 \\ 0 & -1 & 1 \end{pmatrix}.$$

We then have $[T]_\beta = P^{-1}[T]_\alpha P$. ◇

Example 5.3.20. Let $T: \mathbb{R}^3 \to \mathbb{R}^3$ be the linear endomorphism over $\mathbb{R}$ given by

$$T(x, y, z) = (-x, -z, y)$$

and let $S: \mathbb{C}^3 \to \mathbb{C}^3$ be the linear endomorphism over $\mathbb{C}$ given by the same formula above. Determine whether T or S can be diagonalized, and if so find their diagonal forms.

Solution. Let α be the standard $\mathbb{R}$-basis of $\mathbb{R}^3$. Then

$$[T]_\alpha = \begin{pmatrix} -1 & 0 & 0 \\ 0 & 0 & -1 \\ 0 & 1 & 0 \end{pmatrix}.$$

The characteristic polynomial of T is

$$f(t) = -t^3 - t^2 - t - 1 = -(t+1)(t^2+1).$$

This polynomial does not split over $\mathbb{R}$. Hence T is not diagonalizable over $\mathbb{R}$. On the other hand, let α' be the $\mathbb{C}$-basis of $\mathbb{C}^3$, then

$$[S]_{\alpha'} = \begin{pmatrix} -1 & 0 & 0 \\ 0 & 0 & -1 \\ 0 & 1 & 0 \end{pmatrix}.$$

as well. Hence the characteristic polynomial of S is still $f(t)$. Since

$$f(t) = -(t+1)(t+i)(t-i),$$

the characteristic polynomial has three distinct zeros. Thus S is diagonalizable over $\mathbb{C}$, and its diagonal form is

$$\begin{pmatrix} -1 & & \\ & i & \\ & & -i \end{pmatrix}$$

by Proposition 5.3.8.   ◇

Exercises 5.3

1. Determine whether each of the following linear endomorphism is diagonalizable over $\mathbb{R}$.

   (a) $T \colon \mathbb{R}^3 \to \mathbb{R}^3$, $T(x,y,z) = (x + 2y, 2x + y, z)$.

   (b) $T \colon \mathbb{R}^3 \to \mathbb{R}^3$, $T(x,y,z) = (3x - z, y - x + 2z, 4z)$.

   (c) $T \colon \mathbb{R}^4 \to \mathbb{R}^4$, $T(x,y,z,w) = (x + y, x, z + 2w, 2z + w)$.

(d) $T: \mathbb{R}^4 \to \mathbb{R}^4$, $T(x, y, z, w) = (x + y, y, 2z, y + 2z + 2w)$.

(e) $T: \mathbb{R}^2 \to \mathbb{R}^2$, $T(x, y) = (y, -x)$.

2. Suppose A is diagonalizable with the diagonal form

$$\begin{pmatrix} 1 & & & \\ & 2 & & \\ & & -3 & \\ & & & -4 \end{pmatrix}.$$

Is A^t diagonalizable? If so, what is it diagonal form?

3. Let

$$A_1 = \begin{pmatrix} 2 & & & \\ & 2 & & \\ & & 2 & \\ & & & 2 \end{pmatrix}; \quad A_2 = \begin{pmatrix} 2 & & & \\ 1 & 2 & & \\ & & 2 & \\ & & & 2 \end{pmatrix};$$

$$A_3 = \begin{pmatrix} 2 & & & \\ 1 & 2 & & \\ & & 2 & \\ & & 1 & 2 \end{pmatrix}; \quad A_4 = \begin{pmatrix} 2 & & & \\ 1 & 2 & & \\ & & 1 & 2 \\ & & & 2 \end{pmatrix};$$

$$A_5 = \begin{pmatrix} 2 & & & \\ 1 & 2 & & \\ & 1 & 2 & \\ & & 1 & 2 \end{pmatrix}.$$

Note that the characteristic polynomials of the A_i's are all $(t - 2)^4$.

(a) Find the algebraic multiplicities and geometric multiplicities of 2 for each A_i. Determine which A_i's are diagonalizable.

(b) Verify that $(A_i - 2)^4 = 0$ for all i.

(c) For each i, determine the lowest k such that $(A_i - 2)^k = 0$.

4. Let A be a square matrix. Show that A is similar to a upper triangular matrix if and only if the characteristic polynomial splits.

5.4   Powers of a square matrix

There is great advantage for a matrix to be diagonalizable, we will see this in this and the following sections.

How to compute a high power of a square matrix. It is hard work to compute a high power of a square matrix even when it of small size. However, when a square matrix is diagonalizable, it is relatively easy to achieve.

Lemma 5.4.1. *Let A and B be similar square matrices. If $B = P^{-1}AP$ for some invertible matrix P, then*

$$B^n = P^{-1}A^nP$$

for any positive integer n.

Proof. It is clear

$$B^n = \underbrace{(P^{-1}AP)(P^{-1}AP)\cdots(P^{-1}AP)}_{n \text{ times}}$$

$$= P^{-1}A^nP$$

for all n.   □

Example 5.4.2. Let A be the matrix

$$[T]_\alpha = \begin{pmatrix} 1 & 1 & 1 \\ 0 & 1 & 0 \\ 0 & 1 & 2 \end{pmatrix}.$$

in Example 5.3.19. Compute A^{100}.

Solution. From Example 5.3.19, we have that

$$\begin{pmatrix} 1 & & \\ & 1 & \\ & & 2 \end{pmatrix} = P^{-1}AP \quad \text{where} \quad P = \begin{pmatrix} 1 & 0 & 1 \\ 0 & 1 & 0 \\ 0 & -1 & 1 \end{pmatrix},$$

or

$$A = P \begin{pmatrix} 1 & & \\ & 1 & \\ & & 2 \end{pmatrix} P^{-1}.$$

Thus

$$A^{100} = P \begin{pmatrix} 1 & & \\ & 1 & \\ & & 2 \end{pmatrix}^{100} P^{-1}$$

$$= \begin{pmatrix} 1 & 0 & 1 \\ 0 & 1 & 0 \\ 0 & -1 & 1 \end{pmatrix} \begin{pmatrix} 1 & & \\ & 1 & \\ & & 2^{100} \end{pmatrix} \begin{pmatrix} 1 & -1 & -1 \\ 0 & 1 & 0 \\ 0 & 1 & 1 \end{pmatrix}$$

$$= \begin{pmatrix} 1 & 0 & 2^{100} \\ 0 & 1 & 0 \\ 0 & -1 & 2^{100} \end{pmatrix} \begin{pmatrix} 1 & -1 & -1 \\ 0 & 1 & 0 \\ 0 & 1 & 1 \end{pmatrix}$$

$$= \begin{pmatrix} 1 & 2^{100} - 1 & 2^{100} - 1 \\ 0 & 1 & 0 \\ 0 & 2^{100} - 1 & 2^{100} \end{pmatrix}$$

by Lemma 5.4.1. ◇

Example 5.4.3. Let

$$A = \begin{pmatrix} 1/2 & 0 & 0 & 1/2 \\ 0 & 2 & \sqrt{2} & 0 \\ 0 & \sqrt{2} & 3 & 0 \\ 1/2 & 0 & 0 & 1/2 \end{pmatrix}.$$

Find Compute A^{100}.

Solution. The characteristic polynomial of A is

$$f(t) = \det \begin{pmatrix} 1/2 - t & 0 & 0 & 1/2 \\ 0 & 2 - t & \sqrt{2} & 0 \\ 0 & \sqrt{2} & 3 - t & 0 \\ 1/2 & 0 & 0 & 1/2 - t \end{pmatrix}$$

$$= \left(\frac{1}{2} - t \right) \det \begin{pmatrix} 2 - t & \sqrt{2} & 0 \\ \sqrt{2} & 3 - t & 0 \\ 0 & 0 & 1/2 - t \end{pmatrix} - \frac{1}{2} \det \begin{pmatrix} 0 & 0 & 1/2 \\ 2 - t & \sqrt{2} & 0 \\ \sqrt{2} & 3 - t & 0 \end{pmatrix}$$

$$= \left(\frac{1}{2} - t \right)^2 (t^2 - 5t + 4) - \frac{1}{4}(t^2 - 5t + 4)$$

$$= (t-1)(t-4)t(t-1) = t(t-4)(t-1)^2.$$

To verify whether A is diagonalizable, we need to find the geometric multiplicity of the eigenvalue 1. We have

$$\text{nullity}(A-I) = \text{nullity} \begin{pmatrix} -1/2 & 0 & 0 & 1/2 \\ 0 & 1 & \sqrt{2} & 0 \\ 0 & \sqrt{2} & 2 & 0 \\ 1/2 & 0 & 0 & -1/2 \end{pmatrix}$$

$$= 4 - \text{rk} \begin{pmatrix} -1/2 & 0 & 0 & 1/2 \\ 0 & 1 & \sqrt{2} & 0 \\ 0 & \sqrt{2} & 2 & 0 \\ 1/2 & 0 & 0 & -1/2 \end{pmatrix}$$

$$= 4 - 2 = 2.$$

Hence A is diagonalizable with the diagonal form

$$D = \begin{pmatrix} 0 & & & \\ & 1 & & \\ & & 1 & \\ & & & 4 \end{pmatrix}.$$

We next find bases for the following null spaces:

$$\text{Null}(A) = \text{Null} \begin{pmatrix} 1/2 & 0 & 0 & 1/2 \\ 0 & 2 & \sqrt{2} & 0 \\ 0 & \sqrt{2} & 3 & 0 \\ 1/2 & 0 & 0 & 1/2 \end{pmatrix} = \text{Sp} \left(\begin{pmatrix} 1 \\ 0 \\ 0 \\ -1 \end{pmatrix} \right) ;$$

$$\text{Null}(A-I) = \text{Null} \begin{pmatrix} -1/2 & 0 & 0 & 1/2 \\ 0 & 1 & \sqrt{2} & 0 \\ 0 & \sqrt{2} & 2 & 0 \\ 1/2 & 0 & 0 & -1/2 \end{pmatrix}$$

$$= \text{Sp} \left(\begin{pmatrix} 1 \\ 0 \\ 0 \\ 1 \end{pmatrix}, \begin{pmatrix} 0 \\ -\sqrt{2} \\ 1 \\ 0 \end{pmatrix} \right) ;$$

$$\text{Null}\,(A - 4I) = \text{Null}\left(\begin{pmatrix} -7/2 & 0 & 0 & 1/2 \\ 0 & -2 & \sqrt{2} & 0 \\ 0 & \sqrt{2} & -1 & 0 \\ 1/2 & 0 & 0 & -7/2 \end{pmatrix}\right) = \text{Sp}\left(\begin{pmatrix} 0 \\ 1 \\ \sqrt{2} \\ 0 \end{pmatrix}\right).$$

If we choose

$$\beta = \left(\begin{pmatrix} 1 \\ 0 \\ 0 \\ -1 \end{pmatrix}, \begin{pmatrix} 1 \\ 0 \\ 0 \\ 1 \end{pmatrix}, \begin{pmatrix} 0 \\ -\sqrt{2} \\ 1 \\ 0 \end{pmatrix}, \begin{pmatrix} 0 \\ 1 \\ \sqrt{2} \\ 0 \end{pmatrix}\right),$$

then $D = [L_A]_\beta$. Let

$$P = \begin{pmatrix} 1 & 1 & 0 & 0 \\ 0 & 0 & -\sqrt{2} & 1 \\ 0 & 0 & 1 & \sqrt{2} \\ -1 & 1 & 0 & 0 \end{pmatrix}$$

be the base change matrix from β to α. Then $A = PDP^{-1}$.

We perform elementary row operations on

$$\left(\begin{array}{cccc|cccc} 1 & 1 & 0 & 0 & 1 & 0 & 0 & 0 \\ 0 & 0 & -\sqrt{2} & 1 & 0 & 1 & 0 & 0 \\ 0 & 0 & 1 & \sqrt{2} & 0 & 0 & 1 & 0 \\ -1 & 1 & 0 & 0 & 0 & 0 & 0 & 1 \end{array}\right)$$

$$\rightsquigarrow \left(\begin{array}{cccc|cccc} 1 & 1 & 0 & 0 & 1 & 0 & 0 & 0 \\ 0 & 0 & -\sqrt{2} & 1 & 0 & 1 & 0 & 0 \\ 0 & 0 & 1 & \sqrt{2} & 0 & 0 & 1 & 0 \\ 0 & 2 & 0 & 0 & 1 & 0 & 0 & 1 \end{array}\right)$$

$$\rightsquigarrow \left(\begin{array}{cccc|cccc} 1 & 1 & 0 & 0 & 1 & 0 & 0 & 0 \\ 0 & 0 & -\sqrt{2} & 1 & 0 & 1 & 0 & 0 \\ 0 & 0 & 1 & \sqrt{2} & 0 & 0 & 1 & 0 \\ 0 & 1 & 0 & 0 & 1/2 & 0 & 0 & 1/2 \end{array}\right)$$

$$\rightsquigarrow \left(\begin{array}{cccc|cccc} 1 & 0 & 0 & 0 & 1/2 & 0 & 0 & -1/2 \\ 0 & 0 & 0 & 3 & 0 & 1 & \sqrt{2} & 0 \\ 0 & 0 & 1 & \sqrt{2} & 0 & 0 & 1 & 0 \\ 0 & 1 & 0 & 0 & 1/2 & 0 & 0 & 1/2 \end{array}\right)$$

$$\rightsquigarrow \left(\begin{array}{cccc|cccc} 1 & 0 & 0 & 0 & 1/2 & 0 & 0 & -1/2 \\ 0 & 0 & 0 & 1 & 0 & 1/3 & \sqrt{2}/3 & 0 \\ 0 & 0 & 1 & \sqrt{2} & 0 & 0 & 1 & 0 \\ 0 & 1 & 0 & 0 & 1/2 & 0 & 0 & 1/2 \end{array} \right)$$

$$\rightsquigarrow \left(\begin{array}{cccc|cccc} 1 & 0 & 0 & 0 & 1/2 & 0 & 0 & -1/2 \\ 0 & 0 & 0 & 1 & 0 & 1/3 & \sqrt{2}/3 & 0 \\ 0 & 0 & 1 & 0 & 0 & -\sqrt{2}/3 & 1/3 & 0 \\ 0 & 1 & 0 & 0 & 1/2 & 0 & 0 & 1/2 \end{array} \right)$$

$$\rightsquigarrow \left(\begin{array}{cccc|cccc} 1 & 0 & 0 & 0 & 1/2 & 0 & 0 & -1/2 \\ 0 & 1 & 0 & 0 & 1/2 & 0 & 0 & 1/2 \\ 0 & 0 & 1 & 0 & 0 & -\sqrt{2}/3 & 1/3 & 0 \\ 0 & 0 & 0 & 1 & 0 & 1/3 & \sqrt{2}/3 & 0 \end{array} \right).$$

Thus

$$P^{-1} = \frac{1}{6} \begin{pmatrix} 3 & 0 & 0 & -3 \\ 3 & 0 & 0 & 3 \\ 0 & -2\sqrt{2} & 2 & 0 \\ 0 & 2 & 2\sqrt{2} & 0 \end{pmatrix}.$$

We now have

$$A^{100} = (PDP^{-1})^{100} = PD^{100}P^{-1}$$

$$= P \begin{pmatrix} 0 & & & \\ & 1 & & \\ & & 1 & \\ & & & 4^{100} \end{pmatrix} P^{-1}$$

$$= \frac{1}{6} \begin{pmatrix} 3 & 0 & 0 & 3 \\ 0 & 2(4^{100}+2) & 2\sqrt{2}(4^{100}-1) & 0 \\ 0 & 2\sqrt{2}(4^{100}-1) & 4^{101}+2 & 0 \\ 3 & 0 & 0 & 3 \end{pmatrix}.$$

It is possible to calculate A^{100} for A is diagonalizable.   ◇

If A cannot be diagonalized, we can still compute the powers of A if it is relatively simple as we can see in the following example.

Example 5.4.4. Let

$$A = \begin{pmatrix} 2 & 0 & 0 \\ 0 & 2 & 1 \\ 0 & 0 & 2 \end{pmatrix}.$$

We can write $A = 2I_3 + e_{23}$. Observe that

$$I_3 e_{23} = e_{23} = e_{23} I_3 \quad \text{and} \quad (e_{23})^2 = 0.$$

It follows that

$$
\begin{aligned}
A^n &= (2I_3 + e_{23})^n \\
&= (2I_3)^n + \binom{n}{1}(2I_3)^{n-1}e_{23} + \binom{n}{2}(2I_3)^{n-2}e_{23}^2 + \cdots \\
&\quad + \binom{n}{n-1}(2I_3)e_{23}^{n-1} + \binom{n}{n}e_{23}^n \\
&= 2^n I_3 + n2^{n-1}I_3 e_{23} \\
&= 2^n I_3 + n2^{n-1}e_{23} \\
&= \begin{pmatrix} 2^n & & \\ & 2^n & n2^{n-1} \\ & & 2^n \end{pmatrix}.
\end{aligned}
$$

Linear recurrence relations. For a square matrix A, its power A^n can be used to evaluate the general terms of a sequence $\{a_n\}_{n=0}^{\infty}$ defined recursively by

$$a_0 = \alpha,$$
$$a_1 = \beta,$$
$$a_{n+2} = pa_{n+1} + qa_n.$$

It's because the recursive relations may be rewritten in the matrix form

$$(5.1) \qquad \begin{pmatrix} a_{n+1} \\ a_n \end{pmatrix} = \begin{pmatrix} p & q \\ 1 & 0 \end{pmatrix}\begin{pmatrix} a_n \\ a_{n-1} \end{pmatrix} = \begin{pmatrix} p & q \\ 1 & 0 \end{pmatrix}^n \begin{pmatrix} \alpha \\ \beta \end{pmatrix}.$$

Hence if we know how to explicitly express A^n where

$$A = \begin{pmatrix} p & q \\ 1 & 0 \end{pmatrix},$$

we may give the explicit expression for the n-th term a_n in the sequence. We will see this is possible when A is diagonalizable.

We use the well-known Fibonacci sequence to demonstrate this method. The Fibonacci sequence is defined recursively by

$$F_0 = 0, \qquad F_1 = 1, \qquad \text{and} \qquad F_{n+1} = F_n + F_{n-1}, \qquad \text{for } n \geq 1.$$

Consequently, for any positive integer n,

$$\begin{pmatrix} F_{n+1} \\ F_n \end{pmatrix} = \begin{pmatrix} F_n + F_{n-1} \\ F_n \end{pmatrix} = \begin{pmatrix} 1 & 1 \\ 1 & 0 \end{pmatrix} \begin{pmatrix} F_n \\ F_{n-1} \end{pmatrix}$$

and hence

$$(5.2) \qquad \begin{pmatrix} F_{n+1} \\ F_n \end{pmatrix} = \begin{pmatrix} 1 & 1 \\ 1 & 0 \end{pmatrix}^n \begin{pmatrix} F_1 \\ F_0 \end{pmatrix} = \begin{pmatrix} 1 & 1 \\ 1 & 0 \end{pmatrix}^n \begin{pmatrix} 1 \\ 0 \end{pmatrix}.$$

We need to find A^n where

$$A = \begin{pmatrix} 1 & 1 \\ 1 & 0 \end{pmatrix}.$$

To do this we first find the characteristic polynomial of A

$$\det \begin{pmatrix} 1 - t & 1 \\ 1 & -t \end{pmatrix} = t^2 - t - 1 = \left(t - \frac{1 + \sqrt{5}}{2} \right) \left(t - \frac{1 - \sqrt{5}}{2} \right).$$

Thus A is diagonalizable with

$$\begin{pmatrix} \dfrac{1 + \sqrt{5}}{2} & \\ & \dfrac{1 - \sqrt{5}}{2} \end{pmatrix}$$

as the diagonal form. The eigenspaces are

$$E_{(1+\sqrt{5})/2} = \text{Null} \left(\begin{pmatrix} \dfrac{1 - \sqrt{5}}{2} & 1 \\ 1 & \dfrac{-1 - \sqrt{5}}{2} \end{pmatrix} \right) = \text{Sp} \left(\begin{pmatrix} 1 \\ \dfrac{-1 + \sqrt{5}}{2} \end{pmatrix} \right),$$

$$E_{(1-\sqrt{5})/2} = \text{Null} \left(\begin{pmatrix} \dfrac{1 + \sqrt{5}}{2} & 1 \\ 1 & \dfrac{-1 + \sqrt{5}}{2} \end{pmatrix} \right) = \text{Sp} \left(\begin{pmatrix} 1 \\ \dfrac{-1 - \sqrt{5}}{2} \end{pmatrix} \right).$$

Let β be the ordered basis

$$\left(\begin{pmatrix} 1 \\ \dfrac{-1+\sqrt{5}}{2} \end{pmatrix}, \begin{pmatrix} 1 \\ \dfrac{-1-\sqrt{5}}{2} \end{pmatrix} \right).$$

Then

$$[L_A]_\beta = \begin{pmatrix} \dfrac{1+\sqrt{5}}{2} & \\ & \dfrac{1-\sqrt{5}}{2} \end{pmatrix}.$$

Let P be the base change matrix from β to α. We have

$$P = \begin{pmatrix} 1 & 1 \\ \dfrac{-1+\sqrt{5}}{2} & \dfrac{-1-\sqrt{5}}{2} \end{pmatrix}$$

and

$$A = P \begin{pmatrix} \dfrac{1+\sqrt{5}}{2} & \\ & \dfrac{1-\sqrt{5}}{2} \end{pmatrix} P^{-1}.$$

We now compute P^{-1}. We may use Exercise 3 in §4.3 to find

$$P^{-1} = \frac{1}{\sqrt{5}} \begin{pmatrix} \dfrac{1+\sqrt{5}}{2} & 1 \\ \dfrac{-1+\sqrt{5}}{2} & -1 \end{pmatrix}.$$

Let $\varphi = (1+\sqrt{5})/2$. With this notation we have that

$$A = P \begin{pmatrix} \varphi & \\ & -\varphi^{-1} \end{pmatrix} P^{-1}.$$

Furthermore,

$$P = \begin{pmatrix} 1 & 1 \\ \varphi^{-1} & -\varphi \end{pmatrix} \quad \text{and} \quad P^{-1} = \frac{1}{\sqrt{5}} \begin{pmatrix} \varphi & 1 \\ \varphi^{-1} & -1 \end{pmatrix}.$$

Hence, by Lemma 5.4.1, we have

$$A^n = \frac{1}{\sqrt{5}} \begin{pmatrix} 1 & 1 \\ \varphi^{-1} & -\varphi \end{pmatrix} \begin{pmatrix} \varphi & \\ & -\varphi^{-1} \end{pmatrix}^n \begin{pmatrix} \varphi & 1 \\ \varphi^{-1} & -1 \end{pmatrix}$$

$$= \frac{1}{\sqrt{5}} \begin{pmatrix} 1 & 1 \\ \varphi^{-1} & -\varphi \end{pmatrix} \begin{pmatrix} \varphi^n & \\ & (-1)^n \varphi^{-n} \end{pmatrix} \begin{pmatrix} \varphi & 1 \\ \varphi^{-1} & -1 \end{pmatrix}$$

$$= \frac{1}{\sqrt{5}} \begin{pmatrix} 1 & 1 \\ \varphi^{-1} & -\varphi \end{pmatrix} \begin{pmatrix} \varphi^{n+1} & \varphi^n \\ (-1)^n \varphi^{-n-1} & (-1)^{n+1} \varphi^{-n} \end{pmatrix}$$

$$= \frac{1}{\sqrt{5}} \begin{pmatrix} * & * \\ \varphi^n + (-1)^{n+1} \varphi^{-n} & \varphi^{n-1} + (-1)^n \varphi^{-n+1} \end{pmatrix}$$

$$(5.3) \qquad = \frac{1}{\sqrt{5}} \begin{pmatrix} * & * \\ \varphi^n - (-\varphi^{-1})^n & \varphi^{n-1} - (-\varphi^{-1})^{n-1} \end{pmatrix}$$

We leave the first row unchecked for it will not be needed. Since

$$\begin{pmatrix} F_{n+1} \\ F_n \end{pmatrix} = \frac{1}{\sqrt{5}} \begin{pmatrix} * & * \\ \varphi^n - (-\varphi^{-1})^n & \varphi^{n-1} - (-\varphi^{-1})^{n-1} \end{pmatrix} \begin{pmatrix} 1 \\ 0 \end{pmatrix},$$

we have

$$F_n = \varphi^n - (-\varphi^{-1})^n.$$

Or more explicitly, the Fibonacci numbers is expressed as

$$F_n = \frac{1}{\sqrt{5}} \left[\left(\frac{1+\sqrt{5}}{2} \right)^n - \left(\frac{1-\sqrt{5}}{2} \right)^n \right]$$

for $n \geq 1$. This is known as *Binet formula*, discovered in 1843 by French mathematician Jacques-Philippe-Marie Binet (1786–1855).

Example 5.4.5. Let

$$A_n = \begin{pmatrix} 1 & -1 & & & \\ 1 & 1 & -1 & & \\ & 1 & 1 & \ddots & \\ & & \ddots & \ddots & -1 \\ & & & 1 & 1 \end{pmatrix}_{n \times n}.$$

for $n \geq 3$. The is the matrix with 1 on the diagonal, 1 on the lower sub-diagonal, and -1 on the upper sub-diagonal. Find $\det A_n$.

Solution. Even though in the problem, the sequence of matrices are defined for $n \geq 3$, following the spirit of the definition, it is natural to define

$$A_1 = \begin{pmatrix} 1 \end{pmatrix} \quad \text{and} \quad A_2 = \begin{pmatrix} 1 & -1 \\ 1 & 1 \end{pmatrix}.$$

It is clear $\det A_1 = 1$ and $\det A_2 = 2$. Let $n \geq 3$. We may use cofactor expansion on the first column to find

$$\det A_n = 1 \cdot (A_n)_{11} - 1 \cdot (A_n)_{21}$$

where $(A_n)_{11}$ and $(A_n)_{21}$ are the $(1,1)$- and $(2,1)$-cofactors of A_n. Note that the $(1,1)$-minor of A_n is A_{n-1}, while the $(2,1)$-minor of A_n is

$$\begin{pmatrix} -1 & 0 & \cdots & 0 \\ 0 & & & \\ \vdots & & A_{n-2} & \\ 0 & & & \end{pmatrix}.$$

Thus $\det A_n = \det A_{n-1} - (-1)\det A_{n-2} = \det A_{n-1} + \det A_{n-2}$. Note that

$$\det A_1 = F_2, \quad \text{and} \quad \det A_2 = F_3$$

where $\{F_n\}_{n=0}^{\infty}$ is the Fibonacci sequence we discussed above. Hence we have

$$\det A_n = F_{n+1} \quad \text{for } n \geq 1.$$

W conclude that

$$F_n = \frac{1}{\sqrt{5}} \left[\left(\frac{1 + \sqrt{5}}{2} \right)^{n+1} - \left(\frac{1 - \sqrt{5}}{2} \right)^{n+1} \right]$$

for $n \geq 1$.   ◇

We give one more example.

Example 5.4.6. Find the sequence $\{a_n\}_{n=0}^{\infty}$ defined by

$$a_0 = 2, \quad a_1 = 3 \quad \text{and} \quad a_{n+2} = a_{n+1} + 6a_n \quad \text{for } n \geq 0.$$

Solution. By definition, we have

$$\begin{pmatrix} a_{n+1} \\ a_n \end{pmatrix} = \begin{pmatrix} 1 & 6 \\ 1 & 0 \end{pmatrix} \begin{pmatrix} a_n \\ a_{n+1} \end{pmatrix}.$$

Hence

$$\begin{pmatrix} a_{n+1} \\ a_n \end{pmatrix} = A^n \begin{pmatrix} 3 \\ 2 \end{pmatrix}$$

where

$$A = \begin{pmatrix} 1 & 6 \\ 1 & 0 \end{pmatrix}.$$

The characteristic polynomial of A is

$$\det \begin{pmatrix} 1-t & 6 \\ 1 & -t \end{pmatrix} = t^2 - t - 6 = (t-3)(t+2).$$

The eigenvalues are 3 and -2, and the corresponding eigenvectors are

$$\begin{pmatrix} 3 \\ 1 \end{pmatrix} \quad \text{and} \quad \begin{pmatrix} -2 \\ 1 \end{pmatrix}.$$

Let

$$P = \begin{pmatrix} 3 & -2 \\ 1 & 1 \end{pmatrix}.$$

Then

$$A = P \begin{pmatrix} 3 & 0 \\ 0 & -2 \end{pmatrix} P^{-1}.$$

We have

$$
\begin{aligned}
A^n &= P \begin{pmatrix} 3 & 0 \\ 0 & -2 \end{pmatrix}^n P^{-1} \\
&= \frac{1}{5} \begin{pmatrix} 3 & -2 \\ 1 & 1 \end{pmatrix} \begin{pmatrix} 3^n & 0 \\ 0 & (-2)^n \end{pmatrix} \begin{pmatrix} 1 & 2 \\ -1 & 3 \end{pmatrix} \\
&= \frac{1}{5} \begin{pmatrix} 3^{n+1} - (-2)^{n+1} & 2 \cdot 3^{n+1} + 3(-2)^{n+1} \\ 3^n - (-2)^n & 2 \cdot 3^n + 3(-2)^n \end{pmatrix}.
\end{aligned}
$$

Thus $a_n = \frac{7}{5} \cdot 3^n + \frac{3}{5}(-2)^n$ for $n \geq 1$.   ◇

More on Fibonacci numbers. There is a second method to find the expression for the Fibonacci numbers. Remember that $(1, \varphi^{-1})$ is an eigenvector with eigenvalue $\varphi = (1 + \sqrt{5})/2$ for A, and $(1, -\varphi)$ is an eigenvector with eigenvalue $-\varphi^{-1} = (1 - \sqrt{5})/2$ for A. A routine calculation will show that

$$\begin{pmatrix} 1 \\ 0 \end{pmatrix} = \frac{\varphi}{\sqrt{5}} \begin{pmatrix} 1 \\ \varphi^{-1} \end{pmatrix} + \frac{\varphi^{-1}}{\sqrt{5}} \begin{pmatrix} 1 \\ -\varphi \end{pmatrix}.$$

Thus

$$\begin{pmatrix} F_{n+1} \\ F_n \end{pmatrix} = A^n \begin{pmatrix} 1 \\ 0 \end{pmatrix}$$

$$= A^n \left[\frac{\varphi}{\sqrt{5}} \begin{pmatrix} 1 \\ \varphi^{-1} \end{pmatrix} + \frac{\varphi^{-1}}{\sqrt{5}} \begin{pmatrix} 1 \\ -\varphi \end{pmatrix} \right]$$

$$= \frac{\varphi}{\sqrt{5}} A^n \begin{pmatrix} 1 \\ \varphi^{-1} \end{pmatrix} + \frac{\varphi^{-1}}{\sqrt{5}} A^n \begin{pmatrix} 1 \\ -\varphi \end{pmatrix}$$

$$= \frac{\varphi}{\sqrt{5}} \varphi^n \begin{pmatrix} 1 \\ \varphi^{-1} \end{pmatrix} + \frac{\varphi^{-1}}{\sqrt{5}} (-\varphi^{-1})^n \begin{pmatrix} 1 \\ -\varphi \end{pmatrix}$$

$$= \frac{1}{\sqrt{5}} \begin{pmatrix} \varphi^{n+1} - (-\varphi^{-1})^{n+1} \\ \varphi^n - (-\varphi^{-1})^n \end{pmatrix}.$$

We obtain the Binet formula again:

$$F_n = \frac{1}{\sqrt{5}} \left[\varphi^n - (-\varphi^{-1})^n \right].$$

There is yet a third method to find the Binet formula. We may use a generating function to find an expression for the general term of a sequence. A **generating function** for the sequence $\{a_n\}_{n=0}^{\infty}$ is the formal power series

$$a_0 + a_1 t + a_2 t^2 + a_3 t^3 + \cdots + a_n t^n + \cdots = \sum_{n=0}^{\infty} a_n t^n.$$

We proceed to establish the Binet formula. Let

$$Q(t) = F_0 + F_1 t + F_2 t^2 + \cdots + F_n t^n + \cdots$$

be the generating function for the Fibonacci sequence. Then

$$Q(t) = t + t^2 + \sum_{n=0}^{\infty} F_{n+2} t^{n+2}$$

$$= t + t^2 + \sum_{n=0}^{\infty} (F_{n+1} + F_n) t^{n+2}$$

$$= t + t^2 + t \sum_{n=0}^{\infty} F_{n+1} t^{n+1} + t^2 \sum_{n=0}^{\infty} F_n t^n$$

$$= t + t^2 + t(Q(t) - t) + t^2 Q(t)$$
$$= t + tQ(t) + t^2 Q(t).$$

Thus

$$Q(t) = \frac{t}{1 - t - t^2} = t\left(\frac{1}{1 - t - t^2}\right).$$

We now use partial fraction decomposition. It is routine to find

$$\frac{1}{1 - t - t^2} = \frac{1}{(1 - \varphi t)(1 + \varphi^{-1} t)}$$
$$= \frac{1}{\sqrt{5}}\left(\frac{\varphi}{1 - \varphi t} + \frac{\varphi^{-1}}{1 + \varphi^{-1} t}\right)$$

where $\varphi = (1 + \sqrt{5})/2$. Note that

$$\frac{1}{1 - \varphi t} = 1 + \varphi t + \varphi^2 t^2 + \cdots + \varphi^n t^n + \cdots,$$
$$\frac{1}{1 + \varphi^{-1} t} = 1 - \varphi^{-1} t + \varphi^{-2} t^2 + \cdots + (-\varphi^{-1})^n t^n + \cdots.$$

Hence

$$Q(t) = \frac{t}{\sqrt{5}}\left(\sum_{n=0}^{\infty} \varphi^{n+1} t^n + \sum_{n=0}^{\infty} -(-\varphi^{-1})^{n+1} t^n\right)$$
$$= \frac{1}{\sqrt{5}} \sum_{n=0}^{\infty} \left(\varphi^{n+1} - (-\varphi^{-1})^{n+1}\right) t^{n+1}$$
$$= \frac{1}{\sqrt{5}} \sum_{n=1}^{\infty} \left(\varphi^n - (-\varphi^{-1})^n\right) t^n.$$

We conclude that

$$F_n = \frac{1}{\sqrt{5}}\left[\varphi^n - (-\varphi^{-1})^n\right] \quad \text{for} \quad n \geq 1.$$

We have obtained Binet formula again.

Just for fun, let's mention an important property of Fibonacci sequence. Let's consider another Fibonacci sequence $\{f_n\}_{n=0}^{\infty}$ with $f_0 = 1$ and $f_1 = 0$. Observe that the first few terms of $\{f_n\}_{n=0}^{\infty}$ are

$$1, 0, 1, 1, 2, 3, 5, \ldots.$$

It is easy to see that for $n \geq 1$,

$$f_n = F_{n-1}$$

where $\{F_n\}_{n=0}^{\infty}$ is the Fibonacci sequence with $F_0 = 0$ and $F_1 = 1$. From (5.1), we have

$$\begin{pmatrix} F_{n+1} \\ F_n \end{pmatrix} = \begin{pmatrix} 1 & 1 \\ 1 & 0 \end{pmatrix}^n \begin{pmatrix} 1 \\ 0 \end{pmatrix},$$

$$\begin{pmatrix} F_n \\ F_{n-1} \end{pmatrix} = \begin{pmatrix} f_{n+1} \\ f_n \end{pmatrix} = \begin{pmatrix} 1 & 1 \\ 1 & 0 \end{pmatrix}^n \begin{pmatrix} 0 \\ 1 \end{pmatrix}.$$

Thus, we have the matrix identity

$$\begin{pmatrix} F_{n+1} & F_n \\ F_n & F_{n-1} \end{pmatrix} = \begin{pmatrix} 1 & 1 \\ 1 & 0 \end{pmatrix}^n \begin{pmatrix} 1 & 0 \\ 0 & 1 \end{pmatrix} = \begin{pmatrix} 1 & 1 \\ 1 & 0 \end{pmatrix}^n.$$

Taking determinant on both sides, we have

$$F_{n+1}F_{n-1} - F_n^2 = (-1)^n.$$

This is called the *Cassini's identity* .

Markov chains. Before we close this section, we give an application in Statistics using powers of a matrix. We give a brief discussion on Markov chains.

A **stochastic matrix** is a square matrix whose entries are probabilities and the entries in each column add up to 1. For example,

$$\begin{pmatrix} 0.8 & 0.3 \\ 0.2 & 0.7 \end{pmatrix}, \quad \begin{pmatrix} 0.5 & 0.3 & 0.1 \\ 0.25 & 0.4 & 0.3 \\ 0.25 & 0.3 & 0.6 \end{pmatrix}, \quad \begin{pmatrix} 1 & 0 & 1/3 \\ 0 & 1/2 & 1/3 \\ 0 & 1/2 & 1/3 \end{pmatrix}.$$

are stochastic matrices.

Let P be an $n \times n$ stochastic matrix. Suppose given an $n \times 1$ column vector X_0 as an initial vector. We define a **Markov chain** as a sequence of vectors $\{X_n\}$ given by

$$X_1 = PX_0, \; X_2 = PX_1 = P^2X_0, \dots, \; X_{n+1} = PX_n = P^{n+1}X_0.$$

A Markov chain depends only on the current state rather than a more historical description.

We now build a hypothetical model to describe population movement in Metropolitan Taipei. Metropolitan Taipei includes Taipei City and New Taipei City, two contiguous cities in the north of Taiwan. Suppose the total population within the Metropolitan area stay stable within the foreseeable years. Furthermore, suppose the residents in the Metropolitan move within the area, while the number of people moving out is negligible. In 2012, the population in Taipei and in New Taipei City are 2 millions and 3 millions respectively. This information is represented by an initial vector

$$X_0 = \begin{pmatrix} 2 \\ 3 \end{pmatrix}$$

where the unit is millions of people. Let

$$P = \begin{pmatrix} 0.9 & 0.15 \\ 0.1 & 0.85 \end{pmatrix}$$

be the matrix of transition for the period of one year. This matrix says two things:

- the probability of a resident of Taipei city who stayed was 0.9, and the probability of a resident of Taipei City moving to New Taipei City was 0.1;

- the probability of a resident of New Taipei City moving to Taipei City was 0.15, and the probability of a resident who stayed was 0.85.

With 2012 as the initial year, we can predict the population distribution X_1 in 2013 as

$$X_1 = PX_0 = \begin{pmatrix} 0.9 & 0.15 \\ 0.1 & 0.85 \end{pmatrix} \begin{pmatrix} 2 \\ 3 \end{pmatrix} = \begin{pmatrix} 2.25 \\ 2.75 \end{pmatrix}.$$

In 2014, two years later, the population distribution X_2 is given by

$$X_2 = PX_1 = \begin{pmatrix} 0.9 & 0.15 \\ 0.1 & 0.85 \end{pmatrix} \begin{pmatrix} 2.25 \\ 2.75 \end{pmatrix} = \begin{pmatrix} 2.4375 \\ 2.5625 \end{pmatrix}.$$

Now we are going to ask two questions concerning the long-term population distribution.

- Does $\lim_{n \to \infty} X_n$ exist?

- Does $\lim_{n \to \infty} P^n$ exist?

Assume the answer to the first question is affirmative. Let X be the limit. Then from the general term

$$X_{n+1} = PX_n$$

we have

$$X = PX.$$

Therefore X must be an eigenvector of P corresponding to the eigenvalue 1. Let $X = (a, b)^t$. Then

$$\begin{pmatrix} -0.1 & 0.15 \\ 0.1 & -0.15 \end{pmatrix} \begin{pmatrix} a \\ b \end{pmatrix} = \begin{pmatrix} 0 \\ 0 \end{pmatrix} \quad \text{and} \quad a + b = 5.$$

The solution is $a = 3$ and $b = 2$. Eventually, the population in Taipei City is 3 million and the population in New Taipei City is 2 million.

To answer the second question, we diagonalize P first. The characteristic polynomial of P is

$$\det(P - tI) = t^2 - 1.75t + 0.75 = (t - 1)(t - 0.75).$$

It is routine to find eigenvectors corresponding to the two distinct eigenvalues. The vector $(0.6,\ 0.4)$ is an eigenvector corresponding to 1 and the vector $(-1,\ 1)$ is an eigenvector corresponding to 0.75. Let

$$\beta = \left(\begin{pmatrix} 0.6 \\ 0.4 \end{pmatrix}, \begin{pmatrix} -1 \\ 1 \end{pmatrix} \right)$$

Then

$$Q = \begin{pmatrix} 0.6 & -1 \\ 0.4 & 1 \end{pmatrix}$$

is the base change matrix from β to the standard basis of $\mathbb{R}^2$ and

$$P = Q \begin{pmatrix} 1 & 0 \\ 0 & 0.75 \end{pmatrix} Q^{-1}.$$

It follows that

$$\lim_{n \to \infty} P^n = \lim_{n \to \infty} Q \begin{pmatrix} 1 & 0 \\ 0 & 0.75 \end{pmatrix}^n Q^{-1}$$

$$= \lim_{n \to \infty} Q \begin{pmatrix} 1 & 0 \\ 0 & (0.75)^n \end{pmatrix} Q^{-1}$$

$$= Q \begin{pmatrix} 1 & 0 \\ 0 & 0 \end{pmatrix} Q^{-1}$$

$$= \begin{pmatrix} 0.6 & 0.6 \\ 0.4 & 0.4 \end{pmatrix}.$$

We say the transition matrix P of a Markov chain is **regular** if for some power of P all the entries are positive numbers. A regular transition matrix defines a **regular Markov chain**.

The model we just discussed is a regular Markov chain. Regular Markov chains give information above the long-term behavior as stated in the following theorem.

Theorem 5.4.7. *Let P and X_0 be the transition matrix and the initial vector of a regular Markov chain. Then*

(1) *the limit*

$$\lim_{n \to \infty} X_n = \lim_{n \to \infty} P^n X_0 = X$$

exists and it satisfies

$$PX = X.$$

(2) *the limit*

$$\lim_{n \to \infty} P^n = Q$$

is a stochastic matrix. The columns of Q are the same, each being an eigenvector of P corresponding to the eigenvalue 1.

Example 5.4.8. Let A be the stochastic matrix

$$\begin{pmatrix} 0.75 & 0.25 \\ 0.25 & 0.75 \end{pmatrix}.$$

Find $\lim_{n \to \infty} A^n$.

Solution. The characteristic polynomial of A is

$$\det(A - tI) = \det \begin{pmatrix} 0.75 - t & 0.25 \\ 0.25 & 0.75 - t \end{pmatrix} = t^2 - 1.5t + 0.5$$

and the eigenvalues of A are 1 and 0.5.

An eigenvector corresponding to the eigenvalue 1 is

$$\begin{pmatrix} 0.5 \\ 0.5 \end{pmatrix}.$$

Therefore

$$\lim_{n \to \infty} A^n = \begin{pmatrix} 0.5 & 0.5 \\ 0.5 & 0.5 \end{pmatrix}.$$

◇

Exercises 5.4

1. Let

$$A_1 = \begin{pmatrix} 2 & & & \\ & 2 & & \\ & & 2 & \\ & & & 2 \end{pmatrix}; \qquad A_2 = \begin{pmatrix} 2 & & & \\ 1 & 2 & & \\ & & 2 & \\ & & & 2 \end{pmatrix};$$

$$A_3 = \begin{pmatrix} 2 & & & \\ 1 & 2 & & \\ & & 2 & \\ & & 1 & 2 \end{pmatrix}; \qquad A_4 = \begin{pmatrix} 2 & & & \\ 1 & 2 & & \\ & 1 & 2 & \\ & & & 2 \end{pmatrix};$$

$$A_5 = \begin{pmatrix} 2 & & & \\ 1 & 2 & & \\ & 1 & 2 & \\ & & 1 & 2 \end{pmatrix}.$$

Note that the characteristic polynomials of the A_i's are all $(t - 2)^4$.

(a) Find the algebraic multiplicities and geometric multiplicities of 2 for each A_i. Determine which A_i's are diagonalizable.

(b) Verify that $(A_i - 2)^4 = 0$ for all i.

(c) For each i, determine the lowest k such that $(A_i - 2)^k = 0$.

2. Suppose A is a diagonalizable matrix. Show that A^n is diagonalizable for all $n \in \mathbb{Z}_+$. Conversely, if A^n is diagonalizable for some integer $n \geq 2$, is A also diagonalizable?

3. Find $\begin{pmatrix} 0 & 1 & 1 \\ 1 & 0 & 1 \\ 1 & 1 & 0 \end{pmatrix}^{100}$.

4. Let $\{F_n\}_{n=0}^{\infty}$ be the Fibonacci sequence with $F_0 = 0$ and $F_1 = 1$.

(a) Show that
$$\lim_{n \to \infty} \frac{F_{n+1}}{F_n} = \varphi = \frac{1 + \sqrt{5}}{2}.$$

This is called the **golden ratio**, which appears in some patterns in nature and is used by artists and architects.

(b) Let $\psi = -\varphi^{-1}$ be another zero of the polynomial $t^2 - t - 1$. Show that

$$\varphi^n = \varphi^{n-1} + \varphi^{n-2} \qquad \text{and} \qquad \psi^n = \psi^{n-1} + \psi^{n-2}$$

for $n \geq 2$.

5. Find a close form expression for the Fibonacci numbers with arbitrarily given F_0 and F_1.

(a) Find an eigenvector for each eigenvalue of A.

(b) Express $(1, 1)$ as a linear combination of the eigenvectors of found in (a). Use this to find the closed form of Fibonacci numbers.

6. Define the sequence $\{a_n\}_{n=1}^{\infty}$ inductively by letting

$$a_1 = 1, \qquad a_2 = 1, \qquad \text{and} \qquad a_n = a_{n-1} + 2a_{n-2} \quad \text{for } n \geq 3.$$

Find a closed form formula for this sequence.

7. Find an expression for a_n of the sequence $\{a_n\}_{n=0}^{\infty}$ defined recursively by

$$a_0 = 5, \quad a_1 = 13 \quad \text{and} \quad a_{n+1} = 4a_n + 5a_{n-1}.$$

8. Find an expression for a_n of the sequence $\{a_n\}_{n=0}^{\infty}$ defined recursively by

$$a_0 = 1, \quad a_1 = 4 \quad \text{and} \quad a_{n+1} = 4a_n - 4a_{n-1}.$$

9. Find the generating function for the sequence $\{a_n\}_{n=0}^{\infty}$ defined recursively by

$$a_0 = 1, \quad a_1 = 2 \quad \text{and} \quad a_{n+1} = 2a_n - 3a_{n-1}.$$

10. The generating function $G(t)$ for the sequence $\{a_n\}_{n=0}^{\infty}$ is given by

$$G(t) = \frac{1}{(1-t)^3}.$$

Find a_n.

11. Let P be the transition matrix

$$\begin{pmatrix} 0.6 & 0.13 \\ 0.4 & 0.87 \end{pmatrix}.$$

Find an eigenvector of P corresponding to the eigenvalue $t = 1$ of P. Also find the limit $\lim_{n\to\infty} P^n$.

12. Let P be the transition matrix

$$\begin{pmatrix} 0.8 & 0.15 & 0.05 \\ 0.15 & 0.7 & 0.15 \\ 0.05 & 0.15 & 0.8 \end{pmatrix}.$$

Find the limit $\lim_{n\to\infty} P^n$.

5.5 Differential equations

In this section, we discuss how to use matrices to help us solve Differential Equations. Throughout this section, we assume all matrices are with real entries.

Solving a linear system of first-order differential equations. First, let's look at systems of linear differential equations.

Example 5.5.1. Let's solve the system of differential equations

$$
\begin{aligned}
x' &= 3x - y + z \\
y' &= 8x - 3y + 2z \\
z' &= 6x - 3y + 2z
\end{aligned}
$$

where $x = x(t)$, $y = y(t)$ and $z = z(t)$ are differentiable real-valued function of the real variable t.

The system may be rewritten in the matrix form

$$
\begin{pmatrix} x' \\ y' \\ z' \end{pmatrix} = A \begin{pmatrix} x \\ y \\ z \end{pmatrix} \quad \text{where} \quad A = \begin{pmatrix} 3 & -1 & 1 \\ 8 & -3 & 2 \\ 6 & -3 & 2 \end{pmatrix}.
$$

First we diagonalize A. The characteristic polynomial of A is

$$
\begin{aligned}
\det &\begin{pmatrix} 3-t & -1 & 1 \\ 8 & -3-t & 2 \\ 6 & -3 & 2-t \end{pmatrix} \\
&= -t^3 + 2t^2 + t - 2 \\
&= -(t-2)(t+1)(t-1).
\end{aligned}
$$

The eigenvalues of A are ± 1 and 2. The eigenspaces are

$$
E_1 = \text{Null} \left(\begin{pmatrix} 2 & -1 & 1 \\ 8 & -4 & 2 \\ 6 & -3 & 1 \end{pmatrix} \right) = \text{Sp} \left(\begin{pmatrix} 1 \\ 2 \\ 0 \end{pmatrix} \right),
$$

$$
E_{-1} = \text{Null} \left(\begin{pmatrix} 4 & -1 & 1 \\ 8 & -2 & 2 \\ 6 & -3 & 3 \end{pmatrix} \right) = \text{Sp} \left(\begin{pmatrix} 0 \\ -1 \\ -1 \end{pmatrix} \right),
$$

$$
E_2 = \text{Null} \left(\begin{pmatrix} 1 & -1 & 1 \\ 8 & -5 & 2 \\ 6 & -3 & 0 \end{pmatrix} \right) = \text{Sp} \left(\begin{pmatrix} 1 \\ 2 \\ 1 \end{pmatrix} \right).
$$

Thus

$$\beta = \left(\begin{pmatrix} 1 \\ 2 \\ 0 \end{pmatrix}, \begin{pmatrix} 0 \\ -1 \\ -1 \end{pmatrix}, \begin{pmatrix} 1 \\ 2 \\ 1 \end{pmatrix} \right)$$

is an ordered basis such that

$$[L_A]_\beta = \begin{pmatrix} 1 & & \\ & -1 & \\ & & 2 \end{pmatrix}.$$

Let

$$P = \begin{pmatrix} 1 & 0 & 1 \\ 2 & -1 & 2 \\ 0 & -1 & 1 \end{pmatrix}$$

be the basis change matrix from β to the standard basis α of $\mathbb{R}^3$. Then

$$P^{-1}AP = \begin{pmatrix} 1 & & \\ & -1 & \\ & & 2 \end{pmatrix}.$$

We now make a change of variables by letting

$$\begin{pmatrix} u \\ v \\ w \end{pmatrix} = P^{-1} \begin{pmatrix} x \\ y \\ z \end{pmatrix}.$$

Note that

$$\begin{pmatrix} u' \\ v' \\ w' \end{pmatrix} = P^{-1} \begin{pmatrix} x' \\ y' \\ z' \end{pmatrix}$$

since $(ax + by + cz)' = ax' + by' + cz'$ for any scalars $a, b, c \in \mathbb{R}$. Thus

$$\begin{pmatrix} u' \\ v' \\ w' \end{pmatrix} = P^{-1} \begin{pmatrix} x' \\ y' \\ z' \end{pmatrix} = P^{-1}A \begin{pmatrix} x \\ y \\ z \end{pmatrix}$$

$$= P^{-1}APP^{-1} \begin{pmatrix} x \\ y \\ z \end{pmatrix}$$

$$= \begin{pmatrix} 1 & & \\ & -1 & \\ & & 2 \end{pmatrix} \begin{pmatrix} u \\ v \\ w \end{pmatrix}.$$

Thus the real-valued functions u, v and w satisfy the system of differential equations

$$\begin{aligned} u' &= u, \\ v' &= -v, \\ w' &= 2w. \end{aligned}$$

The solution for this system is easy:

$$\begin{aligned} u &= A_1 e^t, \\ v &= A_2 e^{-t}, \\ w &= A_3 e^{2t} \end{aligned}$$

for arbitrary constants A_1, A_2 and A_3. The solution of the original system is now clearly

$$\begin{pmatrix} x \\ y \\ z \end{pmatrix} = P \begin{pmatrix} u \\ v \\ w \end{pmatrix} = \begin{pmatrix} 1 & 0 & 1 \\ 2 & -1 & 2 \\ 0 & -1 & 1 \end{pmatrix} \begin{pmatrix} A_1 e^t \\ A_2 e^{-t} \\ A_3 e^{2t} \end{pmatrix}.$$

We now conclude that

$$\begin{aligned} x(t) &= A_1 e^t & & + A_3 e^{2t}, \\ y(t) &= 2A_1 e^t - A_2 e^{-t} & & + 2A_3 e^{2t}, \\ z(t) &= & -A_2 e^{-t} & + A_3 e^{2t}. \end{aligned}$$

If initial conditions are given for x, y and z, we can also find A_1, A_2 and A_3 for sure.

Exponential of a square matrix.   We may also use the exponential of a matrix to help us solve differential equations. We first discuss how to compute the exponential of a matrix.

For any square matrix over $\mathbb{R}$, we may define the matrix e^A to be

$$e^A = I + A + \frac{A^2}{2!} + \cdots + \frac{A^n}{n!} + \cdots = \sum_{n=0}^{\infty} \frac{A^n}{n!}.$$

The entries of e^A are power series over $\mathbb{R}$. If all the entries converge, we have a well-defined matrix.

Example 5.5.2. Let

$$A = \begin{pmatrix} 0 & 1 & 0 \\ 0 & 0 & 1 \\ 0 & 0 & 0 \end{pmatrix}.$$

It is clear that

$$A^2 = \begin{pmatrix} 0 & 0 & 1 \\ 0 & 0 & 0 \\ 0 & 0 & 0 \end{pmatrix} \quad \text{and} \quad A^n = \mathbf{0} \qquad \text{for } n \geq 3.$$

It follows that

$$e^A = I + A + \frac{A^2}{2!} = \begin{pmatrix} 1 & 1 & 1/2 \\ 0 & 1 & 1 \\ 0 & 0 & 1 \end{pmatrix}.$$

Suppose A and B are similar. There is an invertible matrix P such that $A = PBP^{-1}$. Then

$$\begin{aligned} e^A &= \sum_{n=0}^{\infty} \frac{A^n}{n!} = \sum_{n=0}^{\infty} \frac{1}{n!}(PBP^{-1})^n \\ &= \sum_{n=0}^{\infty} \frac{1}{n!} PB^n P^{-1} = P\left(\sum_{n=0}^{\infty} \frac{B^n}{n!}\right) P^{-1} \\ &= Pe^B P^{-1}. \end{aligned}$$

Thus we have the following result.

Proposition 5.5.3. *Let A and B be square matrices and let P be an invertible matrix of the same size such that $A = PBP^{-1}$. Then*

$$e^A = Pe^B P^{-1}$$

when e^A and e^B are both convergent.

Hence e^A is easy to find when A can be diagonalized.

Proposition 5.5.4. *Let D be the square matrix*

$$\begin{pmatrix} \lambda_1 & & & \\ & \lambda_2 & & \\ & & \ddots & \\ & & & \lambda_n \end{pmatrix}.$$

Then

$$e^D = \begin{pmatrix} e^{\lambda_1} & & & \\ & e^{\lambda_2} & & \\ & & \ddots & \\ & & & e^{\lambda_n} \end{pmatrix}.$$

Proof. By definition we have

$$e^D = \sum_{n=0}^{\infty} \frac{1}{n!} \begin{pmatrix} \lambda_1 & & & \\ & \lambda_2 & & \\ & & \ddots & \\ & & & \lambda_n \end{pmatrix}^n$$

$$= \sum_{n=0}^{\infty} \frac{1}{n!} \begin{pmatrix} \lambda_1^n & & & \\ & \lambda_2^n & & \\ & & \ddots & \\ & & & \lambda_n^n \end{pmatrix}$$

$$= \begin{pmatrix} \sum_{n=0}^{\infty} \dfrac{\lambda_1^n}{n!} & & & \\ & \sum_{n=0}^{\infty} \dfrac{\lambda_2^n}{n!} & & \\ & & \ddots & \\ & & & \sum_{n=0}^{\infty} \dfrac{\lambda_n^n}{n!} \end{pmatrix}$$

$$= \begin{pmatrix} e^{\lambda_1} & & & \\ & e^{\lambda_2} & & \\ & & \ddots & \\ & & & e^{\lambda_n} \end{pmatrix}.$$

This completes the proof. □

Example 5.5.5. Find e^A if

$$A = \begin{pmatrix} 1 & -2 \\ -2 & 1 \end{pmatrix}.$$

Solution. The characteristic polynomial of A is

$$\det(A - tI) = \det \begin{pmatrix} 1 - t & -2 \\ -2 & 1 - t \end{pmatrix}$$
$$= (t - 1)^2 - 4$$
$$= (t + 1)(t - 3).$$

The eigenvalues of A are -1 and 3. By Proposition 5.3.8, A is diagonalizable. We have

$$E_{-1} = \mathrm{Null}\left(\begin{pmatrix} 2 & -2 \\ -2 & 2 \end{pmatrix} \right) = \mathrm{Sp}\left(\begin{pmatrix} 1 \\ 1 \end{pmatrix} \right),$$

$$E_3 = \mathrm{Null}\left(\begin{pmatrix} -2 & -2 \\ -2 & -2 \end{pmatrix} \right) = \mathrm{Sp}\left(\begin{pmatrix} -1 \\ 1 \end{pmatrix} \right).$$

Thus we may use the base change matrix

$$P = \frac{1}{\sqrt{2}} \begin{pmatrix} 1 & -1 \\ 1 & 1 \end{pmatrix}$$

to obtain $A = PDP^{-1}$, where

$$D = \begin{pmatrix} -1 & 0 \\ 0 & 3 \end{pmatrix}.$$

Hence

$$e^A = P^{-1} e^D P$$
$$= \frac{1}{\sqrt{2}} \begin{pmatrix} 1 & -1 \\ 1 & 1 \end{pmatrix} \begin{pmatrix} e^{-1} & 0 \\ 0 & e^3 \end{pmatrix} \frac{1}{\sqrt{2}} \begin{pmatrix} 1 & 1 \\ -1 & 1 \end{pmatrix}$$
$$= \frac{1}{2} \begin{pmatrix} e^{-1} + e^3 & e^{-1} - e^3 \\ e^{-1} - e^3 & e^{-1} + e^3 \end{pmatrix}$$

by Propositions 5.5.3 and 5.5.4.   ◇

Even when a square matrix is not diagonalizable, it is still sometimes possible to evaluate e^A.

Example 5.5.6. Evaluate e^{tA} if

$$A = \begin{pmatrix} 0 & 1 \\ -1 & 0 \end{pmatrix}.$$

Solution. A direct calculation shows that

$$A^{4k} = \begin{pmatrix} 1 & 0 \\ 0 & 1 \end{pmatrix}, \qquad k = 0, 1, 2, 3, \ldots;$$

$$A^{4k+1} = \begin{pmatrix} 0 & 1 \\ -1 & 0 \end{pmatrix}, \qquad k = 0, 1, 2, 3, \ldots;$$

$$A^{4k+2} = \begin{pmatrix} -1 & 0 \\ 0 & -1 \end{pmatrix}, \qquad k = 0, 1, 2, 3, \ldots;$$

$$A^{4k+3} = \begin{pmatrix} 0 & -1 \\ 1 & 0 \end{pmatrix}, \qquad k = 0, 1, 2, 3, \ldots.$$

Let

$$e^{tA} = \begin{pmatrix} a & b \\ c & d \end{pmatrix}.$$

It is easy to see that

$$a = 1 - \frac{t^2}{2!} + \frac{t^4}{4!} - \frac{t^6}{6!} + \cdots + \frac{(-1)^n t^{2n}}{(2n)!} + \cdots = \cos t;$$

$$b = t - \frac{t^3}{3!} + \frac{t^5}{5!} - \frac{t^7}{7!} + \cdots + \frac{(-1)^n t^{2n+1}}{(2n+1)!} + \cdots = \sin t;$$

$$c = -b = -\sin t;$$

$$d = a = \cos t.$$

We conclude that

$$e^{tA} = \begin{pmatrix} \cos t & \sin t \\ -\sin t & \cos t \end{pmatrix}.$$

The matrix A is not diagonalizable. However, since there is a pattern for the entries of A^n, it is still possible to find e^{tA} explicitly.   ◇

Remember we have the identity

$$e^x e^y = e^{x+y}$$

for all real numbers x and y. In proving this identity, it only requires that x and y commute with each other. Hence if we replace x and y by two matrices A and B which commute with each other, the same identity should still hold. Thus we have the following lemma.

Lemma 5.5.7. *Let A and B be two square matrices of the same size such that $AB = BA$. Then*

$$e^A e^B = e^{A+B}.$$

Example 5.5.8. Let

$$A = \begin{pmatrix} 2 & 1 \\ 0 & 2 \end{pmatrix}.$$

Then $A = 2I + N$ where

$$N = \begin{pmatrix} 0 & 1 \\ 0 & 0 \end{pmatrix}$$

is a nilpotent matrix. Note that A is not diagonalizable.

Observe that I commutes with N. (The matrix I commutes with everything.) Also observe that $N^2 = \mathbf{0}$. Consequently,

$$e^{tA} = e^{2tI+tN} = e^{2tI} e^{tN} = e^{2tI}(I + tN)$$

$$= \begin{pmatrix} e^{2t} & 0 \\ 0 & e^{2t} \end{pmatrix} \begin{pmatrix} 1 & t \\ 0 & 1 \end{pmatrix}$$

$$= \begin{pmatrix} e^{2t} & te^{2t} \\ 0 & e^{2t} \end{pmatrix}.$$

Solving linear higher-order differential equations. Why does the exponential of a matrix help us solve differential equations? Observe that

$$\frac{d}{dt} e^{tA} = \frac{d}{dt} \left(I + tA + \frac{t^2}{2!} A^2 + \cdots + \frac{t^n}{n!} A^n + \frac{t^{n+1}}{(n+1)!} A^{n+1} + \cdots \right)$$

$$= A + tA^2 + \frac{t^2}{2!} A^3 + \cdots + \frac{t^{n-1}}{(n-1)!} A^n + \frac{t^n}{n!} A^{n+1} + \cdots$$

$$= Ae^{tA}.$$

Most of the linear differential equations may be expressed as

$$\frac{d}{dt}u(t) = Au(t)$$

where $u(t)$ is a vector function. By evaluating e^{tA}, we obtain a solution to this differential equation.

Example 5.5.9. The second-order differential equation

$$y''(t) + y(t) = 0$$

may be rewritten as

$$\frac{d}{dt}\begin{pmatrix} y(t) \\ y'(t) \end{pmatrix} = \begin{pmatrix} y'(t) \\ y''(t) \end{pmatrix} = \begin{pmatrix} y'(t) \\ -y(t) \end{pmatrix}.$$

Hence if we let

$$u(t) = \begin{pmatrix} y(t) \\ y'(t) \end{pmatrix} \quad \text{and} \quad A = \begin{pmatrix} 0 & 1 \\ -1 & 0 \end{pmatrix},$$

the differential equation in (5.5.9) may be expressed as

$$\frac{d}{dt}u(t) = Au(t).$$

In Example 5.5.6, we have shown that

$$e^{tA} = \begin{pmatrix} \cos t & \sin t \\ -\sin t & \cos t \end{pmatrix}$$

satisfies

$$\frac{d}{dt}e^{tA} = Ae^{tA}.$$

Hence, we may choose

$$u(t) = \begin{pmatrix} \cos t \\ -\sin t \end{pmatrix} \quad \text{or} \quad u(t) = \begin{pmatrix} \sin t \\ \cos t \end{pmatrix}.$$

The functions $y(t) = \cos t$ or $y(t) = \sin t$ are both solutions to the differential equation $y'' + y = 0$. This also implies that

$$y(t) = C_1 \cos t + C_2 \sin t$$

is a solution for arbitrary real numbers C_1 and C_2.

Example 5.5.10. The differential equation

$$y'''(t) - 2y''(t) - y'(t) + 2y(t) = 0$$

may be expressed as

$$\frac{d}{dt}u(t) = \frac{d}{dt}\begin{pmatrix} y(t) \\ y'(t) \\ y''(t) \end{pmatrix} = \begin{pmatrix} y'(t) \\ y''(t) \\ y'''(t) \end{pmatrix}$$

$$= \begin{pmatrix} y'(t) \\ y''(t) \\ -2y(t) + y'(t) + 2y''(t) \end{pmatrix}$$

$$= A\begin{pmatrix} y(t) \\ y'(t) \\ y''(t) \end{pmatrix}$$

where

$$A = \begin{pmatrix} 0 & 1 & 0 \\ 0 & 0 & 1 \\ -2 & 1 & 2 \end{pmatrix}.$$

The matrix A has three distinct eigenvalues -1, 1 and 2. Thus A is a diagonalizable. We may find an invertible matrix P so that

$$D = \begin{pmatrix} -1 & & \\ & 1 & \\ & & 2 \end{pmatrix} = P^{-1}AP.$$

We then have

$$\frac{d}{dt}u(t) = PAP^{-1}u(t)$$

$$\implies P^{-1}\frac{d}{dt}u(t) = \frac{d}{dt}P^{-1}u(t) = D(P^{-1}u(t)).$$

Since we know that

$$\frac{de^{tD}}{dt} = e^{tD} = \begin{pmatrix} e^{-t} & & \\ & e^t & \\ & & e^{2t} \end{pmatrix},$$

its columns give solutions for $P^{-1}u(t)$. Equivalently, the columns of

(5.4)
$$P \begin{pmatrix} e^{-t} & & \\ & e^t & \\ & & e^{2t} \end{pmatrix}$$

give solutions to $u(t)$. The entries of the first row in (5.4) are linear combinations of e^{-t}, e^t and e^{2t}. Observe that these are indeed solutions to the given differential equation. We conclude that the general solution for the differential equation is

$$C_1 e^{-t} + C_2 e^t + C_3 e^{2t}$$

where C_1, C_2 and C_3 are arbitrary real numbers.

Example 5.5.11. Solve the differential equation

$$y''(t) - 4y'(t) + 4y(t) = 0.$$

Solution. Let
$$u(t) = \begin{pmatrix} y(t) \\ y'(t) \end{pmatrix}.$$

It is easy to see that
$$\frac{d}{dt}u(t) = Au(t)$$

where
$$A = \begin{pmatrix} 0 & 1 \\ -4 & 4 \end{pmatrix}.$$

The characteristic polynomial of A is $t^2 - 4t + 4$, and 2 is the only eigenvalue of A. The vector $(1, 2)$ is an eigenvector corresponding to 2. Choose

$$P = \begin{pmatrix} 1 & -2 \\ 2 & 1 \end{pmatrix}.$$

Then

$$P^{-1}AP = \frac{1}{5} \begin{pmatrix} 1 & 2 \\ -2 & 1 \end{pmatrix} \begin{pmatrix} 0 & 1 \\ -4 & 4 \end{pmatrix} \begin{pmatrix} 1 & -2 \\ 2 & 1 \end{pmatrix} = \begin{pmatrix} 2 & 5 \\ 0 & 2 \end{pmatrix}.$$

Hence

$$P^{-1}P^{tA}P = \begin{pmatrix} e^{2t} & 5te^{2t} \\ 0 & e^{2t} \end{pmatrix}.$$

As in the discussion in Example 5.5.10, we conclude that the general solution for the given differential equation is

$$C_1 e^{2t} + C_2 t e^{2t}$$

for arbitrary real numbers C_1 and C_2.   ◇

Exercises 5.5

1. Solve the system of linear differential equations

$$x' = 3x + 2y,$$
$$y' = 6x - y.$$

where $x = x(t)$ and $y = y(t)$ are differentiable real-valued function of the real variable t.

2. Solve the system of differential equations

$$x' = 3x \qquad - z$$
$$y' = \qquad 2y$$
$$z' = -x \qquad + 3z$$

where $x = x(t)$, $y = y(t)$ and $z = z(t)$ are differentiable real-valued function of the real variable t.

3. Find e^A if

$$A = \begin{pmatrix} 5 & -3 \\ 3 & 1 \end{pmatrix}.$$

4. Find e^A if
$$A = \begin{pmatrix} 1 & 2 & -2 \\ 1 & 2 & 1 \\ -1 & -1 & 0 \end{pmatrix}.$$

5. Solve the linear differential equation
$$y'''(t) - 3y''(t) - y'(t) + 3y(t) = 0.$$

6. Solve the linear differential equation
$$y'''(t) - 3y''(t) + 3y'(t) - y(t) = 0.$$

Review Exercises for Chapter 5

1. Let A be a square matrix. Show that A is similar to a upper triangular matrix if and only if the characteristic polynomial splits.

2. Is the matrix
$$\begin{pmatrix} 0 & 1 & 1 \\ 1 & 0 & 1 \\ 0 & 1 & 0 \end{pmatrix}$$
diagonalizable over $\mathbb{Z}_2$?

3. Let A and B be square matrices of the same size.

   (a) If at least one of A or B is invertible, show that AB and BA are similar.

   (b) Is the conclusion of (a) still valid if neither A nor B is invertible.

4. Let $A \in M_{m \times n}(\mathbb{R})$ and $B \in M_{n \times m}(\mathbb{R})$.

   (a) If λ is an eigenvalue of AB, show that λ is also an eigenvalue of BA.

   (b) If $\lambda \neq 0$ is an eigenvalue of AB with algebraic multiplicity k, show that λ is also an eigenvalue of BA with algebraic multiplicity k.

(c) Let $m \leq n$. Show that the characteristic polynomial of BA equals $(-1)^{n-m}t^{n-m}$ times the characteristic polynomial of AB.

CHAPTER 6

Canonical Forms

In this chapter we want to "describe" the answer to the last classification problem. We say "describe" because at this point, we will not be able to prove the whole theory rigourously.

For any linear endomorphism T we will find a matrix representing T which is as simple as possible. For any square matrix we will find a similar matrix which is as simple as possible. We will call these very simple matrices the **canonical forms** of the corresponding linear endomorphisms or matrices.

6.1   Cayley-Hamilton theorem

Before we tackle the classification problem regarding linear endomorphisms, we first describe the significance of the characteristic polynomial to
a matrix (or a linear endomorphism). The process will give us an overview
on how to construct the canonical form of a matrix in general.

Invariant subspaces.   So far we have seen that not all square matrices
are diagonalizable. In this section we *begin* to explore what happens in the
general case. To diagonalize a matrix, we need to find enough eigenvectors
for a linear endomorphism or a square matrix. Let v be an eigenvector with
eigenvalue λ of the linear endomorphism T. We can see that

$$T(\mathrm{Sp}\,(v)) = \mathrm{Sp}\,(T(v)) = \mathrm{Sp}\,(\lambda v) \subseteq \mathrm{Sp}\,(v)\,.$$

The linear endomorphism T may be restricted to a linear endomorphism on
the subspace $\mathrm{Sp}\,(v)$. Basically, to study a diagonalizable linear endomorphism, we may instead study these "smaller" linear endomorphisms on the
various one-dimensional subspaces. In general, we can study a linear endomorphism by studying the subspaces which are mapped into themselves.

Definition 6.1.1. Let T be a linear endomorphism on a vector space V.
A subspace W of V is called a T-**invariant subspace** of V if $T(W) \subseteq W$.

Example 6.1.2. Let T be a linear endomorphism on a vector space V.
The following are some examples of invariant subspaces of T.

(1) The space V: $T(V) \subseteq V$ by definition.

(2) The trivial subspace $\{O\}$: $T(O) = O \subseteq \{O\}$.

(3) The range of T: $T(\mathrm{Range}\,(T)) \subseteq T(V) = \mathrm{Range}\,(T)$.

(4) The null space of T: $T(\mathrm{Null}\,(T)) = \{O\} \subseteq \mathrm{Null}\,(T)$.

(5) The span $\mathrm{Sp}\,(v)$ where v is an eigenvector of T: Let $T(v) = \lambda v$. Then
$T(av) = a\lambda v \in \mathrm{Sp}\,(v)$ for all scalars a.

(6) The eigen space E_λ where λ is eigenvalue of T: Let $v \in E_\lambda$, then
$T(v) = \lambda v \in \mathrm{Sp}\,(v) \subseteq E_\lambda$.

Let T be an linear endomorphism on V. When W is a T-invariant subspace, we may restrict the domain and codomain of T to W. We will use $T|_W$ to denote the linear endomorphism

$$T|_W : \quad W \quad \longrightarrow \quad W$$
$$w \quad \longmapsto \quad w.$$

Example 6.1.3. Let T be the linear endomorphism on $\mathbb{R}^3$ defined by

$$T(a, b, c) = (a + b, b + c, 0).$$

(a) Find $[T]_{(e_1, e_2, e_3)}$.

(b) Is the x-axis $W_1 = \{(a, 0, 0) : a \in \mathbb{R}\}$ T-invariant? If yes, find $[T|_{W_1}]_{(e_1)}$.

(c) Is the xy-plane $W_2 = \{(a, b, 0) : a, b \in \mathbb{R}\}$ T-invariant? If yes, find $[T|_{W_2}]_{(e_1, e_2)}$.

(d) Is the xz-plane $W_3 = \{(a, 0, c) : a, b \in \mathbb{R}\}$ T-invariant? If yes, find $[T|_{W_3}]_{(e_1, e_3)}$.

(e) If the subspace W_4 generated by e_1 and $e_2 - e_3$ T-invariant? If yes, find $[T|_{W_4}]_{(e_1, e_2 - e_3)}$.

Solution. (a) Since

$$T(e_1) = (1, 0, 0),$$
$$T(e_2) = (1, 1, 0),$$
$$T(e_3) = (0, 1, 0),$$

we have

$$[T]_{(e_1, e_2, e_3)} = \begin{pmatrix} 1 & 1 & 0 \\ 0 & 1 & 1 \\ 0 & 0 & 0 \end{pmatrix}.$$

(b) For all $a \in \mathbb{R}$, we have $T(a, 0, 0) = (a, 0, 0) \in W_1$. Thus W_1 is T-invariant. By definition of T, we have $T(e_1) = e_1$. It follows that

$$[T|_{W_1}]_{(e_1)} = \begin{pmatrix} 1 \end{pmatrix}.$$

(c) For $a, b \in \mathbb{R}$, we have $T(a, b, 0) = (a + b, b, 0) \in W_2$. This says that W_2 is T-invariant. Since

$$T(e_1) = e_1,$$
$$T(e_2) = (1, 1, 0) = e_1 + e_2,$$

we have

$$[T|_{W_2}]_{(e_1, e_2)} = \begin{pmatrix} 1 & 1 \\ 0 & 1 \end{pmatrix}.$$

(d) The vector $(1, 0, 1) \in W_3$. However, $T(1, 0, 1) = (1, 1, 0) \notin W_3$. Thus W_3 is not T-invariant.

(e) Since

$$T(W_4) = \mathrm{Sp}\left(T(e_1), T(e_2 - e_3)\right),$$

to check that W_4 is T-invariant, it suffices to check whether $T(e_1)$ and $T(e_2 - e_3)$ are in W_4. Note that

$$T(e_1) = e_1 \in W_4,$$
$$T(e_2 - e_3) = (1, 0, 0) = e_1 \in W_4.$$

We now know that W_4 is T-invariant. We also know that

$$[T|_{W_4}]_{(e_1, e_2 - e_3)} = \begin{pmatrix} 1 & 1 \\ 0 & 0 \end{pmatrix}.$$

The reader is encouraged to find other T-invariant subspaces of $\mathbb{R}^3$.   ◇

Eventually, to study a linear endomorphism T on V we will break V into T-invariant subspaces. This simplifies the original problem.

The characteristic polynomial of T restricted to a T-invariant subspace. Now let's discuss the advantage of T-invariant subspaces. Let T be a linear endomorphism on a vector space V of dimension n. Suppose we can find a T-invariant subspace W of V. Find an ordered basis $\gamma = (v_1, \ldots, v_m)$ for W. Extend γ to an ordered basis $\beta = (v_1, \ldots, v_m, v_{m+1}, \ldots, v_n)$ for V. We then have

$$A = [T]_\beta = \begin{pmatrix} B & C \\ 0 & D \end{pmatrix}$$

where $B = [T|_W]_\gamma$ is an $m \times m$ matrix and D is an $(n - m) \times (n - m)$ matrix. This produces may zero entries in the matrix representing T. This is definitely preferred than another matrix with few zero entries. By Exercise 8 in §4.3, the characteristic polynomial of T is

$$f(t) = \det(A - tI_n) = \det \begin{pmatrix} B - tI_m & C \\ 0 & D - tI_{n-m} \end{pmatrix}$$

$$= \det(B - tI_{n-m}) \det(D - tI_{n-m})$$

$$= g(t) \det(D - tI_{n-m})$$

where $g(t)$ is the characteristic polynomial of $T|_W$. We now have the following result.

Proposition 6.1.4. *Let T be a linear endomorphism on a finite dimensional vector space V and let W be a T-invariant subspace of V. The characteristic polynomial of $T|_W$ is a factor of the characteristic polynomial of T.*

Example 6.1.5. Let T, W_1 and W_2 be as in Example 6.1.3. Verify the previous proposition by finding the characteristic polynomials of T, $T|_{W_1}$, $T|_{W_2}$ and $T|_{W_4}$.

Solution. The characteristic polynomial of T is

$$\det \begin{pmatrix} 1 - t & 1 & \\ & 1 - t & 1 \\ & & -t \end{pmatrix} = -t(t - 1)^2.$$

The characteristic polynomial of $T|_{W_1}$ is $-(t - 1)$. The characteristic polynomial of $T|_{W_2}$ is

$$\det \begin{pmatrix} 1 - t & 1 \\ & 1 - t \end{pmatrix} = (t - 1)^2.$$

The characteristic polynomial of $T|_{W_4}$ is

$$\det \begin{pmatrix} 1 - t & 1 \\ 0 & -t \end{pmatrix} = t(t - 1).$$

As one can see, they all behave as is described in Proposition 6.1.4.   ◇

The T-cyclic subspaces. The following definition provides a way to construct T-invariant subspaces.

Definition 6.1.6. Let T be a linear endomorphism on a vector space V and let $v \in V$. The subspace

$$\mathrm{Sp}\left(\{v, T(v), T^2(v), \dots\}\right) = \mathrm{Sp}\left(\{T^i(v) : i \in \mathbb{N}\}\right)$$

is called the T-**cyclic subspace of** V **generated by** v.

Lemma 6.1.7. *Let T be an linear endomorphism on V. The T-cyclic subspace of V generated by v is T-invariant.*

Proof. Let W be the T-cyclic subspace of V. For an element $w \in W$, there are scalars $a_0, a_1, \dots, a_N$ such that

$$w = a_0 v + a_1 T(v) + a_2 T^2(v) + \cdots + a_N T^N(v).$$

The image of w under T is

$$\begin{aligned}
T(w) &= T(a_0 v + a_1 T(v) + a_2 T^2(v) + \cdots + a_N T^N(v)) \\
&= a_0 T(v) + a_1 T^2(v) + a_2 T^3(v) + \cdots + a_N T^{N+1}(v).
\end{aligned}$$

This is still an element in W. Hence W is T-invariant. $\qquad\square$

Let $v \in V$ and let W be the T-cyclic subspace generated by v. If V is a finite dimensional vector space, then so is W. The infinite subset

$$\left\{v,\ T(v),\ T^2(v),\ T^3(v), \dots\right\}$$

must be linearly dependent. We can find a nonnegative integer m such that

$$v,\ T(v),\ T^2(v),\ \dots,\ T^{m-1}(v)$$

are linearly independent but

$$T^m(v) \in \mathrm{Sp}\left(v,\ T(v),\ T^2(v),\ \dots,\ T^{m-1}(v)\right).$$

Let a_i be scalars such that

$$(6.1) \qquad T^m(v) = a_0 v + a_1 T(v) + a_2 T^2(v) + \cdots + a_{m-1} T^{m-1}(v).$$

We proceed to show that $W = \mathrm{Sp}\left(v,\ T(v),\ T^2(v),\ \ldots,\ T^{m-1}(v)\right)$. It suffices to show that $T^k(v) \in \mathrm{Sp}\left(v,\ T(v),\ T^2(v),\ \ldots,\ T^{m-1}(v)\right)$ for all k. This is clearly true for $k = 1, \ldots, m$ and we may assume that $k \geq m$.

Assume $T^k(v) \in \mathrm{Sp}\left(v,\ T(v),\ T^2(v),\ \ldots,\ T^{m-1}(v)\right)$. If in particular,

$$T^k(v) = b_0 v + b_1 T(v) + b_2 T^2(v) + \cdots + b_{m-1} T^{m-1}(v),$$

then we have

$$
\begin{aligned}
T^{k+1}(v) &= T(T^k(v)) = T(b_0 v + b_1 T(v) + b_2 T^2(v) + \cdots + b_{m-1} T^{m-1}(v)) \\
&= b_0 T(v) + b_1 T^2(v) + b_2 T^3(v) + \cdots + b_{m-1} T^m(v) \\
&\in \mathrm{Sp}\left(v,\ T(v),\ T^2(v),\ \ldots,\ T^{m-1}(v)\right).
\end{aligned}
$$

This shows that every generator of W is in $\mathrm{Sp}\left(v,\ T(v), T^2(v),\ \ldots,\ T^{m-1}(v)\right)$ and thus

$$W = \mathrm{Sp}\left(v,\ T(v),\ T^2(v),\ \ldots,\ T^{m-1}(v)\right).$$

Since these generators are chosen to be linearly independent, we see that

$$\{v,\ T(v),\ T^2(v),\ \ldots,\ T^{m-1}(v)\}$$

forms a basis for W. Furthermore, $\dim W = m$. We summarize the observation in the following proposition.

Proposition 6.1.8. *Let T be a linear endomorphism on a finite dimensional vector space V and let v be a nonzero vector in V. Let W be the T-cyclic subspace generated by v. Let $\dim W = m$. Then*

$$\{v,\ T(v),\ T^2(v),\ \ldots,\ T^{m-1}(v)\}$$

is a basis for W. Moreover, m is also the least positive integer k such that

$$T^k(v) \in \mathrm{Sp}\left(v,\ T(v),\ T^2(v),\ \ldots,\ T^{k-1}(v)\right).$$

Example 6.1.9. Let $T \colon \mathbb{R}^3 \to \mathbb{R}^3$ be the linear endomorphism defined by

$$T(a, b, c) = (-b + c, a + c, c).$$

(a) Find a basis for the T-cyclic subspace W_1 generated by e_1.

(b) Find a basis for the T-cyclic subspace W_2 generated by $e_2 + e_3$.

(c) Find a basis for the T-cyclic subspace W_3 generated by e_3.

Solution. (a) The set $\{e_1, T(e_1) = e_2\}$ is obviously linearly independent. The next vector $T^2(e_1) = T(e_2) = -e_1$ is in $\text{Sp}(e_1, e_2)$. Thus by Proposition 6.1.8, $\{e_1, e_2\}$ is a basis for W_1.

(b) We compute $T(e_2 + e_3) = e_2 + e_3 \in \text{Sp}(e_2 + e_3)$. Hence $\{e_2 + e_3\}$ is a basis for W_2.

(c) The set $\{e_3, T(e_3) = e_1 + e_2 + e_3, T^2(e_3) = 2e_2 + e_3\}$ is easily seen to be linearly independent. Hence it is a basis for $\mathbb{R}^3$. The T-cyclic subspace generated by e_3 is $\mathbb{R}^3$.   ◇

Example 6.1.10. Let T be an linear endomorphism on an n-dimensional real vector space V with n distinct eigenvalues. We will show that there exists $v \in V$ such that $v, T(v), T^2(v), \ldots, T^{n-1}(v)$ form a basis for V over $\mathbb{R}$. This shows that V is T-cyclic.

Let $\lambda_1, \lambda_2, \ldots, \lambda_n$ be the distinct eigenvalues of T. For each i, find an eigenvector v_i for λ_i. Let $v = v_1 + v_2 + \cdots + v_n$. We claim that $S = \{v, T(v), T^2(v), \ldots, T^{n-1}(v)\}$ forms a basis for V. It suffices to show that S is linearly independent. Remember that by Corollary 5.3.7, $\beta = (v_1, v_2, \ldots, v_n)$ is an ordered basis for V. We have that

$$
\begin{aligned}
v = {}& v_1 + & v_2 + \cdots + & \quad v_n, \\
T(v) = {}& \lambda_1 v_1 + & \lambda_2 v_2 + \cdots + & \quad \lambda_n v_n, \\
T^2(v) = {}& \lambda_1^2 v_1 + & \lambda_2^2 v_2 + \cdots + & \quad \lambda_n^2 v_n, \\
& & \vdots & \\
T^{n-1}(v) = {}& \lambda_1^{n-1} v_1 + & \lambda_2^{n-1} v_2 + \cdots + & \lambda_n^{n-1} v_n.
\end{aligned}
$$

The coordinate of $T^i(v)$ with respect to β is

$$(\lambda_1^i, \lambda_2^i, \ldots, \lambda_n^i) \quad \text{for} \quad i = 0, 1, \ldots, n-1.$$

By Proposition 3.5.6, β is a basis if and only if the matrix

$$
A = \begin{pmatrix}
1 & \lambda_1 & \lambda_1^2 & \cdots & \lambda_1^{n-1} \\
1 & \lambda_2 & \lambda_2^2 & \cdots & \lambda_2^{n-1} \\
\vdots & \vdots & \vdots & \ddots & \vdots \\
1 & \lambda_n & \lambda_n^2 & \cdots & \lambda_n^{n-1}
\end{pmatrix}
$$

is invertible. By Proposition 4.3.5, we only need to check that $\det A \neq 0$. By Exercise 4 in §4.3, we have

$$\det A = \prod_{i<j}(\lambda_j - \lambda_i).$$

Since the eigenvalues are assumed to be distinct, $\det A \neq 0$. This completes the claim.

Since we assume that $T\colon V \to V$ is a linear endomorphism, it makes sense to talk about T, T^2, T^3. Moreover, when

$$f(t) = a_m t^m + a_{m-1}t^{m-1} + \cdots + a_1 t + a_0$$

is a polynomial with the scalars a_i as coefficients, it makes sense to talk about $f(T)$. We leave it as an exercise to show that

$$f(T) = a^m T^m + a_{m-1}T^{m-1} + \cdots + a_1 T + a_0 1_V$$

is also a linear endomorphism. When $f(t) = g(t)h(t)$,

$$f(T) = g(T) \circ h(T) = h(T) \circ g(T).$$

See Exercise 1.

Cayley-Hamilton Theorem.   We can now use the basis in Proposition 6.1.8 to help us find the characteristic polynomial of $T|_W$ where W is the T-cyclic subspace generated by v as in our previous discussion. We might as well assume that $v \neq O$ and hence $m \geq 1$. With respect to the base given in Proposition 6.1.8 the matrix of $T|_W$ is

(6.2)
$$\begin{pmatrix} 0 & 0 & \ldots & 0 & 0 & a_0 \\ 1 & 0 & \ldots & 0 & 0 & a_1 \\ 0 & 1 & \cdots & 0 & 0 & a_2 \\ \vdots & \vdots & \ddots & \vdots & \vdots & \vdots \\ 0 & 0 & \cdots & 1 & 0 & a_{m-2} \\ 0 & 0 & \ldots & 0 & 1 & a_{m-1} \end{pmatrix}_{m \times m}$$

using identity (6.1). We have the following proposition.

Proposition 6.1.11. *Let T be a linear endomorphism on a finite dimensional vector space V and let W be the T-cyclic subspace generated by $v \in V$. Let $\dim W = m$. Suppose*

$$T^m(v) = a_0 v + a_1 T(v) + a_2 T^2(v) + \cdots + a_{m-1} T^{m-1}(v)$$

where the a_i's are scalars. Then the characteristic polynomial of $T|_W$ is

$$(-1)^m (t^m - a_{m-1} t^{m-1} - \cdots - a_2 t^2 - a_1 t - a_0).$$

Proof. To complete the proof we need to find $\det A_m$ where

$$A_m = \begin{pmatrix}
-t & 0 & \cdots & 0 & 0 & a_0 \\
1 & -t & \cdots & 0 & 0 & a_1 \\
0 & 1 & \cdots & 0 & 0 & a_2 \\
\vdots & \vdots & \ddots & \vdots & \vdots & \vdots \\
0 & 0 & \cdots & 1 & -t & a_{m-2} \\
0 & 0 & \cdots & 0 & 1 & a_{m-1} - t
\end{pmatrix}.$$

When $m = 1$, $T(v) = a_0 v$. The characteristic polynomial is

$$\det(a_0 - t) = -(t - a_0).$$

The proposition is true. We now assume $m > 1$, and assume the induction hypothesis for $m - 1$. Then by using the cofactor expansion on the first column we have

$\det A_m$

$$= -t \det \begin{pmatrix}
-t & 0 & \cdots & 0 & 0 & a_1 \\
1 & -t & \cdots & 0 & 0 & a_2 \\
0 & 1 & \cdots & 0 & 0 & a_3 \\
\vdots & \vdots & \ddots & \vdots & \vdots & \vdots \\
0 & 0 & \cdots & 1 & -t & a_{m-2} \\
0 & 0 & \cdots & 0 & 1 & a_{m-1} - t
\end{pmatrix}_{(m-1) \times (m-1)}$$

$$- \det \begin{pmatrix} 0 & 0 & \cdots & 0 & 0 & a_0 \\ 1 & -t & \cdots & 0 & 0 & a_2 \\ 0 & 1 & \cdots & 0 & 0 & a_3 \\ \vdots & \vdots & \ddots & \vdots & \vdots & \vdots \\ 0 & 0 & \cdots & 1 & -t & a_{m-2} \\ 0 & 0 & \cdots & 0 & 1 & a_{m-1} - t \end{pmatrix}_{(m-1) \times (m-1)}$$

$$= -t \left((-1)^{m-1} (t^{m-1} - a_{m-1} t^{m-2} - a_{m-2} t^{m-3} - \cdots - a_2 t - a_1) \right)$$

$$- (-1)^m a_0 \det \begin{pmatrix} 1 & -t & 0 & \cdots & \cdots & 0 \\ 0 & 1 & -t & \cdots & \cdots & 0 \\ 0 & 0 & 1 & \ddots & \cdots & \vdots \\ \vdots & \vdots & \vdots & \ddots & \ddots & \vdots \\ 0 & 0 & 0 & \cdots & 1 & -t \\ 0 & 0 & 0 & \cdots & 0 & 1 \end{pmatrix}_{(m-2) \times (m-2)}$$

$$= (-1)^m \left(t^m - a_{m-1} t^{m-2} - a_{m-2} t^{m-2} - \cdots - a_2 t^2 - a_1 t \right) + (-1)^m (-a_0)$$

$$= (-1)^m \left(t - a_{m-1} t^{m-1} - a_{m-2} t^{m-2} - \cdots - a_2 t^2 - a_1 t - a_0 \right).$$

This completes the proof. $\qquad\qquad\square$

Example 6.1.12. Let's revisit the linear endomorphism T in Example 6.1.9. Consider the ordered basis $\beta = (e_3, \ T(e_3) = e_1 + e_2 + e_3, \ T^2(e_3) = 2e_2 + e_3)$ for $W_3 = \mathbb{R}^3$. Note that

$$T^3(e_3) = T(2e_2 + e_3) = -e_1 + e_2 + e_3 = e_3 - (e_1 + e_2 + e_3) + (2e_2 + e_3)$$
$$= e_3 - T(e_3) + T^2(e_3).$$

The matrix

$$[T]_\beta = \begin{pmatrix} 0 & 0 & 1 \\ 1 & 0 & -1 \\ 0 & 1 & 1 \end{pmatrix}.$$

The characteristic polynomial of T is $f(t) = -(t^3 - t^2 + t - 1)$. Note that

$$f(T)(e_3) = -(T^3(e_3) - T^2(e_3) + T(e_3) - e_3) = O;$$
$$f(T)(T(e_3)) = -(T^3(T(e_3)) - T^2(T(e_3)) + T(T(e_3)) - T(e_3))$$
$$= -T(T^3(e_3) - T^2(e_3) + T(e_3) - e_3) = -T(O) = O;$$

$$f(T)(T^2(e_3)) = -(T^3(T^2(e_3)) - T^2(T^2(e_3)) + T(T^2(e_3)) - T^2(e_3))$$
$$= -T^2(T^3(e_3) - T^2(e_3) + T(e_3) - e_3) = -T^2(O) = O.$$

Since $f(T)$ sends a basis to the trivial vector, we have that $f(T)$ is the zero endomorphism on $\mathbb{R}^3$.

Proposition 6.1.13. *Under the same assumption on T, V and W in Proposition 6.1.11, let $f(t)$ be the characteristic polynomial of $T|_W$. Then $f(T)|_W = \mathbf{0}|_W$.*

Proof. By assumption, v satisfies $f(T)(v) = 0$. For all $i = 0, 1, 2, \ldots, m-1$,

$$f(T)(T(v))$$
$$= (-1)^m \left(T^m - a_{m-1}T^{m-1} - \cdots - a_2T^2 - a_1T - a_0 \right) (T^i(v))$$
$$= (-1)^m \left(T^{m+i}(v) - a_{m-1}T^{m-1+i}(v) - \cdots - a_2T^{2+i}(v) \right.$$
$$\left. - a_1T^{1+i}(v) - a_0T^i(v) \right)$$
$$= T^i \left((-1)^m(T^m(v) - a_{m-1}T^{m-1}(v) - \cdots - a_2T^2(v) - a_1T(v) - a_0v) \right)$$
$$= T^i(O) = O.$$

Thus $f(T)$ sends a basis for V to the trivial vector. This implies that $f(T) = \mathbf{0}|_W$. $\qquad\square$

We will state the following fact without proof. We recommend a course in Abstract Algebra (on Module Theory) to get all the insights and techniques to handle this part.

Let T be a linear endomorphism on a finite dimensional vector space V. In general, it is always possible (but not always easy) to find $v_1, \ldots, v_s$ and corresponding $m_1, \ldots, m_s$ such that for each k

$$v_k, \ T(v_k), \ \ldots, \ T^{m_k-1}(v_k)$$

form a basis for W_k, the T-cyclic subspace generated by v_k, and that

$$(6.3) \quad v_1, \ T(v_1), \ \ldots, \ T^{m_1-1}(v_1), \ v_2, \ T(v_2), \ \ldots, \ T^{m_2-1}(v_2), \ldots$$
$$\ldots, \ v_s, \ T(v_s), \ \ldots, \ T^{m_s-1}(v_s)$$

form a basis for V. The significance here is that the matrix with respect to this basis is

(6.4)
$$\begin{pmatrix} [T|_{W_1}] & & & \\ & [T|_{W_2}] & & \\ & & \ddots & \\ & & & [T|_{W_s}] \end{pmatrix}$$

where each $[T|_{W_k}]$ is of the form in (6.2). If the T-cyclic subspaces are "extremely" well-chosen,[1] this matrix is called the *rational canonical form*, or simply the *rational form* of T. Every linear endomorphism and every square matrix has a rational canonical form, and the rational canonical form may be used to classify the linear endomorphisms. However, we will not go into details in discussing this type of canonical form. It is beyond the scope of this course.

Although we cannot prove our claim above at this point, we give an example to demonstrate the point.

Example 6.1.14. Let's re-revisit the linear endomorphism T in Example 6.1.9. The bases provided for W_1 and W_2 together constitute the ordered basis $\gamma = (e_1, \ e_2, \ e_2 + e_3)$ for $\mathbb{R}^3$. The matrix

$$[T]_\gamma = A = \left(\begin{array}{cc|c} 0 & -1 & 0 \\ 1 & 0 & 0 \\ \hline 0 & 0 & 1 \end{array} \right).$$

This matrix in fact consists of two matrices

$$B = [T|_{W_1}]_{(e_1,e_2)} = \begin{pmatrix} 0 & -1 \\ 1 & 0 \end{pmatrix} \quad \text{and} \quad C = [T|_{W_2}]_{(e_2+e_3)} = \begin{pmatrix} 1 \end{pmatrix}.$$

The characteristic polynomial of B is $t^2 + 1$ and the characteristic polynomial of C is $-(t-1)$. The characteristic polynomial of A is

$$f(t) = -(t^2 + 1)(t - 1).$$

---

[1] It is required that the characteristic polynomial of $[T|_{W_k}]$ be a factor of the characteristic polynomial of $[T|_{W_{k+1}}]$ for each k.

This complies with the result in Example 6.1.12. Remember that $B^2 + 1 = \mathbf{0}_2$ and $C - I_1 = \mathbf{0}_1$. Hence

$$
f(A) = \left(\begin{array}{c|c} f(B) & \\ \hline & f(C) \end{array} \right)
$$

$$
= \left(\begin{array}{c|c} (B^2 + I_2)(B - I_2) & \\ \hline & (C^2 + I_1)(C - I_1) \end{array} \right)
$$

$$
= \left(\begin{array}{c|c} \mathbf{0}_2 & \\ \hline & \mathbf{0}_1 \end{array} \right).
$$

Theorem 6.1.15 (Cayley-Hamilton Theorem). *Let T be a linear endomorphism on a finite dimensional vector space V and let $f(t)$ be the characteristic polynomial of T. Then $f(T) = \mathbf{0}_V$ is the zero transformation on V. In other words, T satisfies the characteristic equation.*

Proof. With a well-chosen basis as in (6.3), the matrix of a linear endomorphism is of the form in (6.4). Let $f_i(t)$ be the characteristic polynomial of $T|_{W_i}$ and let the characteristic polynomial of T be $f(t)$. Then

$$
f(t) = f_1(t) f_2(t) \cdots f_s(t).
$$

For each i, $f(t) = g_i(t) f_i(t)$ where $g_i = f_1 \cdots \widehat{f_i} \cdots f_s$. (The $\widehat{}$ indicates that the term is taken out from the product.) By Proposition 6.1.13, $f_i(T)(w) = O$ for all $w \in W_i$. Moreover,

$$
f(T)(w) = g_i(T)(f_i(T)(w)) = g_i(T)(O) = O
$$

for all i. Thus the $f(T)$ sends every basis element in (6.3) to O. We conclude that $f(T) = \mathbf{0}_V$. $\square$

Example 6.1.16. Let

$$
A = \begin{pmatrix} 1 & 2 \\ -2 & 1 \end{pmatrix}.
$$

(a) Show that the characteristic polynomial of A is $t^2 - 2t + 5$.

(b) Verify directly that $A^2 - 2A + 5I_2 = 0$.

(c) Compute $A^5 + 2A^4 - 3A^3 + 10A^2 - 20I$.

Solution. (a) The characteristic polynomial of A is

$$\det \begin{pmatrix} 1-t & 2 \\ -2 & 1-t \end{pmatrix} = (t-1)^2 + 4 = t^2 - 2t + 5.$$

(b) We verify that

$$A^2 - 2A + 5I$$

$$= \begin{pmatrix} 1 & 2 \\ -2 & 1 \end{pmatrix} \begin{pmatrix} 1 & 2 \\ -2 & 1 \end{pmatrix} - 2 \begin{pmatrix} 1 & 2 \\ -2 & 1 \end{pmatrix} + \begin{pmatrix} 5 & 0 \\ 0 & 5 \end{pmatrix}$$

$$= \begin{pmatrix} -3 & 4 \\ -4 & -3 \end{pmatrix} + \begin{pmatrix} -2 & -4 \\ 4 & -2 \end{pmatrix} + \begin{pmatrix} 5 & 0 \\ 0 & 5 \end{pmatrix}$$

$$= \begin{pmatrix} 0 & 0 \\ 0 & 0 \end{pmatrix}.$$

(c) We may use the relation

$$t^5 + 2t^4 - 3t^3 + 10t^2 - 20$$
$$= (t^2 - 2t + 5)(t^3 + 4t^2 - 10) - 20t + 30$$

to simplify computation. We have

$$A^5 + 2A^4 - 3A^3 + 10A^2 - 20I$$
$$= (A^2 - 2A + 5I)(A^3 + 4A^2 - 10I) - 20A + 30I$$
$$= -20A + 30I$$
$$= \begin{pmatrix} -20 & -40 \\ 40 & -20 \end{pmatrix} + \begin{pmatrix} 30 & 0 \\ 0 & 30 \end{pmatrix}$$
$$= \begin{pmatrix} 10 & -40 \\ 40 & 10 \end{pmatrix}.$$

As one can see, Cayley-Hamilton Theorem greatly simplifies the process. ◇

Remember that the constant term of the characteristic polynomial of a square matrix is its determinant. Hence a square matrix is invertible if and only if the constant term of its characteristic polynomial is nonzero. This point can also be demonstrated by the following example.

Example 6.1.17. Let

$$A = \begin{pmatrix} 1 & 0 & 1 \\ 2 & 1 & 2 \\ 0 & 4 & 6 \end{pmatrix}.$$

Show that A is invertible. Find A^{-1} using the characteristic equation.

Solution. The characteristic polynomial of A is

$$\det \begin{pmatrix} 1-t & 0 & 1 \\ 2 & 1-t & 2 \\ 0 & 4 & 6-t \end{pmatrix} = -t^3 + 8t^2 - 5t + 6.$$

From Cayley-Hamilton Theorem we have

$$- A^3 + 8A^2 - 5A + 6I = 0$$
$$\implies 6I = A^3 - 8A^2 + 5A$$
$$= A(A^2 - 8A + 5I).$$

We conclude that $A^{-1} = (A^2 - 8A + 5I)/6.$ ◇

Minimal polynomials. In the proof of Cayley-Hamilton Theorem, we used the fact that the characteristic polynomial of T is a multiple of all the characteristic polynomial of $T|_{W_i}$. The result of Cayley-Hamilton Theorem would have applied if we had had chosen a common multiple of the characteristic polynomials of $T|_{W_i}$ for all i. There is a chance we may find a polynomial $g(t)$ with degree lower than the degree of the characteristic polynomial of T such that $g(T) = \mathbf{0}$. Thus, we may refine Cayley-Hamilton Theorem as shown below.

Theorem 6.1.18. *Let T be a linear endomorphism on a n-dimensional vector space V. There is a polynomial $g(t)$ over the scalar field of V such that $g(T) = \mathbf{0}_V$ and for any polynomial $h(t)$ with $h(T) = \mathbf{0}_V$, $g(t)$ is a factor of $h(t)$.*

The polynomial $g(t)$ is a factor of the characteristic polynomial of T and hence has degree $\leq n$. Moreover, the prime factors of $g(t)$ are exactly those of the characteristic polynomial of T.

Proof. Choose a basis for V as in (6.3) so that the matrix of T is of the form in (6.4). Let $g_i(t)$ be the characteristic polynomial of $T|_{W_i}$. We claim that for any polynomial $h(t)$ such that $h(T|_{W_i}) = \mathbf{0}_{W_i}$, we have that $g_i(t)$ is a factor of $h(t)$.

Let
$$h(t) = g_i(t)q(t) + r(t), \quad \deg r(t) < \deg g_i(t) = m_i.$$

Then
$$h(T|_{W_i}) = g_i(T|_{W_i}) \circ q(T|_{W_i}) + r(T|_{W_i})$$
$$= r(T|_{W_i}).$$

If $r(t) \neq 0$, let $r(t) = b_k t^k + b_{k-1} t^{k-1} + \cdots + b_1 t + b_0$ with $k < m_i$ and $b_k \neq 0$. It implies that

$$b_k T^k(v_i) + b_{k-1} T^{k-1}(v_i) + \cdots + b_1 T(v_i) + b_0 v_i = O,$$

contradicting to the fact that $\{v_i, T(v_i), \ldots, T^{m_k}(v_i)\}$ is linearly independent. Thus $r(t)$ is the zero polynomial and $g_i(t)$ is a factor of $h(t)$.

Choose $g(t)$ to be the least common multiple of $g_1(t), \ldots, g_s(t)$. Then for each i, $g(t) = q_i(t)g_i(t)$. We have

$$g(T)|_{W_i} = g(T|_{W_i}) = q_i(T|_{W_i}) \circ g_i(T|_{W_i}) = \mathbf{0}_{W_i}$$

for all i. The linear endomorphism $g(T)$ sends all basis elements to the trivial vector. Hence, $g(T) = \mathbf{0}_V$.

Let $h(t)$ be a polynomial such that $h(T) = \mathbf{0}_V$. Since $h(T)|_{W_i} = h(T|_{W_i}) = \mathbf{0}_{W_i}$ for all i. By our previous claim, $g_i(t)$ is a factor of $h(t)$ for all i. Thus $h(t)$ is a common multiple of all the $g_i(t)$'s. We have that $g(t)$ is a factor of $h(t)$.

Remember that the characteristic polynomial of T is

$$f(t) = g_1(t)g_2(t) \cdots g_s(t).$$

The least common multiple $g(t)$ is a factor of $f(t)$. Moreover, $f(t)$ and $g(t)$ share the same set of prime factors. $\square$

Definition 6.1.19. If the leading coefficient (the coefficient of the highest degree monomial) of a nonzero polynomial is 1, it is called a **monic** polynomial. If we adjust the polynomial $g(t)$ found in Theorem 6.1.18 so that it becomes monic, the polynomial $g(t)$ is called the **minimal polynomial** of T.

Example 6.1.20. Consider the linear endomorphism T in Example 6.1.14. Its characteristic polynomial is $-(t^2+1)(t-1)$ and its minimal polynomial is $(t^2+1)(t-1)$. Remember that the leading coefficient of the minimal polynomial must be 1!

Example 6.1.21. The characteristic polynomial of I_3 is $(1-t)^3$ while its minimal polynomial is $t-1$. The characteristic polynomial and the minimal polynomial of a linear endomorphism or of a matrix are not always the same.

Example 6.1.22. Let
$$A = \begin{pmatrix} 0 & 1 & 0 \\ 0 & 0 & 0 \\ 0 & 0 & 0 \end{pmatrix}.$$

The characteristic polynomial of A is $-t^3$. By Theorem 6.1.18 the minimal polynomial of A could be t, t^2 or t^3. We check that

$$A \neq \mathbf{0}_3,$$
$$A^2 = \mathbf{0}_3.$$

We conclude that the minimal polynomial of A is t^2.

Proposition 6.1.23. *If two square matrices are similar to each other, they have the same minimal polynomials.*

Proof. Let $A = P^{-1}BP$ where P is an invertible matrix. Remember that $A^k = (P^{-1}BP)^k = P^{-1}B^kP$. Thus for any polynomial

$$f(t) = \sum_{k=0}^{n} a_k t^k, \quad \text{where the } a_k\text{'s are scalars,}$$

we have

$$f(A) = \sum_{k=0}^{n} a_k A^k = \sum_{k=0}^{n} a_k P^{-1} B^k P$$
$$= \sum_{k=0}^{n} P^{-1} a_k B^k P = P^{-1} \left(\sum_{k=0}^{n} a_k B^k \right) P$$

$$= P^{-1}f(B)P.$$

Similarly, $f(B) = Pf(A)P^{-1}$. This shows that

$$f(A) = 0 \iff f(B) = 0.$$

Thus the minimal polynomials of both matrices are the same. $\qquad\square$

Exercises 6.1

1. Let

$$f(t) = a^m t^m + a_{m-1}t^{m-1} + \cdots + a_1 t + a_0$$

be a polynomial where the a_i's are scalars. Show that

$$f(T) = a^m T^m + a_{m-1}T^{m-1} + \cdots + a_1 T + a_0 1_V$$

is a linear endomorphism. When $f(t) = g(t)h(t)$, show that

$$f(T) = g(T) \circ h(T) = h(T) \circ g(T).$$

2. Let

$$A = \begin{pmatrix} B_1 & & & \\ & B_2 & & \\ & & \ddots & \\ & & & B_r \end{pmatrix}$$

where each B_i is a block of square matrix and let $f(t)$ be a polynomial. Show that

$$f(A) = \begin{pmatrix} f(B_1) & & & \\ & f(B_2) & & \\ & & \ddots & \\ & & & f(B_r) \end{pmatrix}.$$

3. Let T be a linear endomorphism on V. Let W_1 and W_2 be T-invariant subspaces of V such that $W_1 \cap W_2 = \{O\}$.

(a) Let β_1, β_2 be ordered bases for W_1 and W_2 respectively. Show that $\beta = (\beta_1, \beta_2)$ (the elements of β_1 followed by elements of β_2) is an ordered basis for $W_1 + W_2$.

(b) Show that

$$[T]_\beta = \begin{pmatrix} [T]_{\beta_1} & \mathbf{0} \\ \mathbf{0} & [T]_{\beta_2} \end{pmatrix}.$$

4. Let A be the matrix

$$\begin{pmatrix} 1 & 2 \\ -2 & 1 \end{pmatrix}$$

in Example 6.1.16. Show that $f(A)$ is always the sum of a diagonal matrix and a skew-symmetric matrix for any polynomial $f(t)$ with real coefficients.

5. Let T be a linear endomorphism. Which of the following subspaces are T-invariant subspaces?

(a) Subspaces generated by eigenvectors (corresponding to possibly different eigenvalues).

(b) Eigenspaces of T^k for some positive integer k.

(c) The null space of T^k for some positive integer k.

(d) The range of T^k for some positive integer k.

(e) The sum of two T-cyclic subspaces.

(f) The sum of two T-invariant subspaces.

6. A **nilpotent** matrix is a square matrix A such that $A^k = \mathbf{0}$ for some positive integer k.

Let A be a square triangular matrix whose diagonal entries are all 0. Show that A is nilpotent.

7. Let

$$A = \begin{pmatrix} 7 & 1 & 2 & 2 \\ 1 & 4 & -1 & -1 \\ -2 & 1 & 5 & -1 \\ 1 & 1 & 2 & 8 \end{pmatrix}.$$

(a) Find the characteristic polynomial of A.

(b) Is A diagonalizable?

(c) Find the L_A-cyclic subspace generated by e_4 and let's call it W. Let $m = \dim W$. Find m and find the matrix of L_A with respect to the ordered basis $(e_4,\ L_A(e_4),\ \ldots,\ L_A^{m-1}(e_4))$.

8. Let

$$A = \begin{pmatrix} 1 & 0 & 0 & 0 \\ 2 & 5 & 6 & 7 \\ 3 & 0 & 8 & 9 \\ 4 & 0 & 0 & 10 \end{pmatrix}.$$

(a) Find the characteristic polynomial of A.

(b) Is A diagonalizable?

(c) Find the L_A-cyclic subspace generated by e_1 and let's call it W. Let $m = \dim W$. Find m and find the matrix of L_A with respect to the ordered basis $(e_1,\ L_A(e_1),\ \ldots,\ L_A^{m-1}(e_1))$.

(d) Can you find a L_A-invariant subspace of dimension 1?

(e) Compute $(A - I)^{10}(A - 5I)^{20}(A - 8I)^{30}(A - 10I)^{40}$.

6.2   Jordan canonical forms

For the rest of this chapter, we describe how to use the *Jordan canonical forms* to classify linear endomorphisms and square matrices.

Jordan blocks and Jordan forms.   Jordan blocks are building blocks of *Jordan forms*. A **Jordan block** of size n and with eigenvalue λ is a square matrix of the form

(6.5)
$$\begin{pmatrix} \lambda & 0 & 0 & \cdots & 0 & 0 & 0 \\ 1 & \lambda & 0 & \cdots & 0 & 0 & 0 \\ 0 & 1 & \lambda & \cdots & 0 & 0 & 0 \\ \vdots & \vdots & \vdots & \ddots & \ddots & \vdots & \vdots \\ 0 & 0 & 0 & \cdots & \lambda & 0 & 0 \\ 0 & 0 & 0 & \cdots & 1 & \lambda & 0 \\ 0 & 0 & 0 & \cdots & 0 & 1 & \lambda \end{pmatrix}_{n \times n}.$$

The determinant and the characteristic polynomial of this Jordan block are λ^n and $(-1)^n(t-\lambda)^n$ respectively.

Example 6.2.1. The matrix

$$\begin{pmatrix} 3 \end{pmatrix}$$

is a Jordan block of size 1 and with eigenvalue 3. The matrix

$$\begin{pmatrix} i & 0 & 0 & 0 \\ 1 & i & 0 & 0 \\ 0 & 1 & i & 0 \\ 0 & 0 & 1 & i \end{pmatrix}$$

is a Jordan block of size 4 and with eigenvalue i. The matrix

$$\begin{pmatrix} 0 & 0 & 0 & 0 & 0 \\ 1 & 0 & 0 & 0 & 0 \\ 0 & 1 & 0 & 0 & 0 \\ 0 & 0 & 1 & 0 & 0 \\ 0 & 0 & 0 & 1 & 0 \end{pmatrix}$$

is a Jordan block of size 5 and with eigenvalue 0.

Let

$$A = \begin{pmatrix} J_1 & \mathbf{0} & \cdots & \mathbf{0} \\ \mathbf{0} & J_2 & \cdots & \mathbf{0} \\ \vdots & \vdots & \ddots & \vdots \\ \mathbf{0} & \mathbf{0} & \cdots & J_k \end{pmatrix}$$

where J_i is the Jordan block with eigenvalue λ_i of size m_i for each i. Then

$$\det A = \lambda_1^{m_1} \lambda_2^{m_2} \cdots \lambda_k^{m_k}$$

and the characteristic polynomial of A is

$$(-1)^{m_1+\cdots+m_k}(t-\lambda_1)^{m_1}(t-\lambda_2)^{m_2}\cdots(t-\lambda_k)^{m_k}.$$

Hence the characteristic polynomial of such a matrix always splits.

We state the following theorem without proof. Its proof is slightly beyond the scope of this textbook.

Theorem 6.2.2. *Let T be a linear endomorphism on a vector space V of dimension n, whose characteristic polynomial splits, say,*

$$f(t) = (-1)^n (t - \lambda_1)^{m_1} (t - \lambda_2)^{m_2} \cdots (t - \lambda_s)^{m_s}$$

where the λ_i's are distinct scalars. Then there is a basis β of V such that

$$[T]_\beta = \begin{pmatrix} J_1 & 0 & \cdots & 0 \\ 0 & J_2 & \cdots & 0 \\ \vdots & \vdots & \ddots & \vdots \\ 0 & 0 & \cdots & J_r \end{pmatrix},$$

*where each J_j is a Jordan block with one of the λ_i as its eigenvalue. For each i, the sum of the sizes of all the Jordan blocks with eigenvalue λ_i appearing in $[T]_\beta$ is equal to m_i. This matrix is called a **Jordan canonical form**, or simply a **Jordan form**, of T (or of $[T]_\beta$).*

Let A be a square matrix whose characteristic polynomial splits. The Jordan form of L_A is also called a Jordan form of A. The matrix A is similar to any of its Jordan form.

Corollary 6.2.3. *A linear endomorphism or a square matrix has a Jordan form if and only if its characteristic polynomial splits.*

Example 6.2.4. The matrices

$$\left(\begin{array}{c|c|c|c} 2 & 0 & 0 & 0 \\ \hline 0 & 2 & 0 & 0 \\ \hline 0 & 1 & 2 & 0 \\ \hline 0 & 0 & 0 & 2 \end{array} \right), \quad \begin{pmatrix} 2 & 0 & 0 & 0 \\ 1 & 2 & 0 & 0 \\ 0 & 1 & 2 & 0 \\ 0 & 0 & 1 & 2 \end{pmatrix} \quad \text{and} \quad \left(\begin{array}{c|c|c|c} 2 & 0 & 0 & 0 \\ 1 & 2 & 0 & 0 \\ 0 & 1 & 2 & 0 \\ \hline 0 & 0 & 0 & 2 \end{array} \right)$$

are all Jordan forms with the characteristic polynomial $(t - 2)^4$.

Example 6.2.5. Both the matrices

$$\left(\begin{array}{c|c|c|c} 3 & 0 & 0 & 0 \\ \hline 0 & 2 & 0 & 0 \\ \hline 0 & 1 & 2 & 0 \\ \hline 0 & 0 & 0 & 1 \end{array} \right) \quad \text{and} \quad \left(\begin{array}{c|c|c|c} 2 & 0 & 0 & 0 \\ \hline 0 & 2 & 0 & 0 \\ \hline 0 & 0 & 1 & 0 \\ \hline 0 & 0 & 0 & 3 \end{array} \right)$$

are Jordan forms with the characteristic polynomial $(t - 1)(t - 2)^2(t - 3)$.

Notice that diagonal matrices are Jordan forms composed of Jordan blocks of size 1.

Example 6.2.6. The characteristic polynomial of the matrix $A = \begin{pmatrix} 0 & -1 \\ 1 & 0 \end{pmatrix}$ is $t^2 + 1$. It does not have a Jordan form over $\mathbb{R}$. However, $\begin{pmatrix} i & 0 \\ 0 & -i \end{pmatrix}$ is a Jordan form of A over $\mathbb{C}$.

Thanks to Fundamental Theorem of Algebra, we have the following result.

Corollary 6.2.7. *Every matrix in $M_n(\mathbb{C})$ (including every matrix in $M_n(\mathbb{R})$) has a Jordan form over $\mathbb{C}$.*

Classification of similar matrices. Let T be a linear endomorphism on V. Suppose that we can find a basis $\beta = (v_1, \ldots, v_m, w_1, \ldots, w_n)$ for V such that $W_1 = \mathrm{Sp}(v_1, \ldots, v_m)$ and $W_2 = \mathrm{Sp}(w_1, \ldots, w_n)$ are T-invariant. This makes

$$[T]_\beta = \begin{pmatrix} B_1 & \\ & B_2 \end{pmatrix},$$

where B_i is the matrix representing $T|_{W_i}$ relative to the bases given above. Clearly, if we let $\beta' = (w_1, \ldots, w_n, v_1, \ldots, v_m)$, we have that

$$[T]_{\beta'} = \begin{pmatrix} B_2 & \\ & B_1 \end{pmatrix}.$$

Thus

$$\begin{pmatrix} B_1 & \\ & B_2 \end{pmatrix} \sim \begin{pmatrix} B_2 & \\ & B_1 \end{pmatrix}.$$

This argument above applies to matrices composed of more blocks. This tells us that if we permute the Jordan blocks in a Jordan form, we still have a similar matrix. Moreover, we have the following result.

Theorem 6.2.8. *Let $J_1, \ldots, J_k$ and $J'_1, \ldots, J'_\ell$ be Jordan blocks. The two Jordan canonical forms*

$$\begin{pmatrix} J_1 & 0 & \cdots & 0 \\ 0 & J_2 & \cdots & 0 \\ \vdots & \vdots & \ddots & \vdots \\ 0 & 0 & \cdots & J_k \end{pmatrix} \quad and \quad \begin{pmatrix} J'_1 & 0 & \cdots & 0 \\ 0 & J'_2 & \cdots & 0 \\ \vdots & \vdots & \ddots & \vdots \\ 0 & 0 & \cdots & J'_\ell \end{pmatrix}$$

are similar if and only if $k = \ell$ and the J_i''s form a permutation of the J_i's.
Thus, we usually refer to these two matrices as the same *Jordan form (up*
to different orderings of the Jordan blocks).

Theorem 6.2.9. *Two square matrices (in $M_n(\mathbb{R})$ or $M_n(\mathbb{C})$) are similar*
if and only if they have the same Jordan canonical form (up to different
orderings of Jordan blocks) over $\mathbb{C}$.

The following theorem is still given without proof.

Two linear endomorphisms on the same finite dimensional vector space
V (over $\mathbb{R}$ or over $\mathbb{C}$) are similar if and only if their matrices are similar
over $\mathbb{C}$.

Example 6.2.10. The two matrices

$$\left(\begin{array}{c|cc} 1 & 0 & 0 \\ \hline 0 & 2 & 0 \\ 0 & 1 & 2 \end{array}\right) \quad \text{and} \quad \left(\begin{array}{cc|c} 2 & 0 & 0 \\ 1 & 2 & 0 \\ \hline 0 & 0 & 1 \end{array}\right)$$

have the same characteristic polynomial and they are also similar. The two
matrices

$$\left(\begin{array}{c|cc} 1 & 0 & 0 \\ \hline 0 & 2 & 0 \\ 0 & 1 & 2 \end{array}\right) \quad \text{and} \quad \left(\begin{array}{c|c|c} 2 & 0 & 0 \\ \hline 0 & 2 & 0 \\ \hline 0 & 0 & 1 \end{array}\right)$$

have the same characteristic polynomial but they are not similar.

Example 6.2.11. Determine whether the two matrices

$$\begin{pmatrix} -1 & 2 \\ -1 & 1 \end{pmatrix} \quad \text{and} \quad \begin{pmatrix} 0 & 1 \\ -1 & 0 \end{pmatrix}$$

are similar.

Solution. The characteristic polynomials of both matrices are

$$t^2 + 1 = (t + i)(t - i).$$

They are both diagonalizable, because they have two distinct eigenvalues.
Hence they are both similar to

$$D = \begin{pmatrix} i & 0 \\ 0 & -i \end{pmatrix}.$$

These two matrices are similar to each other. Note that D is also the Jordan form of both matrices. $\diamond$

Example 6.2.12. Let the characteristic polynomial of a matrix A be

$$f(t) = -(t-1)^2(t-2)^3.$$

Find all possible Jordan forms of A.

Solution. Since the characteristic polynomial of A splits, A has a Jordan form. Without going into more detail, we cannot make sure what its Jordan form is. But, we can list all possibilities. The total size of Jordan blocks with eigenvalue 1 is 2. Thus there might be one of size 2 or 2 blocks of size one. The total size of Jordan blocks with eigenvalue 2 is 3. There might be (i) one of size 3, (ii) one of size 2 and one of size 1, or (iii) 3 blocks of size 1. We may list all possibilities now:

$$\left(\begin{array}{cc|ccc} 1 & & & & \\ 1 & 1 & & & \\ \hline & & 2 & & \\ & & 1 & 2 & \\ & & & 1 & 2 \end{array}\right), \quad \left(\begin{array}{cc|ccc} 1 & & & & \\ & 1 & & & \\ \hline & & 2 & & \\ & & 1 & 2 & \\ & & & 1 & 2 \end{array}\right),$$

$$\left(\begin{array}{cc|cc|c} 1 & & & & \\ 1 & 1 & & & \\ \hline & & 2 & & \\ & & 1 & 2 & \\ \hline & & & & 2 \end{array}\right), \quad \left(\begin{array}{cc|cc|c} 1 & & & & \\ & 1 & & & \\ \hline & & 2 & & \\ & & 1 & 2 & \\ \hline & & & & 2 \end{array}\right),$$

$$\left(\begin{array}{cc|c|c|c} 1 & & & & \\ 1 & 1 & & & \\ \hline & & 2 & & \\ \hline & & & 2 & \\ \hline & & & & 2 \end{array}\right), \quad \left(\begin{array}{cc|c|c|c} 1 & & & & \\ & 1 & & & \\ \hline & & 2 & & \\ \hline & & & 2 & \\ \hline & & & & 2 \end{array}\right).$$

There are altogether six possibilities. $\diamond$

Minimal polynomials and Jordan forms. How does one calculate the minimal polynomial? We start with investigating a Jordan block.

Proposition 6.2.13.   (a) *The minimal polynomial of the Jordan block* (6.5) *is* $(t - \lambda)^n$.

(b) *The minimal polynomial of the Jordan form*

$$J = \begin{pmatrix} J_1 & 0 & \cdots & 0 \\ 0 & J_2 & \cdots & 0 \\ \vdots & \vdots & \ddots & \vdots \\ 0 & 0 & \cdots & J_s \end{pmatrix}$$

is the monic polynomial which is the least common multiple of the minimal polynomials of J_i for $i = 1, \ldots, s$.

(c) *The minimal polynomial of a square matrix is equal to the minimal polynomial of its Jordan form.*

Proof. (a) Let J be the Jordan block in (6.5). The characteristic polynomial of the Jordan block of size n with eigenvalue λ in (6.5) is $(-1)^n (t - \lambda)^n$. Its minimal polynomial must be a factor of $(t - \lambda)^n$. Thus, we only need to show that $(J - \lambda I)^k \neq 0$ for $1 \leq k < n$. This is indeed so. We leave the detail as an exercise. See Exercise 3.

(b) Let $f(t)$ be any polynomial. Then

$$f(J) = \begin{pmatrix} f(J_1) & 0 & \cdots & 0 \\ 0 & f(J_2) & \cdots & 0 \\ \vdots & \vdots & \ddots & \vdots \\ 0 & 0 & \cdots & f(J_s) \end{pmatrix}.$$

If we want $f(J)$ to be the zero matrix, we need $f(J_i)$ to be simultaneously 0 at the same time. Hence the minimal polynomial of J is easily seen to be the monic polynomial which is the least common multiple of the minimal polynomials of all the Jordan blocks.

Part (c) follows from Proposition 6.1.23.   □

Let a be a zero of the polynomial $f(t)$. If the algebraic multiplicity of a is 1, we say a is a **simple** zero of $f(t)$.

Corollary 6.2.14. *A square matrix is diagonalizable if and only if all the zeros of its minimal polynomial are simple.*

Proof. From Theorem 6.2.9, a square matrix is diagonalizable if and only if the Jordan blocks in its Jordan form are all of size 1. If $\lambda_1, \ldots, \lambda_s$ are all the distinct eigenvalues of the given matrix, then its minimal polynomial is $(t - \lambda_1) \cdots (t - \lambda_s)$. All the zeros are simple. □

Example 6.2.15. The minimal polynomials of the matrices in Examples 6.2.4 are $(t - 2)^2$, $(t - 2)^4$ and $(t - 2)^3$ respectively. The minimal polynomials of the matrices in Example 6.2.5 are $(t - 3)(t - 2)^2(t - 1)$ and $(t - 2)(t - 1)(t - 3)$ respectively.

Example 6.2.16. Let A be a matrix of size 5 satisfying $A^2 = A$. Find all possible Jordan forms for A.

Solution. The matrix A satisfies $A^2 - A = \mathbf{0}$. This means that the minimal polynomial $g(t)$ of A is a factor of $t^2 - t$.

Case 1. Assume $g(t) = t^2 - t = t(t - 1)$. This means that the Jordan form has at least one Jordan block with eigenvalue 0 and one Jordan block with eigenvalue 1. Moreover, the Jordan blocks are of size 1. Thus, the Jordan form of A may be

$$
\begin{pmatrix} 0 & & & & \\ & 0 & & & \\ & & 0 & & \\ & & & 0 & \\ & & & & 1 \end{pmatrix}, \quad
\begin{pmatrix} 0 & & & & \\ & 0 & & & \\ & & 0 & & \\ & & & 1 & \\ & & & & 1 \end{pmatrix}, \quad
\begin{pmatrix} 0 & & & & \\ & 0 & & & \\ & & 1 & & \\ & & & 1 & \\ & & & & 1 \end{pmatrix}
$$

and

$$
\begin{pmatrix} 0 & & & & \\ & 1 & & & \\ & & 1 & & \\ & & & 1 & \\ & & & & 1 \end{pmatrix}.
$$

Case 2. Assume $g(t) = t$. The Jordan blocks of A are all with eigenvalue 0 and size 1. The Jordan form of A is

$$
\begin{pmatrix} 0 & & & & \\ & 0 & & & \\ & & 0 & & \\ & & & 0 & \\ & & & & 0 \end{pmatrix}.
$$

Case 3. Assume $g(t) = t - 1$. The Jordan blocks of A are all with eigenvalue 1 and of size 1. The Jordan form of A is

$$\begin{pmatrix} 1 & & & \\ & 1 & & \\ & & \ddots & \\ & & & 1 \end{pmatrix}.$$

There are altogether 6 possible Jordan forms for A.   ◇

Example 6.2.17. Let A be a real matrix whose minimal polynomial is $(t-2)^2(t+3)^2$ and the degree of its characteristic polynomial is 6. Classify A via similarity.

Solution. Let J be the Jordan form of A. From the degree of the characteristic polynomial of A, we know A is of size 6. Moreover, there is a Jordan block with eigenvalue 2 and a Jordan block with eigenvalue -3. From the assumption on the minimal polynomial of A, the largest Jordan block with eigenvalue 2 is of size 2, and the largest Jordan block with eigenvalue -3 is of size 2 as well. There is still a 2×2 space to put in one or two Jordan blocks in J. Hence the possible Jordan forms for A are

$$\left(\begin{array}{cc|cc|cc} 2 & & & & & \\ 1 & 2 & & & & \\ \hline & & -3 & & & \\ & & 1 & -3 & & \\ \hline & & & & 2 & \\ & & & & 1 & 2 \end{array}\right), \quad \left(\begin{array}{cc|cc|c|c} 2 & & & & & \\ 1 & 2 & & & & \\ \hline & & -3 & & & \\ & & 1 & -3 & & \\ \hline & & & & 2 & \\ \hline & & & & & 2 \end{array}\right),$$

$$\left(\begin{array}{cc|cc|cc} 2 & & & & & \\ 1 & 2 & & & & \\ \hline & & -3 & & & \\ & & 1 & -3 & & \\ \hline & & & & -3 & \\ & & & & 1 & -3 \end{array}\right), \quad \left(\begin{array}{cc|cc|c|c} 2 & & & & & \\ 1 & 2 & & & & \\ \hline & & -3 & & & \\ & & 1 & -3 & & \\ \hline & & & & -3 & \\ \hline & & & & & -3 \end{array}\right)$$

and
$$\left(\begin{array}{cc|c|c|c}
2 & & & & \\
1 & 2 & & & \\
\hline
 & & -3 & & \\
 & & 1 & -3 & \\
\hline
 & & & & 2 \\
\hline
 & & & & & -3
\end{array}\right).$$

There are altogether 5 possibilities.   ◇

Example 6.2.18. Let

$$A = \begin{pmatrix} 2 & 2 & -1 \\ -1 & -1 & 1 \\ -1 & -2 & 2 \end{pmatrix}.$$

(a) Find the characteristic polynomial and minimal polynomial of A.

(b) Is A diagonalizable? If not, can you deduce the Jordan form of A?

Solution. (a) The characteristic polynomial of A is

$$-t^3 + 3t^2 - 3t + 1 = -(t-1)^3.$$

The possible minimal polynomial of A must be one of $t - 1$, $(t - 1)^2$ or $(t - 1)^3$. Obviously $A - I \neq 0$. We find

$$(A - I)^2 = \begin{pmatrix} 1 & 2 & -1 \\ -1 & -2 & 1 \\ -1 & -2 & 1 \end{pmatrix}^2 = \mathbf{0}.$$

We conclude that the minimal polynomial of A is $(t - 1)^2$.

(b) From Corollary 6.2.14, A is not diagonalizable. The Jordan form of A has only one eigenvalue 1, and it has a Jordan block of size 2. We conclude that the Jordan form of A must be

$$\left(\begin{array}{cc|c}
1 & & \\
1 & 1 & \\
\hline
 & & 1
\end{array}\right).$$

Fortunately, we do not need more work to deduce this.   ◇

Invariants of similar matrices. At last, we make a summary. Associated with a square matrix there are four important algebraic invariants via similarity. We list them here in degree of significance, from the vaguest to the most decisive:

• *The determinant*: to determine whether a matrix is invertible or not.

• *The characteristic polynomial*: to give the size and the eigenvalues of the matrix. The algebraic multiplicity of each eigenvalue λ gives the total size of the Jordan blocks corresponding to λ.

• *The minimal polynomial*: to give the eigenvalues and an equation for the matrix. The algebraic multiplicity of each eigenvalue λ gives the maximal size of the Jordan blocks corresponding to λ.

• *The canonical form*: to strip the matrix naked so that one knows everything there is to know about the given matrix.

Exercises 6.2

1. Which of the following assertions are true?

   (a) A real 2×2 matrix with a negative determinant is diagonalizable.

   (b) If A is a square matrix such that $A^k = I$ for some positive integer k, then A is diagonalizable.

   (c) If A is a square matrix such that $A^k = \mathbf{0}$ for some positive integer k, then A is diagonalizable.

   (d) If the degree of the minimal polynomial of a square matrix of size n is n, then A is diagonalizable.

2. Let the characteristic polynomial of the matrix A be

$$f(t) = t^3(t-1)^4(t+3).$$

   Find all possible Jordan canonical forms of A.

3. Let J be the Jordan block and with eigenvalue 0 of size n.

   (a) Show that for $k = 1, \ldots, n$ the linear endomorphism L_{J^k} is such that
$$L_{J^k}(e_i) = \begin{cases} e_{i+k}, & \text{if } i = 1, \ldots, n - k; \\ O, & i = n - k + 1, \ldots, n. \end{cases}$$

   (b) For $k = 1, \ldots, n$ show that
$$\text{nullity } J^k = \begin{cases} k, & k = 1, \ldots, n - 1; \\ n, & \text{if } k \geq n \end{cases}$$

     and show that the null space of J^k is $\mathrm{Sp}\,(e_{n-k+1}, \ldots, e_n)$.

   (c) Conclude that $L_{J^k} \neq \mathbf{0}_{\mathbb{R}^n}$ for $k < n$ and $L_{J^k} = \mathbf{0}_{\mathbb{R}^n}$ for $k \geq n$.

4. Let J be a Jordan block and with eigenvalue λ of size n.

   (a) Show that J is an invertible matrix if and only if $\lambda \neq 0$.

   (b) For any $\mu \in \mathbb{R}$, show that the matrix $J - \mu I_n$ is a Jordan block and with eigenvalue $\lambda - \mu$.

5. (a) Let $\lambda \in \mathbb{R}$. Show that

$$\begin{pmatrix} \lambda & 0 & 0 & \cdots & 0 & 0 & 0 \\ 1 & \lambda & 0 & \cdots & 0 & 0 & 0 \\ 0 & 1 & \lambda & \cdots & 0 & 0 & 0 \\ \vdots & \vdots & \vdots & \ddots & \ddots & \vdots & \vdots \\ 0 & 0 & 0 & \cdots & \lambda & 0 & 0 \\ 0 & 0 & 0 & \cdots & 1 & \lambda & 0 \\ 0 & 0 & 0 & \cdots & 0 & 1 & \lambda \end{pmatrix} \sim \begin{pmatrix} \lambda & 1 & 0 & \cdots & 0 & 0 & 0 \\ 0 & \lambda & 1 & \cdots & 0 & 0 & 0 \\ 0 & 0 & \lambda & \cdots & 0 & 0 & 0 \\ \vdots & \vdots & \vdots & \ddots & \ddots & \vdots & \vdots \\ 0 & 0 & 0 & \cdots & \lambda & 1 & 0 \\ 0 & 0 & 0 & \cdots & 0 & \lambda & 1 \\ 0 & 0 & 0 & \cdots & 0 & 0 & \lambda \end{pmatrix}$$

     for any size n.

   (b) Show that every matrix is similar to its transpose.

6. Let $T \colon V \to V$ be a linear endomorphismand let W be a T-invariant subspace of V.

   (a) Show that the minimal polynomial of $T|_W$ is a factor of the minimal polynomial of T.

(b) Show that $T|_W$ is diagonalizable if T is diagonalizable.

7. Suppose A is a square matrix of size $n \geq 3$ such that $A^4 = I$.

   (a) Is A diagonalizable over $\mathbb{R}$?

   (b) Is A diagonalizable over $\mathbb{C}$?

8. Let A be a nilpotent matrix of size 5. Find all possible Jordan form of A.

9. Let A be an idempotent matrix of size 5. Find all possible Jordan form for A.

6.3 How to find Jordan canonical forms

In this section we discuss how to find the Jordan form of a square matrix whose characteristic polynomial splits.

The basis for a Jordan block.   Let's first start with the simple case. Suppose T is a linear endomorphism on V such that

$$[T]_\beta = \begin{pmatrix} \lambda & 0 & 0 & \cdots & 0 & 0 & 0 \\ 1 & \lambda & 0 & \cdots & 0 & 0 & 0 \\ 0 & 1 & \lambda & \cdots & 0 & 0 & 0 \\ \vdots & \vdots & \vdots & \ddots & \ddots & \vdots & \vdots \\ 0 & 0 & 0 & \cdots & \lambda & 0 & 0 \\ 0 & 0 & 0 & \cdots & 1 & \lambda & 0 \\ 0 & 0 & 0 & \cdots & 0 & 1 & \lambda \end{pmatrix}_{n \times n}$$

is a Jordan block for some ordered basis $\beta = \{v_1, \cdots, v_n\}$. We have that

$$[T - \lambda \mathbf{1}_V]_\beta = \begin{pmatrix} 0 & 0 & 0 & \cdots & 0 & 0 & 0 \\ 1 & 0 & 0 & \cdots & 0 & 0 & 0 \\ 0 & 1 & 0 & \cdots & 0 & 0 & 0 \\ \vdots & \vdots & \vdots & \ddots & \ddots & \vdots & \vdots \\ 0 & 0 & 0 & \cdots & 0 & 0 & 0 \\ 0 & 0 & 0 & \cdots & 1 & 0 & 0 \\ 0 & 0 & 0 & \cdots & 0 & 1 & 0 \end{pmatrix}_{n \times n}$$

This says that

$$(T - \lambda \mathbf{1}_V)(v_1) = v_2;$$
$$(T - \lambda \mathbf{1}_V)(v_2) = v_3;$$
$$(T - \lambda \mathbf{1}_V)(v_3) = v_4;$$
$$\vdots$$
$$(T - \lambda \mathbf{1}_V)(v_{n-1}) = v_n;$$
$$(T - \lambda \mathbf{1}_V)(v_n) = O.$$

Observe that v_n is an eigenvector. In fact, the geometric dimension of the eigenspace of λ is 1. Note that

$$(T - \lambda \mathbf{1}_V)^i(v_1) = v_{i+1} \qquad \text{for } i = 1, 2, \ldots, n-1.$$

To retrieve the basis β, it suffices to find v_1. In fact, in this book we would call v_1 a **root** for the Jordan block $[T]_\beta$. We can see that

$$v_1 \in \text{Null}\left((T - \lambda \mathbf{1}_V)^n\right) = V \setminus \text{Null}\left((T - \lambda \mathbf{1}_V)^{n-1}\right).$$

Since nullity $(T - \lambda \mathbf{1}_V)^{n-1} = n - 1$, there is not much choice for v_1. However, the choice of a generator for a Jordan block is by no means unique. Now, let's find an arbitrary vector in

$$\text{Null}\left((T - \lambda \mathbf{1}_V)^n\right) = V \setminus \text{Null}\left((T - \lambda \mathbf{1}_V)^{n-1}\right),$$

we will see that v is also a root for the given Jordan block. Remember that n is determined by the degree of the minimal polynomial of the Jordan block. Let

$$\mathscr{B} = \{v, (T - \lambda \mathbf{1}_V)(v), (T - \lambda \mathbf{1}_V)^2(v), \ldots, (T - \lambda \mathbf{1}_V)^{n-1}(v)\}.$$

We claim that $\mathscr{B}$ is a basis for V. Since there are n elements in $\mathscr{B}$, it suffices to check that $\mathscr{B}$ is linearly independent. If this is not true, there are scalars $a_0, a_1, a_2, \ldots, a_{n-1}$ which are not all zeros such that

$$a_0 v + a_1(T - \lambda \mathbf{1}_V)(v) + a_2(T - \lambda \mathbf{1}_V)^2(v) + \cdots + a_{n-1}(T - \lambda \mathbf{1}_V)^{n-1}(v) = O.$$

This implies that there is a nonzero polynomial $h(t)$ of degree $\leq n-1$ such that $g(T)(v) = 0$. This is a contradiction to the fact that the minimal polynomial of T is $(t - \lambda)^n$.

We have shown that $\mathscr{B}$ is a basis for V. Now let

$$\gamma = (w_0, \ w_1, \ w_2, \ldots, \ w_{n-1})$$

be the ordered basis with

$$\begin{cases} w_0 = v, \quad \text{and} \\ w_i = (T - \lambda 1_V)^i(v) = (T - \lambda 1_V)(w_{i-1}), \quad i = 1, 2, \ldots, n-1. \end{cases}$$

Note that $(T - \lambda 1_V)(w_{n-1}) = O$. This implies that

$$T(w_0) = \lambda w_0 + (T - \lambda 1_V)(w_0) = \lambda w_0 + w_1$$
$$T(w_1) = \lambda w_1 + (T - \lambda 1_V)(w_1) = \lambda w_1 + w_2$$
$$T(w_2) = \lambda w_2 + (T - \lambda 1_V)(w_2) = \lambda w_2 + w_3$$
$$\vdots$$
$$T(w_{n-2}) = \lambda w_{n-2} + (T - \lambda 1_V)(w_{n-2}) = \lambda w_{n-2} + w_{n-1}$$
$$T(w_{n-1}) = \lambda w_{n-1} + (T - \lambda 1_V)(w_{n-1}) = \lambda w_{n-2}$$

Thus $[T]_\gamma$ is also the size n Jordan block with eigenvalue λ.

We demonstrate the process of finding a root using the following example.

Example 6.3.1. Let

$$A = \begin{pmatrix} 0 & 0 & 0 & 1 \\ 0 & 0 & 1 & 0 \\ 1 & 0 & 0 & 0 \\ 0 & 0 & 0 & 0 \end{pmatrix}.$$

(a) Find the characteristic polynomial and the minimal polynomial of A.

(b) Determine the Jordan form J of A.

(c) Find a basis β such that $[L_A]_\beta = J$.

(d) Find P such that $A = PJP^{-1}$.

Solution. (a) The characteristic polynomial of A is

$$
\det \begin{pmatrix} -t & 0 & 0 & 1 \\ 0 & -t & 1 & 0 \\ 1 & 0 & -t & 0 \\ 0 & 0 & 0 & -t \end{pmatrix}
$$

$$
= -t \det \begin{pmatrix} -t & 1 & 0 \\ & -t & 0 \\ & & -t \end{pmatrix} - \det \begin{pmatrix} 0 & -t & 1 \\ 1 & 0 & -t \\ 0 & 0 & 0 \end{pmatrix}
$$

$$
= t^4
$$

by the cofactor expansion using the first row. The only eigenvalue of A is 0.

The minimal polynomial of A is a factor of t^4. We now make a test:

$$
A \neq 0,
$$
$$
A^2 = (e_{31} + e_{23} + e_{14})(e_{31} + e_{23} + e_{14}) = e_{34} + e_{21},
$$
$$
A^3 = (e_{34} + e_{21})(e_{31} + e_{23} + e_{14}) = e_{24},
$$
$$
A^4 = e_{24}(e_{31} + e_{23} + e_{14}) = \mathbf{0}.
$$

We conclude that the minimal polynomial of A is t^4 as well.

(b) From (a), the Jordan form of A is the Jordan block of size 4 with eigenvalue 0:

$$
J = \begin{pmatrix} 0 & & & \\ 1 & 0 & & \\ & 1 & 0 & \\ & & 1 & 0 \end{pmatrix}.
$$

(c) For this, we need to find a root for the Jordan block. The roots are vectors in $\text{Null}\left(A^4\right) \setminus \text{Null}\left(A^3\right) = \mathbb{R}^4 \setminus \text{Null}\left(A^3\right)$. To find $\text{Null}\left(A^3\right)$, we need to solve

$$
\begin{pmatrix} 0 & 0 & 0 & 0 \\ 0 & 0 & 0 & 1 \\ 0 & 0 & 0 & 0 \\ 0 & 0 & 0 & 0 \end{pmatrix} \begin{pmatrix} x \\ y \\ z \\ w \end{pmatrix} = \begin{pmatrix} 0 \\ 0 \\ 0 \\ 0 \end{pmatrix}.
$$

The solution space is

$$
\left\{ \begin{pmatrix} x \\ y \\ z \\ w \end{pmatrix} \in \mathbb{R}^4 : w = 0 \right\} = \mathrm{Sp}\,(e_1, e_2, e_3).
$$

We may choose $v = e_4$ (or any vector whose fourth coordinate is nonzero). Let $\beta = (e_4, Ae_4, A^2 e_4, A^3 e_4)$. Then

$$
[L_A]_\beta = J.
$$

(d) To find the base change matrix P, we compute

$$
Ae_4 = e_1;
$$
$$
A^2 e_4 = Ae_1 = e_3;
$$
$$
A^3 e_4 = Ae_3 = e_2.
$$

Thus

$$
P = \begin{pmatrix} 0 & 1 & 0 & 0 \\ 0 & 0 & 0 & 1 \\ 0 & 0 & 1 & 0 \\ 1 & 0 & 0 & 0 \end{pmatrix}
$$

is the base change matrix from β to the standard basis α of $\mathbb{R}^4$. We verify that

$$
A = [L_A]_\alpha^\alpha = [1_{\mathbb{R}^4}]_\beta^\alpha [L_A]_\beta^\beta [1_{\mathbb{R}^4}]_\alpha^\beta.
$$

Hence $A = PJP^{-1}$. $\diamond$

Example 6.3.2. Let

$$
A = \begin{pmatrix} 1 & 1 & 0 & 1 \\ 0 & 2 & 1 & 0 \\ 1 & 0 & 2 & 0 \\ -1 & 1 & -1 & 3 \end{pmatrix}.
$$

(a) Find the characteristic polynomial and the minimal polynomial of A.

(b) Determine the Jordan form J of A.

(c) Find a basis β such that $[L_A]_\beta = J$.

(d) Find P such that $A = PJP^{-1}$.

Solution. (a) The characteristic polynomial of A is

$$f(t) = \det \begin{pmatrix} 1-t & 1 & 0 & 1 \\ 0 & 2-t & 1 & 0 \\ 1 & 0 & 2-t & 0 \\ -1 & 1 & -1 & 3-t \end{pmatrix}.$$

Using cofactor expansion on the first row, we have that $f(t)$ equals

$$(1-t)\det \begin{pmatrix} 2-t & 1 & 0 \\ 0 & 2-t & 0 \\ 1 & -1 & 3-t \end{pmatrix} - \det \begin{pmatrix} 0 & 1 & 0 \\ 1 & 2-t & 0 \\ -1 & -1 & 3-t \end{pmatrix}$$

$$-\det \begin{pmatrix} 0 & 2-t & 1 \\ 1 & 0 & 2-t \\ -1 & 1 & -1 \end{pmatrix}$$

$$= (1-t)\det \begin{pmatrix} 2-t & 1 & 0 \\ 0 & 2-t & 0 \\ 1 & -1 & 3-t \end{pmatrix} - \det \begin{pmatrix} 0 & 1 & 0 \\ 1 & 2-t & 0 \\ -1 & -1 & 3-t \end{pmatrix}$$

$$-\det \begin{pmatrix} 0 & 2-t & 1 \\ 1 & 0 & 2-t \\ 0 & 1 & 1-t \end{pmatrix}$$

$$= (1-t)(2-t)^2(3-t) + (3-t) + t^2 - 3t + 1$$
$$= (1-t)(2-t)^2(3-t) + t^2 - 4t + 4$$
$$= (t-2)^2 \left[(t-3)(t-1) + 1 \right]$$
$$= (t-2)^4.$$

The minimal polynomial of A is a factor of $f(t)$. So we make the following test:

$$A - 2I = \begin{pmatrix} -1 & 1 & 0 & 1 \\ 0 & 0 & 1 & 0 \\ 1 & 0 & 0 & 0 \\ -1 & 1 & -1 & 1 \end{pmatrix} \neq \mathbf{0};$$

$$(A - 2I)^2 = \begin{pmatrix} 0 & 0 & 0 & 0 \\ 1 & 0 & 0 & 0 \\ -1 & 1 & 0 & 1 \\ -1 & 0 & 0 & 0 \end{pmatrix} \neq \mathbf{0};$$

$$(A - 2I)^3 = \begin{pmatrix} 0 & 0 & 0 & 0 \\ -1 & 1 & 0 & 1 \\ 0 & 0 & 0 & 0 \\ 1 & -1 & 0 & -1 \end{pmatrix} \neq \mathbf{0};$$

$$(A - 2I)^4 = \mathbf{0}.$$

Hence the minimal polynomial of A is $(t - 2)^4$.

(b) Since the characteristic polynomial and the minimal polynomial of A are both $(t - 2)^4$, we know that the Jordan form of A is

$$J = \begin{pmatrix} 2 & & & \\ 1 & 2 & & \\ & 1 & 2 & \\ & & 1 & 2 \end{pmatrix}.$$

(c) We first look for a root for the Jordan block J. For this we need to solve the system of linear equations

$$(A - 2I)^3 \begin{pmatrix} x \\ y \\ z \\ w \end{pmatrix} = \begin{pmatrix} 0 & 0 & 0 & 0 \\ -1 & 1 & 0 & 1 \\ 0 & 0 & 0 & 0 \\ 1 & -1 & 0 & -1 \end{pmatrix} \begin{pmatrix} x \\ y \\ z \\ w \end{pmatrix} = \begin{pmatrix} 0 \\ 0 \\ 0 \\ 0 \end{pmatrix}.$$

The solution space is

$$\text{Null}\left((A - 2I)^3\right) = \{(y + w, y, z, w) : y, z, w \in \mathbb{R}\}$$
$$= \{y(1, 1, 0, 0) + z(0, 0, 1, 0) + w(1, 0, 0, 1) : y, z, w \in \mathbb{R}\}$$
$$= \text{Sp}\left((1, 1, 0, 0), (0, 0, 1, 0), (1, 0, 0, 1)\right).$$

The root v is any vector not in $\text{Null}\left((A - 2I)^3\right)$. We may choose $v = (1, 0, 0, 0)$. We would like to comment again that this is definitely not the only choice for v. Now let

$$\beta = (e_1, (A - 2I)e_1, (A - 2I)^2 e_1, (A - 2I)^3 e_1)$$

$$= \left(\begin{pmatrix} 1 \\ 0 \\ 0 \\ 0 \end{pmatrix}, \begin{pmatrix} -1 \\ 0 \\ 1 \\ -1 \end{pmatrix}, \begin{pmatrix} 0 \\ 1 \\ -1 \\ -1 \end{pmatrix}, \begin{pmatrix} 0 \\ -1 \\ 0 \\ 1 \end{pmatrix} \right).$$

Then $[L_A]_\beta = J$.

(d) Let α be the standard basis of $\mathbb{R}^4$. Then

$$A = [L_A]_\alpha^\alpha = [1_{\mathbb{R}^4}]_\beta^\alpha [L_A]_\beta^\beta [1_{\mathbb{R}^4}]_\alpha^\beta.$$

Let P be the base change matrix from β to α. That is,

$$P = \begin{pmatrix} 1 & -1 & 0 & 0 \\ 0 & 0 & 1 & -1 \\ 0 & 1 & -1 & 0 \\ 0 & -1 & -1 & 1 \end{pmatrix}.$$

Then $A = PJP^{-1}$.   ◇

Jordan forms with a single eigenvalue. Next we study square matrices whose Jordan form might be composed of more than one Jordan blocks. For small-sized square matrices, to determine its Jordan form, sometimes it only needs the characteristic and minimal polynomials.

Example 6.3.3. The characteristic polynomial of

$$A = \begin{pmatrix} 2 & 2 & -1 \\ -1 & -1 & 1 \\ -1 & -2 & 2 \end{pmatrix}$$

is $-(t-1)^3$. Thus the only eigenvalue of A is 1. The minimal polynomial of A must be $(t-1)^k$ for some $1 \le k \le 3$. Clearly $A - I \ne \mathbf{0}$, while some computing reveals that $(A - I)^2 = \mathbf{0}$. This shows that the minimal polynomial of A is $(t-1)^2$. Thus that the size of the maximal Jordan block with eigenvalue 1 of A is 2. This forces

$$J = \left(\begin{array}{c|cc} 1 & 0 & 0 \\ \hline 0 & 1 & 0 \\ 0 & 1 & 1 \end{array} \right)$$

to be the Jordan form of A.

However, the mere knowledge of characteristic polynomial and minimal polynomial cannot always help us determine the Jordan form.

Example 6.3.4. When the characteristic polynomial and the minimal polynomial of the matrix A are $(t-1)^6$ and $(t-1)^2$ respectively, we cannot determine the Jordan form of A.

The only information we have about the Jordan form of A is that

- The size of A is 6.

- The only eigenvalue of A is 1.

- The maximal Jordan block of the Jordan form of A is of size 2.

There are three possibilities:

$$
\left(\begin{array}{cc|cc|cc}
1 & & & & & \\
1 & 1 & & & & \\
\hline
& & 1 & & & \\
& & 1 & 1 & & \\
\hline
& & & & 1 & \\
& & & & 1 & 1
\end{array}\right),
\quad
\left(\begin{array}{cc|cc|c|c}
1 & & & & & \\
1 & 1 & & & & \\
\hline
& & 1 & & & \\
& & 1 & 1 & & \\
\hline
& & & & 1 & \\
\hline
& & & & & 1
\end{array}\right)
$$

and
$$
\left(\begin{array}{cc|c|c|c|c}
1 & & & & & \\
1 & 1 & & & & \\
\hline
& & 1 & & & \\
\hline
& & & 1 & & \\
\hline
& & & & 1 & \\
\hline
& & & & & 1
\end{array}\right).
$$

To determine the Jordan form of A, we need more information.

How do we determine the Jordan forms in general? We shall demonstrate the method with specific examples.

Let T be a linear endomorphism on V whose Jordan form is

$$
\left(\begin{array}{cc|ccc}
2 & 0 & 0 & 0 & 0 \\
1 & 2 & 0 & 0 & 0 \\
\hline
0 & 0 & 2 & 0 & 0 \\
0 & 0 & 1 & 2 & 0 \\
0 & 0 & 0 & 1 & 2
\end{array}\right).
$$

This Jordan form is composed of 2 Jordan blocks corresponding to the same eigenvalue 2 which are of sizes 2 and 3 respectively. Suppose this Jordan form represents T with respect to

$$\beta = \{v_1, v_2, w_1, w_2, w_3\}.$$

Notice that

$$(6.6) \qquad [T - 21_V]_\beta = \left(\begin{array}{cc|ccc} 0 & 0 & 0 & 0 & 0 \\ 1 & 0 & 0 & 0 & 0 \\ \hline 0 & 0 & 0 & 0 & 0 \\ 0 & 0 & 1 & 0 & 0 \\ 0 & 0 & 0 & 1 & 0 \end{array}\right).$$

Hence

$$\begin{cases} \text{nullity } T - 21_V = 2; & \text{Null} (T - 21_V) = \text{Sp} (v_2, w_3); \\ \text{nullity } (T - 21_V)^2 = 4; & \text{Null} ((T - 21_V)^2) = \text{Sp} (v_1, v_2, w_2, w_3); \\ \text{nullity } (T - 21_V)^3 = 5; & \text{Null} ((T - 21_V)^3) = \text{Sp} (v_1, v_2, w_1, w_2, w_3). \end{cases}$$

We can use this to recover the sizes of the Jordan blocks corresponding to 2 in the Jordan form using the nullities of $(T - 21_V)^k$. We can use the **dot diagram** to help us do this. We first put down at the first row two dots, standing for the nullity of $T - 21_V$ (or rather the two vectors v_1 and v_2). At the second row we put down two more dots for the two more vectors v_2 and v_4 in the null space $(T - 21_V)^2$. At the third row we put down one dot for v_5, the new vector in the null space of $(T - 21_V)^3$. The dot diagram for the Jordan form in question is

$$\begin{array}{ll} \bullet \quad \bullet & 2 = \text{nullity } T - 21_V \\ \bullet \quad \bullet & 2 = \text{nullity } (T - 21_V)^2 - \text{nullity } T - 21_V \\ \bullet & 1 = \text{nullity } (T - 21_V)^3 - \text{nullity } (T - 21_V)^2 \end{array}$$

The columns of the dot diagram tells us that the Jordan form consists of two Jordan blocks, one of which is of size 3 and one of which is of size 2. The nullity here does not depend on the matrix used, so even if we don't have the Jordan form of a linear endomorphism we can still recover the Jordan form.

Example 6.3.5. Let T be a linear endomorphism on $\mathbb{R}^{15}$ such that

$$\text{nullity } T - 21_V = 5; \qquad \text{nullity } (T - 21_V)^2 = 8;$$
$$\text{nullity } (T - 21_V)^3 = 11; \quad \text{nullity } (T - 21_V)^4 = 13;$$
$$\text{nullity } (T - 21_V)^5 = 14; \quad \text{nullity } (T - 21_V)^6 = 15.$$

We can see that the only eigenvalue of T is 2. The dot diagram for T is

$$
\begin{array}{ccccc}
\bullet & \bullet & \bullet & \bullet & \bullet \\
\bullet & \bullet & \bullet & & \\
\bullet & \bullet & \bullet & & \\
\bullet & \bullet & & & \\
\bullet & & & & \\
\bullet & & & &
\end{array}
$$

In the Jordan form for T, there are 6 Jordan block with eigenvalue 2, one of size 5, two of size 3, one of size 2 and two of size 1. The Jordan form of T is

$$
\left(
\begin{array}{ccccc|ccc|ccc|cc|c|c}
2 & & & & & & & & & & & & & & \\
1 & 2 & & & & & & & & & & & & & \\
 & 1 & 2 & & & & & & & & & & & & \\
 & & 1 & 2 & & & & & & & & & & & \\
 & & & 1 & 2 & & & & & & & & & & \\
\hline
 & & & & & 2 & & & & & & & & & \\
 & & & & & 1 & 2 & & & & & & & & \\
 & & & & & & 1 & 2 & & & & & & & \\
\hline
 & & & & & & & & 2 & & & & & & \\
 & & & & & & & & 1 & 2 & & & & & \\
 & & & & & & & & & 1 & 2 & & & & \\
\hline
 & & & & & & & & & & & 1 & 2 & & \\
 & & & & & & & & & & & & 1 & 2 & \\
\hline
 & & & & & & & & & & & & & & 2 \\
\hline
 & & & & & & & & & & & & & & & 2
\end{array}
\right).
$$

Once we have determined the Jordan form of a linear endomorphism, we can go ahead to find the basis with respect to which the matrix is a Jordan form. Let's demonstrate the method using (6.6) again. There are two Jordan blocks in (6.6). We need to find the roots v_1 and w_1 for the Jordan blocks. Observe that

$$v_1 \in (T - 2\mathbf{1}_V)^3 \setminus (T - 2\mathbf{1}_V)^2.$$

Note that there are more than one choice for v_1, but any will do. Once we have chosen the v_1, we can use it to find v_2 and v_3. Observe also that

$$w_1 \in (T - 2\mathbf{1}_V)^2 \setminus (T - 2\mathbf{1}_V).$$

Note that we also have $v_2 \in (T - 2\mathbf{1}_V)^2 \setminus (T - 2\mathbf{1}_V)$. Remember to choose w_1 such that w_1 and v_2 are linearly independent. We can then use w_1 to find w_2.

Example 6.3.6. Let A and J be as in Example 6.3.3. Find P such that $J = P^{-1}AP$.

Solution. To do this, we need to find $\text{Null}\,(A - I)$. We need to solve

$$(A - I) \begin{pmatrix} x \\ y \\ z \end{pmatrix} = \begin{pmatrix} 1 & 2 & -1 \\ -1 & -2 & 1 \\ -1 & -2 & 1 \end{pmatrix} \begin{pmatrix} x \\ y \\ z \end{pmatrix} = \begin{pmatrix} 0 \\ 0 \\ 0 \end{pmatrix}.$$

We perform elementary row operations on $A - I$:

$$A - I \rightsquigarrow \begin{pmatrix} 1 & 2 & -1 \\ 0 & 0 & 0 \\ 0 & 0 & 0 \end{pmatrix}.$$

Thus $x + 2y - z = 0$. The solution space is

$$\begin{aligned} \text{Null}\,(A - I) &= \{(-2y + z, y, z) : y, z \in \mathbb{R}\} \\ &= \{y(-2, 1, 0) + z(1, 0, 1) \in \mathbb{R}^3 : y, z \in \mathbb{R}\} \\ &= \text{Sp}\,((-2, 1, 0), (1, 0, 1)). \end{aligned}$$

To find a root for the Jordan block of size 2, we need a vector not in $\text{Null}\,(A - I)$. We may choose $v = (1, 0, 0)$ to obtain a partial basis for the Jordan form

$$\left\{ \begin{pmatrix} 1 \\ 0 \\ 0 \end{pmatrix}, (A - I) \begin{pmatrix} 1 \\ 0 \\ 0 \end{pmatrix} \right\} = \left\{ \begin{pmatrix} 1 \\ 0 \\ 0 \end{pmatrix}, \begin{pmatrix} 1 \\ -1 \\ -1 \end{pmatrix} \right\}.$$

To find a root for the size 1 Jordan block, we need to look in $\text{Null}(A - I)$, but not in $\text{Sp}((1, -1, -1))$. We may choose $w = (1, 0, 1)$ for example. Now

$$\beta = \left(\begin{pmatrix} 1 \\ 0 \\ 0 \end{pmatrix}, \begin{pmatrix} 1 \\ -1 \\ -1 \end{pmatrix}, \begin{pmatrix} 1 \\ 0 \\ 1 \end{pmatrix} \right)$$

is an ordered basis such that

$$[L_A]_\beta^\beta = \left(\begin{array}{cc|c} 1 & & \\ 1 & 1 & \\ \hline & & 1 \end{array} \right).$$

Let α be the standard basis for $\mathbb{R}^3$. Then

$$P = [1_{\mathbb{R}^3}]_\beta^\alpha = \begin{pmatrix} 1 & 1 & 1 \\ 0 & -1 & 0 \\ 0 & -1 & 1 \end{pmatrix}$$

is the base change matrix from β to α and $J = P^{-1}AP$. $\diamond$

Jordan forms with mixed eigenvalues.   Let T be a linear endomorphism on V whose Jordan form is

$$\left(\begin{array}{cc|ccc} 1 & 0 & 0 & 0 & 0 \\ 1 & 1 & 0 & 0 & 0 \\ \hline 0 & 0 & 2 & 0 & 0 \\ 0 & 0 & 1 & 2 & 0 \\ 0 & 0 & 0 & 1 & 2 \end{array} \right).$$

This Jordan form is composed of 2 Jordan blocks corresponding to 1 and 2 respectively. Suppose this Jordan form represents T with respect to $\beta = \{v_1, v_2, w_1, w_2, w_3\}$. Let $W_1 = \text{Sp}(v_1, v_2)$ and $W_2 = \text{Sp}(w_1, w_2, w_3)$. Notice that W_1 and W_2 are T-invariant. Moreover,

$$[T - 1_V]_\beta = \left(\begin{array}{cc|ccc} 0 & 0 & 0 & 0 & 0 \\ 1 & 0 & 0 & 0 & 0 \\ \hline 0 & 0 & 1 & 0 & 0 \\ 0 & 0 & 1 & 1 & 0 \\ 0 & 0 & 0 & 1 & 1 \end{array} \right)$$

and

$$[T - 21_V]_\beta = \left(\begin{array}{ccc|ccc} -1 & 0 & 0 & 0 & 0 \\ 1 & -1 & 0 & 0 & 0 \\ \hline 0 & 0 & 0 & 0 & 0 \\ 0 & 0 & 1 & 0 & 0 \\ 0 & 0 & 0 & 1 & 0 \end{array}\right).$$

Notice that

$$W_1 \cap \text{Null}\left((T - 21_V)^k\right) = \{O\}$$
$$W_2 \cap \text{Null}\left((T - 1_V)^k\right) = \{O\}$$

for all k, while

$$W_1 = \text{Null}\left((T - 1_V)^2\right)$$
$$W_2 = \text{Null}\left((T - 21_V)^3\right).$$

Hence by constructing the dot diagrams for the eigenvalues 1 and 2 we may recover the sizes of the Jordan blocks corresponding to 1 and 2 in the Jordan form.

Example 6.3.7. Let T be a linear endomorphism on a 13-dimensional real vector space such that

$$\begin{aligned} &\text{nullity } T + 1_V = 3; &&\text{nullity } (T + 1_V)^2 = 5; \\ &\text{nullity } (T + 1_V)^3 = 7; \\ &\text{nullity } T - 31_V = 2; &&\text{nullity } (T - 31_V)^2 = 4; \\ &\text{nullity } (T - 31_V)^3 = 5; &&\text{nullity } (T - 31_V)^4 = 6. \end{aligned}$$

(a) Find the Jordan form for T.

(b) Find the characteristic polynomial and the minimal polynomial of T.

Solution. First note that

$$\dim(T + 1_V)^3 + \dim(T - 31_V) = 13,$$

we have already exhausted the eigenvalues of T. They are -1 and 3.

(a) The dot diagram for the eigenvalue -1 is

Hence in the Jordan form of T there are three Jordan blocks with eigenvalue -1. Two are of size 3 and one is of size 1.

The dot diagram for the eigenvalue 3 is

$$\begin{matrix} \bullet & \bullet \\ \bullet & \bullet \\ \bullet & \\ \bullet & \end{matrix}$$

There are two Jordan blocks with eigenvalue 3. One is of size 4 and one is of size 2. We now know the Jordan form of T is

$$\begin{pmatrix}
-1 & & & & & & & & \\
1 & -1 & & & & & & & \\
& 1 & -1 & & & & & & \\
& & & -1 & & & & & \\
& & & 1 & -1 & & & & \\
& & & & 1 & -1 & & & \\
& & & & & & -1 & & \\
& & & & & & & 3 & \\
& & & & & & & 1 & 3 \\
& & & & & & & & 1 & 3 \\
& & & & & & & & & 1 & 3 \\
& & & & & & & & & & & 3 \\
& & & & & & & & & & & 1 & 3
\end{pmatrix}.$$

(b) The characteristic polynomial of T is $-(t+1)^7(t-3)^6$. The minimal polynomial of T is $(t+1)^3(t-3)^4$. ◇

Example 6.3.8. Let

$$A = \begin{pmatrix}
1 & 0 & 0 & 0 & 0 \\
0 & 0 & 2 & 0 & 0 \\
0 & 0 & 0 & 0 & 1 \\
0 & 0 & 0 & 0 & 0 \\
0 & 0 & 0 & 0 & 0
\end{pmatrix}.$$

Find the Jordan form J of A. Find a basis such that the matrix for L_A with respect to this basis is the Jordan form you found. Find an invertible matrix P such that $P^{-1}AP = J$.

Solution. First we need to compute the characteristic polynomial of A. We find

$$f(t) = \det \begin{pmatrix} 1-t & 0 & 0 & 0 & 0 \\ 0 & -t & 2 & 0 & 0 \\ 0 & 0 & -t & 0 & 1 \\ 0 & 0 & 0 & -t & 0 \\ 0 & 0 & 0 & 0 & -t \end{pmatrix}$$

$$= (1-t)\det \begin{pmatrix} -t & 2 & 0 & 0 \\ 0 & -t & 0 & 1 \\ 0 & 0 & -t & 0 \\ 0 & 0 & 0 & -t \end{pmatrix}$$

$$= (1-t)(-t)^4 = t^4(t-1).$$

Thus the eigenvalues of T are 0 and 1.

The Jordan block of 1 is easy. There is only one of size 1. To find its root, we look in

$$\text{Null}(A-I) = \text{Null}\begin{pmatrix} 0 & 0 & 0 & 0 & 0 \\ 0 & -1 & 2 & 0 & 0 \\ 0 & 0 & -1 & 0 & 1 \\ 0 & 0 & 0 & -1 & 0 \\ 0 & 0 & 0 & 0 & -1 \end{pmatrix} = \text{Sp}(e_1).$$

We choose $v = e_1$.

Next, we look for the Jordan blocks with eigenvalue 0. First, we compute the null spaces:

$$\text{Null}(A) = \text{Null}\begin{pmatrix} 1 & 0 & 0 & 0 & 0 \\ 0 & 0 & 2 & 0 & 0 \\ 0 & 0 & 0 & 0 & 1 \\ 0 & 0 & 0 & 0 & 0 \\ 0 & 0 & 0 & 0 & 0 \end{pmatrix} = \text{Sp}(e_2, e_4);$$

$$\text{Null}(A^2) = \text{Null}\begin{pmatrix} 1 & 0 & 0 & 0 & 0 \\ 0 & 0 & 0 & 0 & 1 \\ 0 & 0 & 0 & 0 & 0 \\ 0 & 0 & 0 & 0 & 0 \\ 0 & 0 & 0 & 0 & 0 \end{pmatrix} = \text{Sp}(e_2, e_3, e_4);$$

$$\text{Null}\left(A^3\right) = \text{Null}\left(\begin{pmatrix} 1 & 0 & 0 & 0 & 0 \\ 0 & 0 & 0 & 0 & 0 \\ 0 & 0 & 0 & 0 & 0 \\ 0 & 0 & 0 & 0 & 0 \\ 0 & 0 & 0 & 0 & 0 \end{pmatrix}\right) = \text{Sp}\left(e_2, e_3, e_4, e_5\right).$$

Since the algebraic multiplicity of the eigenvalue 0 is 4, we have reached the limit and there is no need to continue. We have

$$\text{nullity } A = 2,$$
$$\text{nullity } A^2 = 3,$$
$$\text{nullity } A^3 = 4.$$

The dot diagram for the eigenvalue 0 is

$$\begin{array}{cc} \bullet & \bullet \\ \bullet & \\ \bullet & \end{array}$$

For the eigenvalue 0, there are two Jordan blocks, one of size 3 and one of size 1. We conclude the Jordan form of A is

$$J = \begin{pmatrix} 1 & & & & \\ & 0 & & & \\ & 1 & 0 & & \\ & & 1 & 0 & \\ & & & & 0 \end{pmatrix}.$$

To find a basis such that the matrix for L_A is J, first we look in $\text{Null}\,(A - I)$. We choose $u = e_1$. This is a root for the Jordan block with eigenvalue 1. Next, choose a vector in $\text{Null}\left(A^3\right) \setminus \text{Null}\left(A^2\right)$. We may choose $v = e_5$. This is a root for the size 3 Jordan block with eigenvalue 0. Note that

$$v = e_5, \quad Ae_5 = e_3, \quad A^2 e_5 = Ae_3 = 2e_2.$$

In particular, $A^2 e_5 \in \text{Null}\,(A)$. To find a root for the remaining size 1 Jordan block with eigenvalue 0, we need to look in $\text{Null}\,(A) \setminus \text{Sp}\,(2e_2)$. We may choose $w = e_4$. Thus we have found an ordered basis

$$\beta = (e_1, e_5, e_3, 2e_2, e_4)$$

such that $[L_A]_\beta = J$.

Let α be the standard basis of $\mathbb{R}^5$. Since

$$J = [L_A]_\beta = [1_{\mathbb{R}^5}]_\alpha^\beta [L_A]_\alpha^\alpha [1_{\mathbb{R}^5}]_\beta^\alpha,$$

we may choose P to be

$$[1_{\mathbb{R}^5}]_\beta^\alpha = \begin{pmatrix} 1 & 0 & 0 & 0 & 0 \\ 0 & 0 & 0 & 2 & 0 \\ 0 & 0 & 1 & 0 & 0 \\ 0 & 0 & 0 & 0 & 1 \\ 0 & 1 & 0 & 0 & 0 \end{pmatrix},$$

the base change matrix from β to the standard basis of $\mathbb{R}^5$.   ◊

Example 6.3.9. Let

$$A = \begin{pmatrix} 2 & -1 & 0 & 1 \\ 0 & 3 & -1 & 0 \\ 0 & 1 & 1 & 0 \\ 0 & -1 & 0 & 3 \end{pmatrix}.$$

Find an invertible matrix P such that $P^{-1}AP$ is the Jordan form of A.

Solution. We first find the characteristic polynomial of A,

$$f(t) = \det \begin{pmatrix} 2-t & -1 & 0 & 1 \\ 0 & 3-t & -1 & 0 \\ 0 & 1 & 1-t & 0 \\ 0 & -1 & 0 & 3-t \end{pmatrix}$$

$$= (2-t) \det \begin{pmatrix} 3-t & -1 & 0 \\ 1 & 1-t & 0 \\ -1 & 0 & 3-t \end{pmatrix}$$

$$= (2-t)(3-t) \det \begin{pmatrix} 3-t & -1 \\ 1 & 1-t \end{pmatrix}$$

$$= (2-t)(3-t)\left[(3-t)(1-t)+1\right]$$

$$= (t-3)(t-2)^3.$$

The eigenvalues of A are 3 and 2. There is one size 1 Jordan block with eigenvalue 3. We find

$$
\text{Null}\,(A - 3I) = \text{Null}\left(\begin{pmatrix} -1 & -1 & 0 & 1 \\ 0 & 0 & -1 & 0 \\ 0 & 1 & -2 & 0 \\ 0 & -1 & 0 & 0 \end{pmatrix}\right) = \text{Sp}\left(\begin{pmatrix} 1 \\ 0 \\ 0 \\ 1 \end{pmatrix}\right).
$$

Let $u = (1,0,0,1)$. This is a root for the Jordan block with eigenvalue 3. Next, we find the Jordan blocks with eigenvalue 2. We find

$$
\text{Null}\,(A - 2I) = \text{Null}\left(\begin{pmatrix} 0 & -1 & 0 & 1 \\ 0 & 1 & -1 & 0 \\ 0 & 1 & -1 & 0 \\ 0 & -1 & 0 & 1 \end{pmatrix}\right)
$$

$$
= \text{Sp}\left(\begin{pmatrix} 1 \\ 0 \\ 0 \\ 0 \end{pmatrix}, \begin{pmatrix} 0 \\ 1 \\ 1 \\ 1 \end{pmatrix}\right);
$$

$$
\text{Null}\,((A - 2I)^2) = \text{Null}\left(\begin{pmatrix} 0 & -2 & 1 & 1 \\ 0 & 0 & 0 & 0 \\ 0 & 0 & 0 & 0 \\ 0 & -2 & 1 & 1 \end{pmatrix}\right)
$$

$$
= \text{Sp}\left(\begin{pmatrix} 1 \\ 0 \\ 0 \\ 0 \end{pmatrix}, \begin{pmatrix} 0 \\ 1 \\ 2 \\ 0 \end{pmatrix}, \begin{pmatrix} 0 \\ 0 \\ -1 \\ 1 \end{pmatrix}\right).
$$

The dot diagram of the eigenvalue 2 is

$$
\begin{matrix} \bullet & \bullet \\ \bullet & \end{matrix}
$$

There are two Jordan blocks with eigenvalue 2. On is of size 2, and the other one is of size 1. The roots of the size 2 block lie in $\text{Null}\,((A - 2I)^2) \setminus \text{Null}\,(A - 2I)$. We may choose $v = (0,0,-1,1)$. This is a root for the size 2

Jordan block with eigenvalue 2. The basis for this block is

$$\{v, (A - 2I)v\} = \left\{ \begin{pmatrix} 0 \\ 0 \\ -1 \\ 1 \end{pmatrix}, \begin{pmatrix} 1 \\ 1 \\ 1 \\ 1 \end{pmatrix} \right\}.$$

Finally, to find a root for the size 1 Jordan block with eigenvalue 2, we need to look in

$$\text{Null} \, (A - 2I) \setminus \text{Sp} \, ((1, 1, 1, 1)).$$

We may choose $w = e_1$.

To conclude, we may choose

$$\beta = \left(\begin{pmatrix} 1 \\ 0 \\ 0 \\ 1 \end{pmatrix}, \begin{pmatrix} 0 \\ 0 \\ -1 \\ 1 \end{pmatrix}, \begin{pmatrix} 1 \\ 1 \\ 1 \\ 1 \end{pmatrix}, \begin{pmatrix} 1 \\ 0 \\ 0 \\ 0 \end{pmatrix} \right).$$

Then

$$[L_A]_\beta = \begin{pmatrix} 3 & & & \\ \hline & 2 & & \\ & 1 & 2 & \\ \hline & & & 2 \end{pmatrix}$$

is the Jordan form of A.

Let α be the standard basis for $\mathbb{R}^4$. Then

$$[L_A]_\beta = [\mathbf{1}_{\mathbb{R}^4}]_\alpha^\beta [L_A]_\alpha^\alpha [\mathbf{1}_{\mathbb{R}^4}]_\beta^\alpha.$$

We may choose P to be the base change matrix from β to α, i.e.,

$$P = \begin{pmatrix} 1 & 0 & 1 & 1 \\ 0 & 0 & 1 & 0 \\ 0 & -1 & 1 & 0 \\ 1 & 1 & 1 & 0 \end{pmatrix}.$$

We have $[L_A]_\beta = P^{-1}AP$.   ◇

Example 6.3.10. Determine whether

$$A = \begin{pmatrix} 0 & 1 & 0 \\ -1 & 2 & 0 \\ -1 & 1 & 1 \end{pmatrix} \qquad \text{and} \qquad B = \begin{pmatrix} 3 & 1 & 2 \\ -4 & -1 & -4 \\ 0 & 0 & 1 \end{pmatrix}$$

are similar.

Solution. First, we compare the characteristic polynomials. The characteristic polynomial of A is

$$\det \begin{pmatrix} -t & 1 & 0 \\ -1 & 2-t & 0 \\ -1 & 1 & 1-t \end{pmatrix} = (1-t)\det \begin{pmatrix} -t & 1 \\ -1 & 2-t \end{pmatrix} = -(t-1)^3.$$

The characteristic polynomial of B is

$$\det \begin{pmatrix} 3-t & 1 & 2 \\ -4 & 1-t & -4 \\ 0 & 0 & 1-t \end{pmatrix} = (1-t)\det \begin{pmatrix} 3-t & 1 \\ -4 & -1-t \end{pmatrix} = -(t-1)^3.$$

The two matrices have the same characteristic polynomial. We continue to check their minimal polynomials. We check that

$$A - I = \begin{pmatrix} -1 & 1 & 0 \\ -1 & 1 & 0 \\ -1 & 1 & 0 \end{pmatrix} \neq \mathbf{0}; \qquad (A-I)^2 = \mathbf{0};$$

$$B - I = \begin{pmatrix} 2 & 1 & 2 \\ -4 & -2 & -4 \\ 0 & 0 & 0 \end{pmatrix} \neq \mathbf{0}; \qquad (B-I)^2 = \mathbf{0}.$$

Both matrices have the same minimal polynomials $(t-1)^2$. We next compare the Jordan forms.

The maximal Jordan block in both matrices is of size 2. This forces the Jordan form of both matrices to be

$$\left(\begin{array}{cc|c} 1 & & \\ 1 & 1 & \\ \hline & & 1 \end{array} \right).$$

Hence we conclude that $A \sim B$. $\qquad\qquad\diamond$

Exercises 6.3

1. Find the Jordan form of

$$\begin{pmatrix} 1 & 0 & 1 \\ 0 & 0 & 0 \\ 0 & 0 & -1 \end{pmatrix}.$$

2. Let $T\colon \mathbb{R}^3 \to \mathbb{R}^3$ be the linear endomorphism such that

$$\begin{aligned} T(u_1) &= -u_1 - 2u_2 + 6u_3; \\ T(u_2) &= -u_1 \qquad\;\; + 3u_3; \\ T(u_3) &= -u_1 - \;\; u_2 + 4u_3. \end{aligned}$$

Find an ordered basis β for $\mathbb{R}^3$ such that $[T]_\beta$ is a Jordan form.

3. Let

$$A = \begin{pmatrix} 1 & 0 & 0 & 0 \\ 0 & 1 & 0 & 0 \\ -2 & -2 & 0 & 1 \\ -2 & 0 & -1 & 2 \end{pmatrix}.$$

Does A have a Jordan form over $\mathbb{R}$? If yes, find P such that $P^{-1}AP$ is its Jordan form.

4. Let $\omega = \cos\dfrac{2\pi}{3} + i\sin\dfrac{2\pi}{3}$. Show that

$$\begin{pmatrix} 0 & 1 & 0 \\ 0 & 0 & 1 \\ 1 & 0 & 0 \end{pmatrix} \sim \begin{pmatrix} 1 & 0 & 0 \\ 0 & \omega & 0 \\ 0 & 0 & \omega^2 \end{pmatrix}$$

over $\mathbb{C}$.

5. Show that

$$\begin{pmatrix} 0 & 1 & \alpha \\ 0 & 0 & 1 \\ 0 & 0 & 0 \end{pmatrix} \sim \begin{pmatrix} 0 & 1 & 0 \\ 0 & 0 & 1 \\ 0 & 0 & 0 \end{pmatrix}$$

for all $\alpha \in \mathbb{R}$.

6. Find the Jordan form of

$$\begin{pmatrix} 1 & 0 & 0 & 0 & 1 & 0 \\ 0 & 1 & 1 & 1 & 0 & -1 \\ 0 & 0 & 1 & 0 & 0 & 0 \\ 0 & 0 & 0 & 1 & 0 & 0 \\ 0 & 0 & 1 & 0 & 1 & -1 \\ 0 & 0 & 0 & 0 & 0 & 1 \end{pmatrix}$$

over $\mathbb{C}$.

7. Find the Jordan form of

$$\begin{pmatrix} 1 & 1 & \cdots & 1 \\ 1 & 1 & \cdots & 1 \\ \vdots & \vdots & \ddots & \vdots \\ 1 & 1 & \cdots & 1 \end{pmatrix}_{n \times n}.$$

Review Exercises for Chapter 6

1. Let J be the size r Jordan block with λ as the eigenvalue. Find J^n and e^J.

2. We say a linear endomorphism $T\colon V \to V$ is a **projection** if there is a basis $\{v_1, \ldots, v_n\}$ for V and for some r with $0 \le r \le n$ such that

$$T(v_i) = \begin{cases} v_i, & \text{for } 1 \le i \le r, \\ O, & \text{otherwise.} \end{cases}$$

Show that T is a projection if and only if its associated matrix is idempotent. (Hint: Consider the Jordan form of T.)

3. Let A and $B \in M_2(\mathbb{R})$. If $AB = -BA$, show that $(AB)^2 = rI_2$ where r is a nonnegative real number. (Hint: Use Exercises 3 and 4 Review Exercises for Chapter 5 and compare the Jordan forms.)

4. Is it possible to find $A, B \in M_2(\mathbb{R})$ such that $AB - BA = I_2$? How about if we assume the entries are in $\mathbb{Z}_2$?

5. Let A be a square matrix of size n and let $v \in F^n$. show that there is a nonzero polynomial $f(t)$ with coefficients in F such that $f(A)v = 0$.

6. Let A and B be square matrices of size n such that $AB = BA$.

   (a) Show that A and B share at least one common eigenvector. (Hint: Use Exercise 5.)

   (b) Show that there is an invertible matrix P such that both $P^{-1}AP$ and $P^{-1}BP$ are upper triangular.

7. Let $\mathbb{Z}_3 = \{0, 1, 2\}$ with

+	0	1	2
0	0	1	2
1	1	2	0
2	2	0	1

and

·	0	1	2
0	0	0	0
1	0	1	2
2	0	2	1

Let A be a size 3 matrix with entries in $\mathbb{Z}_3$ such that $A^3 = I$. Can you find the Jordan form of A?

CHAPTER 7

Inner Product Spaces

In this chapter we will study inner product spaces. A structure with an inner product admits the notion of **distance** and **angle**, which allows applications to geometry, physics and other disciplines. Our goal in this chapter will be to prove the Spectral Theorem using the additional inner product structure on vector spaces.

Students are familiar with the concept of inner products in $\mathbb{R}^2$ or $\mathbb{R}^3$. We will define and generalize the notion of inner products on all real vector spaces and in all complex vector spaces.

Throughout the whole chapter all vector spaces are over real or over complex numbers.

7.1   Inner product spaces

Real inner product spaces.   We first explain what an inner product is in a real vector space.

Definition 7.1.1. Let V be a real vector space. An **inner product** on V is a function that assigns to every pair of vectors v and w in V a real number

$$V \times V \quad \longrightarrow \quad \mathbb{R}$$
$$(v, \, w) \quad \longmapsto \quad \langle v, \, w \rangle$$

such that the following conditions hold for all u, v and w in V and all a in $\mathbb{R}$:

(i) $\langle u + v, \, w \rangle = \langle u, \, w \rangle + \langle v, \, w \rangle$;

(ii) $\langle av, \, w \rangle = a \langle v, \, w \rangle$;

(iii) $\langle v, \, w \rangle = \langle w, \, v \rangle$;

(iv) $\langle v, \, v \rangle > 0$ if $v \neq O$.

A real vector space endowed with a specific inner product is called a **real inner product space** or simply an **inner product space**.

This definition seems to be different from the usual notion of what a freshman would think of as an inner product in $\mathbb{R}^2$ or in $\mathbb{R}^3$.

Proposition 7.1.2. *Define*

$$\langle v, \, w \rangle = a_1 b_1 + \cdots + a_n b_n,$$

*for $v = (a_1, \ldots, a_n)$ and $w = (b_1, \ldots, b_n)$ in $\mathbb{R}^n$. This is an inner product on $\mathbb{R}^n$. This inner product is called the **standard inner product** on $\mathbb{R}^n$.*

Proof. We verify that this product satisfies all the conditions in Definition 7.1.1.

Let $u = (a_1, \ldots, a_n)$, $v = (b_1, \ldots, b_n)$, $w = (c_1, \ldots, c_n) \in \mathbb{R}^n$, and $a \in \mathbb{R}$. To check (i), we see that

$$\langle u + v, \, w \rangle = \langle (a_1 + b_1, \ldots, a_n + b_n), \, (c_1, \ldots, c_n) \rangle$$

$$= (a_1 + b_1)c_1 + \cdots + (a_n + b_n)c_n$$
$$= a_1c_1 + \cdots + a_nc_n + b_1c_1 + \cdots + b_nc_n$$
$$= \langle u, \ w \rangle + \langle v, \ w \rangle.$$

To check (ii), we have

$$\langle av, \ w \rangle = \langle (ab_1, \ldots, ab_n), \ (c_1, \ldots, c_n) \rangle$$
$$= ab_1c_1 + \cdots + ab_nc_n$$
$$= a(b_1c_1 + \cdots + b_nc_n)$$
$$= a \langle v, \ w \rangle.$$

To check (iii), we have

$$\langle v, \ w \rangle = b_1c_1 + \cdots + b_nc_n$$
$$= c_1b_1 + \cdots + c_nb_n$$
$$= \langle w, \ v \rangle.$$

To check (iv),

$$\langle v, \ v \rangle = a_1^2 + a_2^2 + \cdots + a_n^2 > 0$$

unless $a_1 = \cdots = a_n = 0$. Thus $\langle \cdot, \ \cdot \rangle$ defined here is a linear product. $\square$

The standard inner product is just one of many inner products that one can define in a real vector space. We will look at more examples later.

The following is a list of properties of a real inner product.

Proposition 7.1.3. *Let V be a real inner product space. Then the following statements are true for all u, v and $w \in V$ and all $a \in \mathbb{R}$:*

(a) $\langle u, \ v + w \rangle = \langle u, \ v \rangle + \langle u, \ w \rangle$;

(b) $\langle v, \ aw \rangle = a \langle v, \ w \rangle$;

(c) $\langle v, \ O \rangle = \langle O, \ v \rangle = 0$;

(d) $\langle v, \ v \rangle = 0$ *if and only if* $v = O$;

(e) *If* $\langle u, \ v \rangle = \langle u, \ w \rangle$ *for all* $u \in V$, *then* $v = w$.

Proof. (a) We have

$$
\begin{aligned}
\langle u,\ v+w \rangle &= \langle v+w,\ u \rangle && \text{by (iii)},\\
&= \langle v,\ u \rangle + \langle w,\ u \rangle && \text{by (i)},\\
&= \langle u,\ v \rangle + \langle u,\ w \rangle && \text{by (iii)}.
\end{aligned}
$$

(b) It is also clear that

$$
\begin{aligned}
\langle v,\ aw \rangle &= \langle aw,\ v \rangle && \text{by (iii)},\\
&= a\,\langle w,\ v \rangle && \text{by (ii)},\\
&= a\,\langle v,\ w \rangle && \text{by (iii)}.
\end{aligned}
$$

(c) We first check the value of $\langle O,\ v \rangle$. We have

$$
\begin{aligned}
\langle O,\ v \rangle &= \langle O+O,\ v \rangle\\
&= \langle O,\ v \rangle + \langle O,\ v \rangle && \text{by (i)}.
\end{aligned}
$$

Thus $\langle O,\ v \rangle = 0$. By (iii), we also have $\langle v,\ O \rangle = 0$.

(d) When $v \neq O$, we have $\langle v,\ v \rangle > 0$ by (iv). When $v = O$, we have $\langle O,\ O \rangle = 0$ by (c).

(e) In particular, we have

$$
\langle v-w,\ v \rangle = \langle v-w,\ w \rangle .
$$

It then follows that

$$
\begin{aligned}
\langle v-w,\ v-w \rangle &= \langle v-w,\ v \rangle - \langle v-w,\ w \rangle && \text{by (a) and (b)},\\
&= 0.
\end{aligned}
$$

From (d) we have $v - w = O$. Hence $v = w$. $\qquad\qquad\square$

Example 7.1.4. Let $f(x)$ and $g(x)$ be two vectors in $\mathscr{P}$. Define

$$
\langle f(x),\ g(x) \rangle = \int_0^1 f(t)g(t)\ dt.
$$

(a) Verify that this is an inner product.

(b) Compute $\langle x^m,\ x^n \rangle$ for all $m, n \geq 0$.

(c) Compute $\langle 1 + 2x, \ x^2 + 2x^3 \rangle$.

Solution. We check the four conditions in Definition 8. Let $f, g, h \in \mathscr{P}$ and $a \in \mathbb{R}$.

(a) To check (i), we have that

$$
\langle f(x) + g(x), \ h(x) \rangle = \int_0^1 (f(t) + g(t)) h(t) \, dt
$$
$$
= \int_0^1 f(t) h(t) \, dt + \int_0^1 g(t) h(t) \, dt
$$
$$
= \langle f(x), \ h(x) \rangle + \langle g(x), \ h(x) \rangle.
$$

To check (ii), we have

$$
\langle a f(x), \ g(x) \rangle = \int_0^1 a f(t) g(t) \, dt
$$
$$
= a \int_0^1 f(t) g(t) \, dt
$$
$$
= a \langle f(x), \ g(x) \rangle.
$$

To check (iii), we have

$$
\langle f(x), \ g(x) \rangle = \int_0^1 f(t) g(t) \, dt
$$
$$
= \int_0^1 g(t) f(t) \, dt
$$
$$
= \langle g(x), \ f(x) \rangle.
$$

Finally, to check (iv), we have

$$
\langle f(x), \ f(x) \rangle = \int_0^1 f(t)^2 \, dt > 0
$$

since $f(x)^2 > 0$ except at a finite subset of points in $[0, 1]$ when $f(x)$ is a nonzero polynomial. We have used only basic properties of definite integrals of one variable discussed in a first course on Calculus.

(b) For all $m, n \geq 0$,

$$
\langle x^m, \ x^n \rangle = \int_0^1 t^{m+n} \, dt = \frac{t^{m+n+1}}{m+n+1} \Big|_0^1 = \frac{1}{m+n+1}.
$$

(c) We have

$$\langle 1 + 2x,\ x^2 + 2x^3 \rangle = \langle 1,\ x^2 + 2x^3 \rangle + 2\langle x,\ x^2 + 2x^3 \rangle$$
$$= \langle 1,\ x^2 \rangle + 2\langle 1,\ x^3 \rangle + 2\langle x,\ x^2 \rangle + 4\langle x,\ x^3 \rangle$$
$$= \frac{1}{3} + 2\cdot\frac{1}{4} + 2\cdot\frac{1}{4} + 4\cdot\frac{1}{5}$$
$$= 2\frac{2}{15}$$

using (b). Of course, we may also compute the inner product by directly computing the definite integrals.   ◇

Let $S = \{v_1, \ldots, v_n\}$ be a generating set for a finite dimensional vector space V. Let $v,\ w \in V$ such that

$$v = a_1 v_1 + \cdots + a_n v_n \quad \text{and} \quad w = b_1 v_1 + \cdots + b_n v_n$$

with $a_i, b_i \in \mathbb{R}$. We have that

$$\langle v,\ w \rangle = \left\langle \sum_{i=1}^{n} a_i v_i,\ \sum_{j=1}^{n} b_j v_j \right\rangle$$
$$= \sum_{i=1}^{n} a_i \left\langle v_i,\ \sum_{j=1}^{n} b_j v_j \right\rangle$$
$$= \sum_{i=1}^{n} a_i \left[\sum_{j=1}^{n} b_j \langle v_i,\ v_j \rangle \right]$$
$$= \sum_{i=1}^{n} \sum_{j=1}^{n} a_i b_j \langle v_i,\ v_j \rangle$$
$$= \sum_{i,j=1,\ldots,n} a_i b_j \langle v_i,\ v_j \rangle.$$

Thus we can compute the value of $\langle v,\ w \rangle$ if we have the values of $\langle v_i,\ v_j \rangle$ for all $i, j = 1, \ldots, n$.

Example 7.1.5. For each positive real number $k > 1$, define

$$\langle (a,b),\ (c,d) \rangle_k = kac + ad + bc + bd, \qquad a,b,c,d \in \mathbb{R}$$

Show that $\langle\,\cdot\,,\,\cdot\,\rangle_k$ is an inner product for each k. Hence there are infinitely many ways to define an inner product on $\mathbb{R}^2$. In fact, this is true for any real vector space.

Solution. Let $u = (a_1, a_2)$, $v = (b_1, b_2)$, $w = (c_1, c_2) \in \mathbb{R}^2$ and $a \in \mathbb{R}$. To check (i), we see that

$$\langle u + v, \ w \rangle_k$$
$$= \langle (a_1 + b_1, a_2 + b_2), \ (c_1, c_2) \rangle_k$$
$$= k(a_1 + b_1)c_1 + (a_1 + b_1)c_2 + (a_2 + b_2)c_1 + (a_2 + b_2)c_2$$
$$= ka_1c + a_1c_2 + a_2c_1 + a_2c_2 + kb_1c_1 + b_1c_2 + b_2c_1 + b_2c_2$$
$$= \langle u, \ w \rangle_k + \langle v, \ w \rangle_k .$$

To check (ii), we have that

$$\langle av, \ w \rangle_k = \langle (ab_1, ab_2), \ (c_1, c_2) \rangle_k$$
$$= kab_1c_1 + ab_1c_2 + ab_2c_1 + ab_2c_2$$
$$= a(kb_1c_1 + b_1c_2 + b_2c_1 + b_2c_2)$$
$$= a \langle v, \ w \rangle_k .$$

To check (iii),

$$\langle v, \ w \rangle_k = kb_1c_1 + b_1c_2 + b_2c_1 + b_2c_2$$
$$= kc_1b_1 + c_1b_2 + c_2b_1 + c_2b_2$$
$$= \langle w, \ v \rangle_k .$$

To check (iv), we have

$$\langle v, \ v \rangle_k = kb_1^2 + 2b_1b_2 + b_2^2$$
$$= (b_1 + b_2)^2 + (k - 1)b_1^2 \geq 0$$

since $k > 1$. In order for $\langle v, \ v \rangle_k = 0$, we must have $b_1 + b_2 = b_1 = 0$. It follows that $b_1 = b_2 = 0$. Thus $\langle v, \ v \rangle_k > 0$ if $v \neq O$.   ◇

Complex inner product space. Similarly we can define inner products on a complex vector space. The treatment will be parallel to that of real inner product spaces.

Let $z = a + bi$ where $a, \ b \in \mathbb{R}$. Remember that we use $\bar{z}$ to denote the **conjugate** of z and that $\bar{z} = a - bi$. We will use $|z|$, the absolute value of z, for the value $\sqrt{z\bar{z}} = \sqrt{a^2 + b^2}$. Furthermore, we have $|z| = 0$ if and only if $z = 0$.

Definition 7.1.6. Let V be a complex vector space. An **inner product** on V is a function that assigns to every pair of vectors v and w in V a complex number

$$V \times V \longrightarrow \mathbb{C}$$
$$(v, w) \longmapsto \langle v, w \rangle$$

such that the following conditions hold for all u, v and w in V and all α in $\mathbb{C}$:

(i) $\langle u + v, w \rangle = \langle u, w \rangle + \langle v, w \rangle$;

(ii) $\langle \alpha v, w \rangle = \alpha \langle v, w \rangle$;

(iii) $\langle w, v \rangle = \overline{\langle v, w \rangle}$;

(iv) $\langle v, v \rangle > 0$ if $v \neq O$.

A complex vector space endowed with a specific inner product is called a **complex inner product space** or simply an **inner product space**.

Proposition 7.1.7. *Define*

$$\langle v, w \rangle = \alpha_1 \overline{\beta_1} + \cdots + \alpha_n \overline{\beta_n}$$

*for $v = (\alpha_1, \ldots, \alpha_n)$ and $w = (\beta_1, \ldots, \beta_n)$ in $\mathbb{C}^n$. This is an inner product on $\mathbb{C}^n$. This inner product is called the **standard inner product** on $\mathbb{C}^n$.*

Proof. Let $u = (\alpha_1, \ldots, \alpha_n)$, $v = (\beta_1, \ldots, \beta_n)$, $w = (\gamma_1, \ldots, \gamma_n) \in \mathbb{C}^n$ and $\alpha \in \mathbb{C}$. To check (i) in Definition 7.1.6, we have that

$$\begin{aligned}
\langle u + v, w \rangle &= \langle (\alpha_1 + \beta_1, \ldots, \alpha_n + \beta_n), (\gamma_1, \ldots, \gamma_n) \rangle \\
&= (\alpha_1 + \beta_1)\overline{\gamma_1} + \cdots + (\alpha_n + \beta_n)\overline{\gamma_n} \\
&= \alpha_1 \overline{\gamma_1} + \cdots + \alpha_n \overline{\gamma_n} + \beta_1 \overline{\gamma_1} + \cdots + \beta_m \overline{\gamma_n} \\
&= \langle u, w \rangle + \langle v, w \rangle.
\end{aligned}$$

To check (ii), we have

$$\begin{aligned}
\langle \alpha v, w \rangle &= \langle (\alpha\beta_1, \ldots, \alpha\beta_n), (\gamma_1, \ldots, \gamma_n) \rangle \\
&= \alpha\beta_1 \overline{\gamma_1} + \cdots + \alpha\beta_n \overline{\gamma_n} \\
&= \alpha(\beta_1 \overline{\gamma_1} + \cdots + \beta_n \overline{\gamma_n}) = \alpha \langle v, w \rangle.
\end{aligned}$$

To check (iii), we have

$$
\begin{aligned}
\langle w,\, v\rangle &= \langle (\gamma_1, \ldots, \gamma_n),\, (\beta_1, \ldots, \beta_n)\rangle \\
&= \gamma_1 \overline{\beta_1} + \cdots + \gamma_n \overline{\beta_n} \\
&= \overline{\beta_1 \overline{\gamma_1} + \cdots + \beta_n \overline{\gamma_n}} \\
&= \overline{\langle v,\, w\rangle}.
\end{aligned}
$$

To check (iv), we have

$$
\begin{aligned}
\langle v,\, v\rangle &= \beta_1 \overline{\beta_1} + \cdots + \beta_n \overline{\beta_n} \\
&= |\beta_1|^2 + \cdots + |\beta_n|^2 \\
&\geq 0.
\end{aligned}
$$

Note that

$$
\begin{aligned}
\langle v,\, v\rangle = 0 &\iff |\beta_1| = \cdots = |\beta_n| = 0 \\
&\iff \beta_1 = \cdots = \beta_n = 0 \\
&\iff v = O.
\end{aligned}
$$

We have verified that $\langle\,\cdot\,,\,\cdot\,\rangle$ is an inner product over $\mathbb{C}$. $\quad\square$

Example 7.1.8. Let's practice the standard inner product on $\mathbb{C}^2$. For example,

$$
\begin{aligned}
&\langle (4 - 5i, 3),\, (1 + i, 1 + 2i)\rangle \\
&= (4 - 5i)(1 - i) + 3(1 - 2i) \\
&= 2 - 15i.
\end{aligned}
$$

Definition 7.1.9. Let $A = \left(a_{ij}\right) \in M_{m \times n}(F)$ where $F = \mathbb{R}$ or $\mathbb{C}$. Define the **conjugate transpose** or **adjoint** of A, denoted A^*, to be the $n \times m$ matrix $\left(b_{ij}\right)$ such that $b_{ij} = \overline{a_{ji}}$ for all i, j.

Remark. The terminology "adjoint" has two different meanings. One was used in Chapter 4 when we discussed the determinant of a matrix. The other is defined here in the discussion of the inner product. One can usually decide what "adjoint" refers to from context. When there is confusion, we can always use "conjugate transpose" for "adjoint".

The terminology "conjugate transpose" is adequate. We first replace all the entries in the matrix by their conjugates, we then transpose the matrix. Note that when A is a matrix with real entries we have $A^* = A^t$.

Example 7.1.10. Let

$$A = \begin{pmatrix} i & 1 - 2i & 2 \\ 1 - i & 3 & 3 + 4i \end{pmatrix}.$$

Then

$$A^* = \begin{pmatrix} -i & 1 + i \\ 1 + 2i & 3 \\ 2 & 3 - 4i \end{pmatrix}.$$

Definition 7.1.11. Let $A = (a_{ij})_{n \times n}$ be a square matrix. We can define the **trace** of A, denoted $\operatorname{tr} A$, to be

$$a_{11} + a_{22} + \cdots + a_{nn}.$$

In other words, the trace of a square matrix is defined to be the sum of its diagonal.

Proposition 7.1.12. *For any two matrices A, B in $M_{m \times n}(\mathbb{R})$ (or in $M_{m \times n}(\mathbb{C})$), define*

$$\langle A, B \rangle = \operatorname{tr}(B^* A).$$

*This is an inner product on $M_{m \times n}(\mathbb{R})$ over $\mathbb{R}$ (or on $M_{m \times n}(\mathbb{C})$ over $\mathbb{C}$). This inner product is called the **Frobenius inner product** on $M_{m \times n}(\mathbb{R})$ (or on $M_{m \times n}(\mathbb{C})$).*

We can prove this proposition and other theorems regarding inner products over $\mathbb{R}$ and over $\mathbb{C}$ at the same time. Definition 7.1.6 may be seen as a generalization of Definition 7.1.1. Since when the conditions in Definition 7.1.6 are restricted to $\mathbb{R}$, they coincide with the conditions in Definition 7.1.1. We now check the conditions in Definition 7.1.6 on the product defined here.

We leave the proof of Proposition 7.1.12 as an exercise. See Exercises 4 and 6.

Next we will give another well-known example of complex inner product space.

Example 7.1.13. Consider $\mathscr{H}$, the set of all continuous complex-valued functions defined on the closed interval $[0, 2\pi]$. Let $f \in \mathscr{H}$, we may actually write

$$f = f_1 + if_2,$$

where f_1 and f_2 are continuous real-valued functions defined on $[0, 2\pi]$. Moreover,

$$\int_0^{2\pi} f(t)\,dt = \int_0^{2\pi} f_1(t)\,dt + i \int_0^{2\pi} f_2(t)\,dt.$$

It is routine to verify that

$$\langle f,\ g \rangle = \frac{1}{2\pi} \int_0^{2\pi} f(t)\overline{g(t)}\,dt$$

is an inner product. Thus $\mathscr{H}$ is a complex inner product space.

The following is a list of properties of a complex inner product.

Proposition 7.1.14. *Let V be a complex inner product space. Then the following statements are true for all u, v and $w \in V$ and all $\alpha \in \mathbb{C}$:*

(a) $\langle u,\ v + w \rangle = \langle u,\ v \rangle + \langle u,\ w \rangle$;

(b) $\langle v,\ \alpha w \rangle = \overline{\alpha}\,\langle v,\ w \rangle$;

(c) $\langle v,\ O \rangle = \langle O,\ v \rangle = 0$;

(d) $\langle v,\ v \rangle = 0$ *if and only if $v = O$;*

(e) *If $\langle u,\ v \rangle = \langle u,\ w \rangle$ for all $u \in V$, then $v = w$.*

The proof here is very similar to that of Proposition 7.1.3. We leave it as an exercise. See Exercise 6.

Let $S = \{v_1, \ldots, v_n\}$ be a generating set for the finite dimensional inner product space V. Let $v,\ w \in V$ such that

$$v = \alpha_1 v_1 + \cdots + \alpha_n v_n \quad \text{and} \quad w = \beta_1 v_1 + \cdots + \beta_n v_n$$

with $\alpha_i, \beta_i \in \mathbb{C}$. Then

(7.1) $$\langle v,\ w \rangle = \sum_{i=1}^n \alpha_i \langle v_i,\ w \rangle$$

$$= \sum_{i=1}^{n} \sum_{j=1}^{n} \alpha_i \overline{\beta_j} \langle v_i, \ v_j \rangle$$

$$= \sum_{i,j=1,\ldots,n} \alpha_i \overline{\beta_j} \langle v_i, \ v_j \rangle.$$

Thus we can compute the value of $\langle v, \ w \rangle$ if the values of $\langle v_i, \ v_j \rangle$ are known for all $i, j = 1, \ldots, n$.

Exercises 7.1

1. Let $\langle -, \ - \rangle$ be an inner product on $\mathbb{R}^2$. Suppose that

$$\langle (1,0), \ (1,0) \rangle = 3;$$
$$\langle (1,0), \ (1,1) \rangle = -4;$$
$$\langle (1,1), \ (1,1) \rangle = 9.$$

Compute $\langle (3,5), \ (-1,7) \rangle$.

2. Verify that the standard inner product on $\mathbb{C}^n$ and the Frobenius inner product on $M_{m \times n}(\mathbb{R})$ and on $M_{m \times n}(\mathbb{C})$ are all inner products.

3. Let $a, b, c, d \in \mathbb{R}$. For $\alpha = a + bi$ and $\beta = c + di$ in $\mathbb{C}$, define

$$\langle \alpha, \ \beta \rangle = ac + bd.$$

Is $\langle -, \ - \rangle$ is an inner product on $\mathbb{C}$ over $\mathbb{R}$? Is $\langle -, \ - \rangle$ an inner product on $\mathbb{C}$ over $\mathbb{C}$?

4. Verify the following properties regarding the trace of square matrices for all square matrices A, B and any scalar c:

   (a) $\operatorname{tr} A = \operatorname{tr} A^{\mathrm{t}}$;

   (b) $\operatorname{tr}(AB) = \operatorname{tr}(BA)$;

   (c) $\operatorname{tr}(A + B) = \operatorname{tr} A + \operatorname{tr} B$;

   (d) $\operatorname{tr}(cA) = c \operatorname{tr} A$;

   (e) $\operatorname{tr}(PAP^{-1}) = \operatorname{tr} A$ where P is invertible.

5. Prove Proposition 7.1.12.

6. Prove Proposition 7.1.14.

7. Consider the Frobenius inner product in $M_3(\mathbb{R})$.

   (a) Compute $\langle E_{ij},\ E_{k\ell} \rangle$ for all i, j, k, ℓ.

   (b) Compute $\left\langle \begin{pmatrix} 1 & 2 & 3 \\ 0 & 4 & 0 \\ 0 & 1 & 3 \end{pmatrix}, \begin{pmatrix} 4 & 0 & 1 \\ 2 & 1 & 0 \\ 1 & 1 & 4 \end{pmatrix} \right\rangle$.

8. Let $w(x)$ be a positive-valued continuous function on the interval $[-1, 1]$. For any $p(x), q(x)$ in $\mathscr{P}_n$, define

$$\langle p(x),\ q(x) \rangle = \int_{-1}^{1} p(x)q(x)w(x)\,dx.$$

   Show that $\langle -,\ - \rangle$ is an inner product on $\mathscr{P}_n$.

9. Let $A, B \in M_n(\mathbb{C})$ and $\alpha \in \mathbb{C}$. Prove the following assertions:

   (a) $(A + B)^* = A^* + B^*$;

   (b) $(\alpha A)^* = \overline{\alpha} A^*$;

   (c) $(AB)^* = B^* A^*$;

   (d) $A^{**} = A$;

   (e) $I_n{}^* = I_n$.

7.2   Norm and angle

From now on when we say inner product spaces we mean *real* or *complex* inner product spaces. The field F will stand for either $\mathbb{R}$ or $\mathbb{C}$.

Norm.   Inner products allow us to define the concepts of distance and angles even in spaces where these notions would not have made sense.

Definition 7.2.1. Let V be an inner product space. For $v \in V$, we define the **norm** or **length** of v, denoted $\|v\|$, to be the *nonnegative* real number $\sqrt{\langle v,\ v \rangle}$. Hence the **distance** between v and $w \in V$ is defined to be $\|v - w\|$.

Remark. We use $\| \ \|$ to denote the norm of a vector and we use $| \ |$ to denote the *absolute value* of a scalar.

Definition 7.2.2. A vector whose norm is 1 is called a **unit vector**.

Example 7.2.3. The norm with respect to the standard inner product is the usual Euclidean definition of length. In this case

$$\|v\| = \|(a_1, \ldots, a_n)\| = \left[\sum_{i=1}^{n} |a_i|^2 \right]^{1/2}$$

for $v = (a_1, \ldots, a_n) \in \mathbb{C}^n$.

As one may reasonably expect, the well-known properties of Euclidean length in $\mathbb{R}^3$ hold in general for norms.

Theorem 7.2.4. *Let V be an inner product space. For all v, $w \in V$ and $c \in F$, the following statements are true:*

(a) $\|v\| \geq 0$ *and* $\|v\| = 0$ *if and only if $v = 0$;*

(b) *the distance between v and w is 0 if and only if $v = w$;*

(c) $\|cv\| = |c| \|v\|$;

(d) ***Cauchy-Schwarz Inequality:*** $|\langle v,\ w \rangle| \leq \|v\| \|w\|$;

(e) ***Triangle Inequality:*** $\|v + w\| \leq \|v\| + \|w\|$.

Proof. (a) This is Proposition 5(d) and Proposition 7.1.14(d).
(b) This follows from (a).
(c) We have that

$$\|cv\|^2 = \langle cv,\ cv \rangle = c\bar{c} \langle v,\ v \rangle = |c|^2 \|v\|^2.$$

Taking square root, we have $\|cv\| = |c| \|v\|$ since $\|cv\|$, $|c|$ and $\|v\|$ are all defined to be nonnegative.
(d) Let

$$c = \frac{\langle v,\ w \rangle}{\langle w,\ w \rangle}.$$

By definition, we have that $\langle v - cw,\ v - cw \rangle \geq 0$. Hence

$$\langle v - cw,\ v - cw \rangle = \langle v,\ v - cw \rangle - c \langle w,\ v - cw \rangle$$

$$= \langle v,\ v \rangle - \overline{c} \langle v,\ w \rangle - c \langle w,\ v \rangle + c\overline{c} \langle w,\ w \rangle$$

$$= \|v\|^2 - \frac{\overline{\langle v,\ w \rangle}}{\|w\|^2} \langle v,\ w \rangle - \frac{\langle v,\ w \rangle}{\|w\|^2} \overline{\langle v,\ w \rangle} + |c|^2 \|w\|^2$$

$$= \|v\|^2 - \frac{|\langle v,\ w \rangle|^2}{\|w\|^2} - \frac{|\langle v,\ w \rangle|^2}{\|w\|^2} + \frac{|\langle v,\ w \rangle|^2}{\|w\|^4} \|w\|^2$$

$$= \|v\|^2 - \frac{|\langle v,\ w \rangle|^2}{\|w\|^2} \geq 0.$$

Hence

$$\|v\|^2 \geq \frac{|\langle v,\ w \rangle|^2}{\|w\|^2} \quad \Longrightarrow \quad \|v\|^2 \|w\|^2 \geq |\langle v,\ w \rangle|^2.$$

Taking square roots on both sides, we have $\|v\| \|w\| \geq |\langle v,\ w \rangle|$.

(e) By part (d), we have

$$(\|v\| + \|w\|)^2 = \|v\|^2 + \|w\|^2 + 2 \|v\| \|w\|$$
$$\geq \|v\|^2 + \|w\|^2 + 2 |\langle v,\ w \rangle|.$$

On the other hand,

$$\|v + w\|^2 = \langle v + w,\ v + w \rangle$$
$$= \langle v,\ v + w \rangle + \langle w,\ v + w \rangle$$
$$= \langle v,\ v \rangle + \langle v,\ w \rangle + \langle w,\ v \rangle + \langle w,\ w \rangle$$
$$= \|v\|^2 + \|w\|^2 + \langle v,\ w \rangle + \overline{\langle v,\ w \rangle}.$$

Let $\langle v,\ w \rangle = a + bi$ where $a, b \in \mathbb{R}$. Then

$$2 |\langle v,\ w \rangle| = 2\sqrt{a^2 + b^2} \geq 2 |a|$$
$$\geq 2a = \langle v,\ w \rangle + \overline{\langle v,\ w \rangle}.$$

We have

$$(\|v\| + \|w\|)^2 \geq \|v + w\|^2.$$

Taking square roots on both sides, we have

$$\|v\| + \|w\| \geq \|v + w\|$$

since both $\|v\| + \|w\|$ and $\|v + w\|$ are nonnegative. $\square$

Theorem 7.2.4(d) and (e) for the Euclidean norm can be rewritten as the following two well-known inequalities.

Corollary 7.2.5. *The following two inequalities hold for all $a_1, \ldots, a_n$ and $b_1, \ldots, b_n$ in $\mathbb{R}$ or in $\mathbb{C}$:*

(a) **Triangle Inequality:**

$$\Big[\sum_{i=1}^{n} |a_i + b_i|^2\Big]^{1/2} \leq \Big[\sum_{i=1}^{n} |a_i|^2\Big]^{1/2} + \Big[\sum_{i=1}^{n} |b_i|^2\Big]^{1/2};$$

(b) **Cauchy-Schwarz Inequality:**

$$\Big(\sum_{i=1}^{n} a_i \overline{b_i}\Big)^2 \leq \Big(\sum_{i=1}^{n} |a_i|^2\Big)\Big(\sum_{i=1}^{n} |b_i|^2\Big).$$

The norm is a generalization of the concept of Euclidean length.

Definition 7.2.6. Suppose given a vector space V over $F = \mathbb{R}$ or $\mathbb{C}$. A **norm** on V is a function ν from V to $\mathbb{R}$ such that

(i) $\nu(v) \geq 0$,

(ii) $\nu(av) = |a|\nu(v)$,

(iii) $\nu(v + w) \leq \nu(v) + \nu(w)$ (triangle inequality), and

(iv) if $\nu(v) = 0$ then $v = 0$,

for all $a \in F$ and $v, w \in V$.

Theorem 7.2.4 tells us that the norm defined with respect to a linear product is also a norm in the general sense.

The angle. The Cauchy-Schwarz Inequality has an important significance for inner products over $\mathbb{R}$:

$$-1 \leq \frac{\langle v, w \rangle}{\|v\|\|w\|} \leq 1$$

for all v, w in any real inner product space V. For $v, w \in V$, define the **angle** θ between v and w ($0 \leq \theta < \pi$ with no orientation) by letting

$$\cos\theta = \frac{\langle v, w \rangle}{\|v\|\|w\|}.$$

More generally, we may make the following definition.

Definition 7.2.7. Let V be an inner product space. Vectors $v, w \in V$ are said to be **orthogonal (perpendicular)** to each other if $\langle v, w \rangle = 0$. A subset S of V is orthogonal if any two distinct vectors in S are orthogonal to each other. An **orthonormal** subset is an orthogonal subset consisting of unit vectors.

Lemma 7.2.8. *Let V be an inner product space over $F = \mathbb{R}$ or $\mathbb{C}$ and let $v, w \in V$.*

(a) *The zero vector is orthogonal to all vectors in V.*

(b) *If $v \perp w$ then $av \perp bw$ for all $a, b \in F$.*

Proof. (a) From Proposition 7.1.14(c) we have

$$\langle O, v \rangle = 0 \quad \text{for all } v \in V.$$

(b) Assume $\langle v, w \rangle = 0$. Then

$$\langle av, bw \rangle = a\bar{b} \langle v, w \rangle = 0.$$

We have $av \perp bw$. $\square$

Remarks. (1) If $S = \{v_1, v_2, v_3, \dots\}$ is a orthonormal subset of an inner product space V, then $\langle v_i, v_j \rangle = \delta_{ij}$ for all i, j.

(2) If $S = \{v_1, v_2, v_3, \dots\}$ is an orthogonal subset of V, then we can create an orthonormal subset $\left\{ \dfrac{v_1}{\|v_1\|}, \dfrac{v_2}{\|v_2\|}, \dfrac{v_3}{\|v_3\|}, \dots \right\}$. This process is called **normalizing** S.

Orthogonal and orthonormal bases. When dealing with an inner product space, it is often advantageous to find a basis which is orthonormal.

Definition 7.2.9. In an inner product space, an ordered basis which is orthogonal is called an **orthogonal basis**, and an ordered basis which is orthonormal is called an **orthonormal basis**.

Example 7.2.10. The standard basis for $\mathbb{R}^n$ or $\mathbb{C}^n$ is an orthonormal basis with respect to the standard inner product.

Example 7.2.11. The basis $\mathscr{B} = \{(1,2),\ (2,-1)\}$ is an orthogonal basis for $\mathbb{R}^2$ with respect to the standard inner product. We may normalize it to obtain an orthonormal basis $\mathscr{B}' = \{(1/\sqrt{5}, 2/\sqrt{5}),\ (2/\sqrt{5}, -1/\sqrt{5})\}$.

Proposition 7.2.12. *Let S be an orthogonal subset of an inner product space V consisting of nonzero vectors. Then S is linearly independent.*

Proof. Let $v_1, \ldots, v_n \in S$ and $a_1, \ldots, a_n$ be scalars such that

$$a_1 v_1 + a_2 v_2 + \cdots + a_n v_n = O.$$

We then have

$$\langle a_1 v_1 + \cdots + a_n v_n,\ v_j \rangle = \left\langle \sum_{i=1}^{n} a_i v_i,\ v_j \right\rangle$$

$$= a_j \langle v_j,\ v_j \rangle = 0$$

for all j. By assumption, $v_i \neq 0$, and so $\langle v_j,\ v_j \rangle \neq 0$. It follows that $a_j = 0$ for all j. Thus S is linearly independent. $\qquad\square$

The following results demonstrate the advantage of having an orthogonal or an orthonormal basis.

Proposition 7.2.13. *Let $\{u_1, \ldots, u_n\}$ be an orthonormal basis of the inner product space V. If $v = a_1 u_1 + \cdots + a_n u_n$ and $w = b_1 u_1 + \cdots + b_n u_n$ where $a_i,\ b_j \in F$, then*

$$\langle v,\ w \rangle = a_1 \bar{b}_1 + \cdots + a_n \bar{b}_n.$$

Proof. this follows from the formula in (7.1). $\qquad\square$

Example 7.2.14. Suppose $\{(1,2),\ (4,7)\}$ is an orthonormal basis for the inner product space $V = \mathbb{R}^2$. Find $\langle (2,3),\ (3,4) \rangle$.

Solution. We find

$$(2,3) = -2(1,2) + (4,7),$$
$$(3,4) = -5(1,2) + 2(4,7).$$

Hence $\langle (2,3),\ (3,4) \rangle = 10 + 2 = 12.$ $\qquad\qquad\diamond$

Lemma 7.2.15. *Let S be a generating set for an inner product space. If $\langle v, \, x \rangle = 0$ for all $x \in S$, then $v = O$.*

Proof. From the formula in (7.1), this would imply $\langle v, \, w \rangle = 0$ for all w in $\mathrm{Sp}\,(S)$. We have $v = O$ by Proposition 7.1.14. $\square$

Proposition 7.2.16. *Let $S = \{v_1, v_2, \ldots, v_k\}$ be an orthogonal subset of an inner product space V consisting of nonzero vectors. Then for $v \in \mathrm{Sp}\,(S)$,*

$$v = \sum_{i=1}^{k} \frac{\langle v, \, v_i \rangle}{\langle v_i, \, v_i \rangle} v_i.$$

Proof. For $1 \le j \le k$, we have

$$\left\langle v - \sum_{i=1}^{k} \frac{\langle v, \, v_i \rangle}{\langle v_i, \, v_i \rangle} v_i, \, v_j \right\rangle$$

$$= \langle v, \, v_j \rangle - \sum_{i=1}^{k} \frac{\langle v, \, v_i \rangle}{\langle v_i, \, v_i \rangle} \langle v_i, \, v_j \rangle$$

$$= \langle v, \, v_j \rangle - \frac{\langle v, \, v_j \rangle}{\langle v_j, \, v_j \rangle} \langle v_j, \, v_j \rangle$$

$$= 0$$

by orthogonality of S. We conclude that $v = O$ by Lemma 7.2.15. $\square$

We have the following immediate result.

Corollary 7.2.17. *Let $S = \{v_1, v_2, \ldots, v_k\}$ be an orthonormal subset of of an inner product space V consisting of nonzero vectors. Then*

$$v = \sum_{i=1}^{k} \langle v, \, v_i \rangle v_i$$

for all $v \in \mathrm{Sp}\,(S)$. Furthermore, if $S = \{v_1, v_2, \ldots, v_n\}$ is an orthonormal basis for V, then

$$v = \sum_{i=1}^{n} \langle v, \, v_i \rangle v_i$$

for all $v \in V$.

Proof. The entries in $[T]_\beta$ is chosen so that

$$T(v_j) = \sum_{i=1}^{n} a_{ij} v_j.$$

From Corollary 7.2.17, $a_{ij} = \langle T(v_j),\ v_i \rangle$ for all i, j.   □

Corollary 7.2.18. *Let $\beta = (v_1, v_2, \ldots, v_n)$ be an orthonormal basis for a finite dimensional inner product space V. Let T be a linear endomorphism on V. Then $[T]_\beta = \left(a_{ij} \right)$, where $a_{ij} = \langle T(v_j),\ v_i \rangle$.*

We may define an inner product on V to make any basis of your choice orthonormal.

Gram-Schmidt process.   When the inner product is already given, it still remains to develop a method for constructing orthonormal bases. To construct an orthonormal basis, we may first construct an orthogonal basis and then normalize it.

To find an orthogonal basis, we first take an arbitrary basis, and then we adjust the vectors one by one so that they become orthogonal to each other. The next theorem tells us how to do that.

Theorem 7.2.19 (Gram-Schmidt Process). *Let $S = \{w_1, w_2, \ldots, w_n\}$ be a linearly independent subset of an inner product space V. Define $S' = \{v_1, v_2, \ldots, v_n\}$, where $v_1 = w_1$ and*

$$(7.2) \qquad v_k = w_k - \sum_{i=1}^{k-1} \frac{\langle w_k,\ v_i \rangle}{\langle v_i,\ v_i \rangle} v_i, \qquad \text{for } 2 \le k \le n.$$

Then S' is an orthogonal subset consisting of nonzero vectors such that $\mathrm{Sp}\,(S) = \mathrm{Sp}\,(S')$.

Thus, if we start with a basis for V, we may use the Gram-Schmidt process to transform it into an orthogonal basis. We can then normalize it to obtain an orthonormal basis.

Proof. We prove this theorem in three steps.

Step 1. We show that S' is orthogonal. We do this by showing that for all $k \ge 2$, we have $v_k \perp v_j$ for all $j < k$. First, when $k = 2$,

$$\langle v_2,\ v_1 \rangle = \left\langle w_2 - \frac{\langle w_2,\ v_1 \rangle}{\langle v_1,\ v_1 \rangle} v_1,\ v_1 \right\rangle$$

$$= \langle w_2, \ v_1 \rangle - \left\langle \frac{\langle w_2, \ v_1 \rangle}{\langle v_1, \ v_1 \rangle} v_1, \ v_1 \right\rangle$$

$$= \langle w_2, \ v_1 \rangle - \frac{\langle w_2, \ v_1 \rangle}{\langle v_1, \ v_1 \rangle} \langle v_1, \ v_1 \rangle$$

$$= 0.$$

We now assume $k > 2$. For $j < k$ we have

$$\langle v_k, \ v_j \rangle = \left\langle w_k - \sum_{i=1}^{k-1} \frac{\langle w_k, \ v_i \rangle}{\langle v_i, \ v_i \rangle} v_i, \ v_j \right\rangle$$

$$= \langle w_k, \ v_j \rangle - \sum_{i=1}^{k-1} \frac{\langle w_k, \ v_i \rangle}{\langle v_i, \ v_i \rangle} \langle v_i, \ v_j \rangle$$

$$= \langle w_k, \ v_j \rangle - \frac{\langle w_k, \ v_j \rangle}{\langle v_j, \ v_j \rangle} \langle v_j, \ v_j \rangle = 0$$

since by induction $\langle v_i, \ v_j \rangle = 0$ for $i \neq j$ when $i, j \leq k$. We have completed the induction step and shown that S' is orthogonal.

Step 2. We show that

$$\mathrm{Sp}\,(v_1, \ldots, v_k) = \mathrm{Sp}\,(w_1, \ldots, w_k)$$

for $k = 1, 2, \ldots, n$. Clearly $\mathrm{Sp}\,(v_1) = \mathrm{Sp}\,(w_1)$ since $v_1 = w_1$. Let $k \geq 2$. From (7.2), we have

$$\mathrm{Sp}\,(v_1, \ldots, v_{k-1}, v_k) = \mathrm{Sp}\,(v_1, \ldots, v_{k-1}, w_k).$$

By the induction hypothesis we have

$$\mathrm{Sp}\,(v_1, \ldots, v_{k-1}) = \mathrm{Sp}\,(w_1, \ldots, w_{k-1}).$$

Thus

$$\mathrm{Sp}\,(v_1, \ldots, v_{k-1}, v_k) = \mathrm{Sp}\,(v_1, \ldots, v_{k-1}, w_k) = \mathrm{Sp}\,(w_1, \ldots, w_k).$$

It follows that

$$\mathrm{Sp}\,(v_1, \ldots, v_n) = \mathrm{Sp}\,(w_1, \ldots, w_n).$$

Step 3. Because S is assumed to be linearly independent, $\dim \mathrm{Sp}\,(S) = n$. The set S' is a set of n elements generating $\mathrm{Sp}\,(S)$. It follows that S' is also a basis for $\mathrm{Sp}\,(S)$. Thus S' is linearly independent. The set must consist of nonzero vectors. $\qquad\square$

Example 7.2.20. In $\mathbb{R}^4$ the subset $\{(1,0,1,0),\ (1,1,1,1),\ (0,1,2,1)\}$ is linearly independent. Hence it is a basis for a subspace V in $\mathbb{R}^4$. Use this basis and Gram-Schmidt process to find an orthonormal basis for V.

Solution. We use (7.2) to obtain

$$v_1 = (1,0,1,1);$$

$$v_2 = (1,1,1,1) - \frac{3}{3}(1,0,1,1) = (0,1,0,0);$$

$$v_3 = (0,1,2,1) - \frac{3}{3}(1,0,1,1) - \frac{1}{1}(0,1,0,0) = (-1,0,1,0).$$

The set $S = \{v_1, v_2, v_3\}$ is orthogonal. We normalize S to obtain

$$B = \left\{ \left(\frac{1}{\sqrt{3}}, 0, \frac{1}{\sqrt{3}}, \frac{1}{\sqrt{3}} \right),\ (0,1,0,0),\ \left(-\frac{1}{\sqrt{2}}, 0, \frac{1}{\sqrt{2}}, 0 \right) \right\}.$$

This is an orthonormal basis for V. ◇

Example 7.2.21. Let V be the inner product space $\mathscr{P}_2$ equipped with the inner product

$$\langle f(x),\ g(x) \rangle = \int_{-1}^{1} f(t)g(t)\ dt.$$

Find an orthonormal basis $\mathscr{B}$ for V.

Solution. We start with the standard basis $\{1,\ x,\ x^2\}$ for $\mathscr{P}_2$. We use Gram-Schmidt Process to find an orthogonal basis for V. Let $f_1 = 1$. We find

$$\langle f_1,\ f_1 \rangle = \langle 1,\ 1 \rangle = 2;$$
$$\langle x,\ f_1 \rangle = \langle x,\ 1 \rangle = 0.$$

Thus

$$f_2 = x - \frac{0}{2} \cdot f_1 = x.$$

We further find

$$\langle f_2,\ f_2 \rangle = \langle x,\ x \rangle = \frac{2}{3};$$
$$\langle x^2,\ f_1 \rangle = \langle x^2,\ 1 \rangle = \frac{2}{3};$$
$$\langle x^2,\ f_2 \rangle = \langle x^2,\ x \rangle = 0.$$

We have

$$f_3 = x^2 - \frac{2/3}{2} \cdot 1 - \frac{0}{2/3} \cdot x = x^2 - \frac{1}{3}.$$

The set $S = \left\{1,\ x,\ x^2 - (1/3)\right\}$ is orthogonal. We normalize S to obtain an orthonormal basis. For this, we need to find

$$\langle f_3,\ f_3 \rangle = \left\langle x^2 - \frac{1}{3},\ x^2 - \frac{1}{3} \right\rangle = \frac{8}{45}.$$

Thus

$$B = \left\{ g_1 = \frac{1}{\sqrt{2}},\ g_2 = \frac{\sqrt{3}}{\sqrt{2}}x,\ g_3 = \frac{\sqrt{45}}{\sqrt{8}}\left(x^2 - \frac{1}{3}\right) \right\}$$

is an orthonormal basis for V. ◇

Definition 7.2.22. Let S be a subset of an inner product space V. We define the **orthogonal complement** of S to be

$$S^{\perp} = \{v \in V : \langle v,\ w \rangle = 0 \text{ for all } w \in S\}.$$

Lemma 7.2.23. *Let S be a subset of an inner product space V. Then $S^{\perp}$ is a subspace of V.*

Proof. Since $\langle O,\ v \rangle = 0$ for all $v \in V$, the zero vector O is in $S^{\perp}$. Let $w, w' \in S^{\perp}$. Then $\langle v,\ w \rangle = \langle v,\ w' \rangle = 0$ for all $v \in S$. It follows that

$$\langle v,\ w + w' \rangle = \langle v,\ w \rangle + \langle v,\ w' \rangle = 0$$

for all $v \in S$. Hence $w + w' \in S^{\perp}$. Let a be an arbitrary scalar. We have $\langle v,\ aw \rangle = \bar{a}\langle v,\ w \rangle = 0$ for all $v \in S$. Thus aw is also in $S^{\perp}$. We have shown that $S^{\perp}$ is a subspace of V. □

Example 7.2.24. Find $\{e_1\}^{\perp}$ in the standard inner product space $\mathbb{R}^3$.

Solution. Let $v = (a, b, c) \in \{e\}^{\perp}$. Then

$$\langle (1, 0, 0),\ (a, b, c) \rangle = a = 0.$$

Hence $\{e_1\}^{\perp} = \{(0, b, c) : b, c \in \mathbb{R}\} = \mathrm{Sp}\,(e_2, e_3)$. ◇

Lemma 7.2.25. *Let S be a subset of an inner product space V. Then $S^{\perp} = \mathrm{Sp}\,(S)^{\perp}$.*

Proof. Since $S \subseteq \mathrm{Sp}\,(S)$, we have $\mathrm{Sp}\,(S^{\perp}) \subseteq S^{\perp}$. Conversely, let $w \in S^{\perp}$. Take any $v \in \mathrm{Sp}\,(S)$. Then

$$v = a_1 v_1 + a_2 v_2 + \cdots + a_n v_n$$

where $v_1, \ldots, v_n \in S$ and $a_1, \ldots, a_n$ are scalars. We have

$$\langle v,\ w \rangle = \sum_{i=1}^{n} a_i \langle v_i,\ w \rangle = 0.$$

Hence $w \in \mathrm{Sp}\,(S^{\perp})$. We also have $S^{\perp} \subseteq \mathrm{Sp}\,(S^{\perp})$. □

The next theorem tells us how to find the orthogonal complement of a given subspace in a finite dimensional inner product space.

Proposition 7.2.26. *Let V be a finite dimensional inner product space. We can extend any orthogonal basis $S = \{v_1, \ldots, v_k\}$ to an orthogonal basis*

$$\{v_1, \ldots, v_k, v_{k+1}, \ldots, v_n\}$$

for V. Furthermore we have that

$$\mathrm{Sp}\,(S)^{\perp} = S^{\perp} = \mathrm{Sp}\,(v_{k+1}, \ldots, v_n).$$

Proof. We first extend $S = \{v_1, \ldots, v_k\}$ to an arbitrary basis

$$T = \{v_1, \ldots, v_k, w_{k+1}, \ldots, w_n\}.$$

We use the formula in (7.2) to transform T to an orthogonal basis for V. One can see that in this case, the first k elements in T will remain intact. Hence we will obtain an orthogonal basis

$$\{v_1, \ldots, v_k, v_{k+1}, \ldots, v_n\}$$

for V. We normalize it to obtain an orthonormal basis

$$B = \{u_1, \ldots, u_k, u_{k+1}, \ldots, u_n\}$$

for V. Here $u_i = v_i / \|v_i\|$.

Let $v \in V$. Find scalars $a_1, \ldots, a_n$ such that

$$v = a_1 u_1 + a_2 u_2 + \cdots + a_n u_n.$$

Then $v \in S^\perp$ if and only if $\langle v,\ v_i \rangle = 0$ for $i = 1, 2, \ldots, k$. We have

$$\langle v,\ v_i \rangle = \left\langle \sum_{j=1}^{n} a_j u_j,\ \|v_i\|\, u_i \right\rangle$$

$$= a_i \|v_i\| = 0$$

for all i. Since $\|v_i\| \neq 0$, we have $a_i = 0$ for $i = 1, 2, \ldots, k$. Hence $v \in S^\perp$ if and only if

$$v \in \mathrm{Sp}\,(u_{k+1}, \ldots, u_n) = \mathrm{Sp}\,(v_{k+1}, \ldots, v_n)\,.$$

Thus $\mathrm{Sp}\,\left(S^\perp\right) = S^\perp = \mathrm{Sp}\,(v_{k+1}, \ldots, v_n)$ by Lemma 7.2.25. $\qquad \square$

Corollary 7.2.27. *Let W be a subspace of a finite dimensional inner product space V.*

(a) *Then $V = W \oplus W^\perp$ and $\dim W^\perp = \dim V - \dim W$.*

(b) *In particular, $\mathbf{0}^\perp = V$ and $V^\perp = \mathbf{0}$.*

(c) *If W is a nontrivial proper subspace of V then so is $W^\perp$.*

Exercises 7.2

1. We usually use the notation $e^{i\theta}$ to denote $\cos\theta + i\sin\theta$, where $\theta \in \mathbb{R}$.

   (a) Show that $e^{i\alpha} e^{i\beta} = e^{i(\alpha+\beta)}$ for all α and $\beta \in \mathbb{R}$.

   (b) For any integer n, let $f_n(t) = e^{int}$, where $t \in [0, 2\pi]$. Define $S = \{f_n : n \in \mathbb{Z}\}$. Show that S a is orthonormal subset in $\mathscr{H}$ in Example 7.1.13.

2. Consider the real standard inner product space $\mathbb{R}^3$.

   (a) Verify that $S = \{(1, -2, 1),\ (2, 1, 0),\ (-1, 2, 5)\}$ is an orthogonal basis.

   (b) Normalize S to obtain an orthonormal basis $\mathscr{B}$ for $\mathbb{R}^3$.

   (c) Express $(5, 7, 1)$ as a linear combination of elements in $\mathscr{B}$.

3. (a) Let V be a real inner product space. Prove the **Law of cosines**:

$$\|v - w\|^2 = \|v\|^2 + \|w\|^2 - 2\|v\|\|w\|\cos\theta$$

where $v, w \in V$ and θ is the angle between v and w.

(b) Two vectors v and w in a *real* inner product space are orthogonal if and only if
$$\|v + w\|^2 = \|v\|^2 + \|w\|^2.$$

(c) Show that (b) becomes false if "real" is changed to "complex".

(d) Let v and w be vectors in a real inner product space. Show that $v - w$ and $v + w$ are orthogonal if $\|v\| = \|w\|$. Discuss the corresponding statement for complex spaces.

(e) Let v and w be vectors in an inner product space. Show that

$$\|v + w\|^2 + \|v - w\|^2 = 2\|v\|^2 + 2\|w\|^2.$$

4. Use the Gram-Schmidt process to transform the following basis

$$\{(1, i, i),\ (0, i, 1),\ (0, 0, 1)\}$$

for $\mathbb{C}^3$ into an orthonormal basis for $\mathbb{C}^3$.

5. Find an orthonormal basis in $M_{2\times 3}(\mathbb{R})$ and in $M_{2\times 3}(\mathbb{C})$ with respect to the Frobenius inner product.

6. Let $M_{m\times n}(\mathbb{R})$ be the inner product space equipped with the Frobenius inner product. Show that the standard basis

$$\alpha = (e_{11}, \ldots, e_{1n}, e_{21}, \ldots, e_{2n}, \ldots, e_{m1}, \ldots, e_{mn})$$

is an orthonormal basis.

7. Let $\langle \cdot, \cdot \rangle \colon \mathbb{R}^2 \times \mathbb{R}^2 \to \mathbb{R}$ be defined by

$$\langle (x, y),\ (u, v) \rangle = 5xu - 2(xv + yu) + yv.$$

(a) Show that $\langle \cdot, \cdot \rangle$ is an inner product.

(b) Determine the lengths of the vectors

$$(1,0), \quad (0,1), \quad (1,3), \quad (-1,2).$$

(c) Find an orthonormal basis for this inner product.

(d) Draw a picture of the unit circle around the origin.

8. Let S be a subset of finite dimensional inner product space V. Show that $S = S^{\perp\perp}$ if and only if S is a subspace of V.

7.3 The spectral theorem

The adjoint of a linear endomorphism. Let V be an arbitrary vector space over the field F. An F-linear transformation from V to F is also called a **linear functional**. We first show that linear functionals defined on a finite dimensional inner product space are very special.

For the rest of this section V stands for a *finite dimensional* inner product space over $F = \mathbb{R}$ or $\mathbb{C}$ unless otherwise noted.

Lemma 7.3.1. *Let w be a fixed vector in V. The function $g_w \colon V \to F$ defined by $g_w(v) = \langle v, w \rangle$ for all $v \in V$ is a linear functional.*

Proof. Let $v, v' \in V$ and $a \in F$. Then

$$g_w(v + v') = \langle v + v', w \rangle$$
$$= \langle v, w \rangle + \langle v', w \rangle = g_w(v) + g_w(v')$$

and

$$g_w(av) = \langle av, w \rangle = a \langle v, w \rangle = a g_w(v).$$

Thus g_w is a linear functional. □

Interestingly, the converse to this lemma is also true.

Example 7.3.2. Consider the standard inner product space $\mathbb{R}^2$. The function $g \colon \mathbb{R}^2 \to \mathbb{R}$ given by $g(a, b) = 2a + b$ is a linear functional. In fact, $g(a, b) = \langle (a, b), (2, 1) \rangle$.

Theorem 7.3.3. *Let* $g \colon V \to F$ *be a linear functional. There exists a unique* $w \in V$ *such that* $g(v) = \langle v, \ w \rangle$ *for all* $v \in V$.

Proof. Let $\beta = \{v_1, \ldots, v_n\}$ be an orthonormal basis for V and let

$$w = \sum_{i=1}^{n} \overline{g(v_i)} v_i.$$

For any v in V, find scalars $a_1, \ldots, a_n$ such that $v = \sum_{1}^{n} a_i v_i$. Then

$$
\begin{aligned}
\langle v, \ w \rangle &= \left\langle \sum_{i=1}^{n} a_i v_i, \ \sum_{i=1}^{n} \overline{g(v_i)} v_i \right\rangle \\
&= \sum_{i=1}^{n} a_i \overline{\overline{g(v_i)}} = \sum_{i=1}^{n} a_i g(v_i) \\
&= g \Big(\sum_{i=1}^{n} a_i v_i \Big) = g(v)
\end{aligned}
$$

for all $v \in V$. $\qquad\square$

Example 7.3.4. Consider the standard inner product space $\mathbb{C}^3$. Let

$$
\begin{aligned}
g \colon \quad & \mathbb{C}^3 && \longrightarrow && \mathbb{C} \\
& (\alpha, \ \beta, \ \gamma) && \longmapsto && -i\alpha + (3 - 4i)\beta + (-5 + 7i)\gamma
\end{aligned}
$$

Then $g((\alpha, \ \beta, \ \gamma)) = \langle (\alpha, \ \beta, \ \gamma), \ (i, \ 3 + 4i, \ -5 - 7i) \rangle$ is a linear functional.

Let T be a linear endomorphism on V and let $w \in V$. The map $g_w \circ T$ sends $v \in V$ to $\langle T(v), \ w \rangle$. It is clearly a linear functional. Hence we have the following result.

Theorem 7.3.5. *Let* T *be a linear endomorphism on* V. *There exists a unique linear endomorphism* T^* *on* V *such that*

$$\langle T(v), \ w \rangle = \langle v, \ T^*(w) \rangle$$

for all $v, \ w \in V$.

Proof. It remains to show that T^* is a linear endomorphism. Let $w, w' \in V$ and $a \in F$. We have

$$\langle v, \ T^*(w + w') \rangle = \langle T(v), \ w + w' \rangle$$

$$\begin{aligned} &= \langle T(v),\ w \rangle + \langle T(v),\ w' \rangle \\ &= \langle v,\ T^*(w) \rangle + \langle v,\ T^*(w') \rangle \\ &= \langle v,\ T^*(w) + T^*(w') \rangle. \end{aligned}$$

Thus by Proposition 7.1.14(e),

$$T^*(w + w') = T^*(w) + T^*(w').$$

Similarly,

$$\begin{aligned} \langle v,\ T^*(aw) \rangle = \langle T(v),\ aw \rangle &= \overline{a}\,\langle T(v),\ w \rangle \\ &= \overline{a}\,\langle v,\ T^*(w) \rangle = \langle v,\ aT^*(w) \rangle \end{aligned}$$

for all $v \in V$. We also have

$$T^*(aw) = aT^*(w).$$

We conclude that T^* is a linear endomorphism. $\qquad\square$

The linear endomorphism T^* obtained in Theorem 7.3.5 is called the **adjoint** of T. The next theorem tells us how to compute the adjoint of a given linear endomorphism.

Proposition 7.3.6. *Let β be an orthonormal basis for V. If T is a linear endomorphism on V then $[T^*]_\beta = [T]_\beta^*$.*

Proof. Let $\beta = (u_1, \ldots, u_n)$. Let $[T]_\beta = \left(a_{ij} \right)$ and $[T^*]_\beta = \left(b_{ij} \right)$. Then

$$T^*(u_j) = \sum_{i=1}^{n} b_{ij} u_i, \quad j = 1, 2, \ldots, n.$$

We have that

$$\overline{b_{ij}} = \left\langle u_i,\ \sum_{k=1}^{n} b_{kj} u_j \right\rangle = \langle u_i,\ T^*(u_j) \rangle$$

$$= \langle T(u_i),\ u_j \rangle = \left\langle \sum_{k=1}^{n} a_{ki} u_k,\ u_j \right\rangle$$

$$= a_{ij}$$

for all i, j. Thus $[T^*]_\beta$ is the adjoint (or conjugate transpose) of $[T]_\beta$. $\qquad\square$

Example 7.3.7. The standard ordered basis $\alpha = (e_1, e_2, \ldots, e_n)$ in an orthonormal basis in the standard inner product space F^n. Let $A \in M_n(F)$. Then $(L_A)^* = L_{A^*}$ since both the matrix relative to α are A^*.

Example 7.3.8. Let T be the linear endomorphism on the standard inner product space $\mathbb{C}^2$ given by

$$T(z_1, z_2) = ((3i + 1)z_1 - (2i)z_2, 5z_1 + (6 + 7i)z_2).$$

Compute T^*.

Solution. Let α be the standard basis for $\mathbb{C}^2$. Then

$$[T]_\alpha = \begin{pmatrix} 3i + 1 & -2i \\ 5 & 6 + 7i \end{pmatrix}.$$

It follows that

$$[T^*]_\alpha = \begin{pmatrix} 1 - 3i & 5 \\ 2i & 6 - 7i \end{pmatrix}$$

by Proposition 7.3.6. We find

$$\begin{pmatrix} 1 - 3i & 5 \\ 2i & 6 - 7i \end{pmatrix} \begin{pmatrix} z_1 \\ z_2 \end{pmatrix} = \begin{pmatrix} (1 - 3i)z_1 + 5z_2 \\ 2iz_1 + (6 - 7i)z_2 \end{pmatrix}.$$

We have $T^*(z_1, z_2) = ((1 - 3i)z_1 + 5z_2, 2iz_1 + (6 - 7i)z_2)$.   ◇

From Proposition 7.3.6 and Exercise 9 in §7.1 we have the following basic results for the adjoint of a linear endomorphism.

Corollary 7.3.9. *Let S and T be linear endomorphisms on V. Then the following assertions are true:*

(a) $(S + T)^* = S^* + T^*$;

(b) $(\alpha T)^* = \overline{\alpha} T^*$ *for all* $\alpha \in F$;

(c) $(S \circ T)^* = T^* \circ S^*$;

(d) $T^{**} = T$;

(e) $1_V^* = 1_V$.

Proposition 7.3.10. *Let T be a linear endomorphism on V. If T has an eigenvector of eigenvalue λ then T^* has an eigenvector of eigenvalue $\bar{\lambda}$.*

Proof. Since λ is an eigenvalue of T, the characteristic polynomial of T is $f(t) = (t - \lambda)g(t)$. Since the characteristic polynomial of a matrix and of its transpose are the same, we have that the characteristic polynomial of T^* is $\overline{f(t)} = (t - \bar{\lambda})\overline{g(t)}$ from Proposition 7.3.6. Hence $\bar{\lambda}$ is an eigenvalue of T^*. $\square$

Normal operators. Now suppose the finite dimensional inner product space V has an orthonormal basis β consisting of eigenvectors of a linear endomorphism T, then $[T]_\beta$ is a diagonal matrix. This implies that $[T^*]_\beta = [T]_\beta^*$ is also a diagonal matrix. Because diagonal matrices commute with each other, we should have that $TT^* = T^*T$. Now let's study this condition.

Definition 7.3.11. Let T be a linear endomorphism on V. We say that T is **normal** if $TT^* = T^*T$. A real or complex square matrix A is **normal** if $AA^* = A^*A$.

A linear endomorphism which is normal is often called a **normal operator**.

If a linear endomorphism T is diagonalizable relative to an orthonormal basis then T is normal. Now the question is that whether the converse is also true. In other words, would a normal linear endomorphism always possess an orthonormal basis consisting of eigenvectors? The answer is unfortunately negative!

Example 7.3.12. Let $T \colon \mathbb{R}^2 \to \mathbb{R}^2$ be the counterclockwise rotation by θ where $\theta \neq k\pi$ for any $k \in \mathbb{Z}$. The matrix representing T in the standard basis is

$$A = \begin{pmatrix} \cos\theta & -\sin\theta \\ \sin\theta & \cos\theta \end{pmatrix}.$$

Verify whether T is normal and whether T has any eigenvector.

Solution. Since

$$AA^* = AA^{\mathrm{t}} = I_2 = A^{\mathrm{t}}A = A^*A,$$

the linear endomorphism T is normal. The characteristic polynomial of T is $t^2 - 2\cos\theta\, t + 1$. The discriminant of this polynomial is $\Delta = 4\cos^2\theta - 4 < 0$ for we assume that $\theta \neq k\pi$ for any $k \in \mathbb{Z}$. Thus this polynomial has no zero in $\mathbb{R}$. In other words, T has no eigenvectors.   ◇

Example 7.3.13. Any real *skew-symmetric* matrix A ($A^t = -A$) is normal. However, no nonzero real skew-symmetric matrices are diagonalizable over $\mathbb{R}$ (see Problem 13).

Nevertheless, the normality condition does suffice for an orthonormal basis consisting of eigenvectors to exist for *complex* inner product spaces.

Lemma 7.3.14. *Let T be a normal operator on V. The following statements are true.*

(a) *The norms $\|T(v)\| = \|T^*(v)\|$ for all $v \in V$.*

(b) *The linear endomorphism $T - c\mathbf{1}_V$ is normal for all $c \in F$.*

(c) *If v is an eigenvector for T with eigenvalue λ, then v is also an eigenvector for T^* of eigenvalue $\overline{\lambda}$.*

(d) *If λ_1 and λ_2 are distinct eigenvalues of T with corresponding eigenvectors v_1 and v_2 respectively, then $v_1 \perp v_2$.*

Proof. (a) We compare the norms

$$\|T(v)\|^2 = \langle T(v),\, T(v) \rangle$$
$$= \langle v,\, T^*(v) \rangle = \langle v,\, TT^*(v) \rangle$$
$$= \langle v,\, (T^*)^*T^*(v) \rangle = \langle T^*(v),\, T^*(v) \rangle$$
$$= \|T^*(v)\|^2 .$$

Taking square roots on both sides, we have $\|T(v)\| = \|T^*(v)\|$.

(b) Using Corollary 7.3.9, we have

$$(T - c\mathbf{1}_V)(T - c\mathbf{1}_V)^*$$
$$= (T - c\mathbf{1}_V)(T^* - \overline{c}\mathbf{1}_V^*)$$
$$= (T - c\mathbf{1}_V)(T^* - \overline{c}\mathbf{1}_V)$$

$$= TT^* - \bar{c}T - cT^* + c\bar{c}\mathbf{1}_V$$
$$= T^*T - cT^* - \bar{c}T + c\bar{c}\mathbf{1}_V$$
$$= (T^* - \bar{c}\mathbf{1}_V)(T - c\mathbf{1}_V)$$
$$= (T^* - \bar{c}\mathbf{1}_V^*)(T - c\mathbf{1}_V)$$
$$= (T - c\mathbf{1}_V)^*(T - c\mathbf{1}_V).$$

Hence $T - c\mathbf{1}_V$ is normal.

(c) Since v is an eigenvector of T with eigenvalue λ, we have

$$(T - \lambda\mathbf{1}_V)(v) = O.$$

From (a) and (b) we have

$$\left\|(T^* - \bar{\lambda}\mathbf{1}_V)(v)\right\| = \|(T - \lambda\mathbf{1}_V)(v)\| = 0.$$

Hence $(T^* - \bar{\lambda}\mathbf{1}_V)(v) = O$. That is, $T^*(v) = \bar{\lambda}v$.

(d) We have

$$\lambda_1 \langle v_1, \ v_2 \rangle = \langle \lambda_1 v_1, \ v_2 \rangle = \langle T(v_1), \ v_2 \rangle$$
$$= \langle v_1, \ T^*(v_2) \rangle = \langle v_1, \ \bar{\lambda}_2 v_2 \rangle$$
$$= \lambda_2 \langle v_1, \ v_2 \rangle$$

using (c). It follows that $(\lambda_1 - \lambda_2) \langle v_1, \ v_2 \rangle = 0$. By assumption $\lambda_1 \neq \lambda_2$, we have $\langle v_1, \ v_2 \rangle = 0$, that is, $v_1 \perp v_2$. $\qquad\square$

Lemma 7.3.15. *Let T be a normal operator on V and let λ be an eigenvalue of T. Then*

$$\begin{cases} T(E_\lambda) \subseteq E_\lambda; \\ T^*(E_\lambda) \subseteq E_\lambda; \end{cases} \quad and \quad \begin{cases} T(E_\lambda^\perp) \subseteq E_\lambda^\perp; \\ T^*(E_\lambda^\perp) \subseteq E_\lambda^\perp. \end{cases}$$

Proof. If $v \in E_\lambda$, then $T(v) = \lambda v \in E_\lambda$. This implies that $T(E_\lambda) \subseteq E_\lambda$. From Lemma 7.3.14(c) we also have $T^*(v) = \bar{\lambda}v \in E_\lambda$. This also implies $T^*(E_\lambda) \subseteq E_\lambda$.

On the other hand, let $v \in E_\lambda^\perp$. Then for all $w \in E_\lambda$,

$$\langle T(v), \ w \rangle = \langle v, \ T^*(w) \rangle = \langle v, \ \bar{\lambda}w \rangle = 0$$

by Proposition 7.3.14(c). It follows that $T(v) \in E_\lambda^\perp$. Similarly,

$$\langle w, \ T^*(v) \rangle = \langle T(w), \ v \rangle = \langle \lambda w, \ v \rangle = 0$$

for all $w \in E_\lambda$. We conclude that $T^*(v) \in E_\lambda^\perp$. $\square$

Theorem 7.3.16. *Let T be a linear endomorphism on V. Then V has an orthonormal basis consisting of eigenvectors if and only if T is normal and the characteristic polynomial of T splits.*

Proof. It remains to show the "if" part. If $\dim V = 1$, there is nothing to prove. We assume $\dim V \geq 2$. Since the characteristic polynomial of T splits, we may find an eigenvalue λ of T. Find an orthonormal basis $\{v_1, \ldots, v_r\}$ for E_λ. We are done if $E_\lambda = V$. Otherwise, $E_\lambda^\perp$ is a subspace of dimension $< n$. From Lemma 7.3.15 the linear endomorphism T restricted to $E_\lambda^\perp$ is still normal, and its characteristic polynomial still splits, being a factor of the original characteristic polynomial. From the induction hypothesis, we may find an orthonormal basis $\{w_1, \ldots, w_{n-r}\}$ for $E_\lambda^\perp$ consisting of eigenvectors of T (restricted to $E_\lambda^\perp$). The set $B = \{v_1, \ldots, v_r, w_1, \ldots, w_{r-1}\}$ is an orthogonal set consisting of nonzero vectors. This is a linearly independent set by Proposition 7.2.12. Hence B is an orthonormal basis for V. Moreover, B consists of eigenvectors. $\square$

Corollary 7.3.17. *Let T be a linear endomorphism on a finite dimensional complex inner product space V. Then V has an orthonormal basis consisting of eigenvectors if and only if T is normal. Thus any normal operator or normal matrix is diagonalizable over $\mathbb{C}$.*

Proof. Any nonzero polynomial splits over $\mathbb{C}$. The result follows from Theorem 7.3.16. $\square$

The result of Corollary 7.3.17 does not extend to infinite dimensional complex inner product spaces.

Self-adjoint endomorphisms.   From Theorem 7.3.16 we can see that for real inner product spaces, normality does not necessarily imply diagonalizability. The difficulty lies in that the characteristic polynomial might not split over $\mathbb{R}$. Next we would like to replace normality by a stronger condition so that even real inner product spaces can have an orthonormal basis of eigenvectors.

Let T be a linear endomorphism on V. Suppose V has an orthonormal basis $\beta = \{v_1, \ldots, v_n\}$ consisting of eigenvectors of T. Let the eigenvalue of v_i be λ_i. By Lemma 7.3.14(c), β is also an orthonormal basis for T^*, where v_i is an eigenvector of eigenvalue $\overline{\lambda}_i$. Thus

$$[T]_\beta = \begin{pmatrix} \lambda_1 & & & \\ & \lambda_2 & & \\ & & \ddots & \\ & & & \lambda_n \end{pmatrix} \quad \text{and} \quad [T^*]_\beta = \begin{pmatrix} \overline{\lambda}_1 & & & \\ & \overline{\lambda}_2 & & \\ & & \ddots & \\ & & & \overline{\lambda}_n \end{pmatrix}.$$

Hence, if V is a finite dimensional *real* inner product space which has an orthonormal basis consisting of eigenvectors of T, we have that $T = T^*$.

Definition 7.3.18. Let T be a linear endomorphism on an inner product space V. We say that T is **self-adjoint** indexself-adjoint (or **Hermitian**) if $T = T^*$. A (real or complex) square matrix A is **self-adjoint** (or **Hermitian**) if $A = A^*$.

A self-adjoint linear endomorphism is also called a **self-adjoint operator**.

Obviously, self-adjoint operators are normal since $TT^* = T^2 = T^*T$. Thus, we can apply Theorem 7.3.16 and Corollary 7.3.17 to prove the Main Theorem of this section.

Theorem 7.3.19 (The spectral theorem). *Let T be a linear endomorphism on V. Then T is self-adjoint if and only if there exists an orthonormal basis for V consisting of eigenvectors with real eigenvalues. Thus any Hermitian matrix is similar to a real diagonal matrix.*

Proof. Let $\dim V = n$.

The "if" part: Let β be an ordered basis consisting of eigenvectors. Then $[T]_\beta$ is a real diagonal matrix

$$\begin{pmatrix} \lambda_1 & & \\ & \ddots & \\ & & \lambda_n \end{pmatrix}.$$

This implies that

$$[T^*]_\beta = \begin{pmatrix} \lambda_1 & & \\ & \ddots & \\ & & \lambda_n \end{pmatrix}^* = \begin{pmatrix} \lambda_1 & & \\ & \ddots & \\ & & \lambda_n \end{pmatrix}.$$

We have that $T = T^*$. The linear endomorphism T is self-adjoint.

The "only if" part: It suffices to verify that the theorem is valid for L_A. First, we assume L_A is a self-adjoint operator on $\mathbb{C}^n$. Since L_A is normal, the inner product space $\mathbb{C}^n$ has an orthonormal basis β consisting of eigenvectors by Corollary 7.3.17 . It follows that

$$[L_A]_\beta = \begin{pmatrix} \lambda_1 & & \\ & \ddots & \\ & & \lambda_n \end{pmatrix}, \quad \lambda_1, \ldots, \lambda_n \in \mathbb{C}.$$

Since T is self-adjoint, we have that

$$[T^*]_\beta = \begin{pmatrix} \overline{\lambda_1} & & \\ & \ddots & \\ & & \overline{\lambda_n} \end{pmatrix} = [T]_\beta = \begin{pmatrix} \lambda_1 & & \\ & \ddots & \\ & & \lambda_n \end{pmatrix}.$$

This implies that $\lambda_1 = \overline{\lambda_1}$ for all i. Hence each eigenvalue is a real number. This further implies that the characteristic polynomial of A is $(-1)^n \prod_{i=1}^n (t - \lambda_i)$.

Now, we consider L_A as a linear endomorphism on $\mathbb{R}^n$. We already know that A is normal and the characteristic polynomial of A splits over $\mathbb{R}$. From Theorem 7.3.16, there is an orthonormal basis for $\mathbb{R}^n$ consisting of eigenvectors of L_A. Furthermore, the eigenvalues of L_A are all real. $\square$

Example 7.3.20. Let

$$A = \begin{pmatrix} 1 & 2 \\ 2 & 1 \end{pmatrix}.$$

Find an orthonormal basis β such that $[L_A]_\beta$ is a diagonal matrix.

Solution. Note that A is a symmetric real matrix. Thus A is self-adjoint. The characteristic polynomial of A is

$$\det \begin{pmatrix} 1 - t & 2 \\ 2 & 1 - t \end{pmatrix} = (t - 3)(t + 1).$$

We find a basis for each eigenspace:

$$E_3 = \text{Null}\left(\begin{pmatrix} -2 & 2 \\ 2 & -2 \end{pmatrix}\right) = \text{Sp}\left(\begin{pmatrix} 1 \\ 1 \end{pmatrix}\right),$$

$$E_{-1} = \text{Null}\left(\begin{pmatrix} 2 & 2 \\ 2 & 2 \end{pmatrix}\right) = \text{Sp}\left(\begin{pmatrix} 1 \\ -1 \end{pmatrix}\right).$$

The set

$$\left\{\begin{pmatrix} 1 \\ 1 \end{pmatrix}, \begin{pmatrix} 1 \\ -1 \end{pmatrix}\right\}$$

is an orthogonal basis for $\mathbb{R}^2$. We normalize it to obtain an ordered orthonormal basis

$$\beta = \left(\begin{pmatrix} 1/\sqrt{2} \\ 1/\sqrt{2} \end{pmatrix}, \begin{pmatrix} 1/\sqrt{2} \\ -1/\sqrt{2} \end{pmatrix}\right).$$

It then follows that

$$[L_A]_\beta = \begin{pmatrix} 3 & \\ & -1 \end{pmatrix}$$

is a diagonalization for A. ◇

Example 7.3.21. From Exercise 6 in §7.2 we know that the standard basis is an orthonormal basis for the inner product space $M_2(\mathbb{R})$ equipped with the Frobenius inner product. Let T be the $\mathbb{R}$-linear endomorphism

$$\begin{array}{rccc} T: & M_2(\mathbb{R}) & \longrightarrow & M_2(\mathbb{R}) \\ & A & \longmapsto & A^t \end{array}$$

Check that T is self-adjoint. Find an orthonormal basis β such that $[T]_\beta$ is diagonal.

Solution. We know that the standard basis

$$\alpha = (e_{11}, \ e_{12}, \ e_{21}, \ e_{22})$$

is an orthonormal basis for $M_2(\mathbb{R})$. We find

$$T(e_{11}) = e_{11}, \quad T(e_{12}) = e_{21},$$
$$T(e_{21}) = e_{12}, \quad T(e_{22}) = e_{22}.$$

Hence

$$[T]_\alpha = \begin{pmatrix} 1 & 0 & 0 & 0 \\ 0 & 0 & 1 & 0 \\ 0 & 1 & 0 & 0 \\ 0 & 0 & 0 & 1 \end{pmatrix}$$

is a symmetric matrix. We have that T is self-adjoint.

The characteristic polynomial of T is readily seen to be $(t+1)(t-1)^3$. We find a basis for each eigenspaces:

$$E_1 = \mathrm{Sp}\,(e_{11},\; e_{22},\; e_{12}+e_{21})\,,$$
$$E_{-1} = \mathrm{Sp}\,(e_{12}-e_{21})\,.$$

We also find

$$\|e_{11}\| = \|e_{22}\| = 1, \quad \|e_{12}+e_{21}\| = \sqrt{2} \quad \text{and} \quad \|e_{12}-e_{21}\| = \sqrt{2}.$$

Thus

$$\beta = \left(e_{11},\; e_{22},\; \frac{e_{12}+e_{21}}{\sqrt{2}},\; \frac{e_{12}-e_{21}}{\sqrt{2}} \right)$$

is an orthonormal basis for $M_2(\mathbb{R})$ such that

$$[T]_\beta = \begin{pmatrix} 1 & & & \\ & 1 & & \\ & & 1 & \\ & & & -1 \end{pmatrix}$$

is diagonal.   ◇

We summarize the most important results for real and complex matrices below.

- A normal matrix is always diagonalizable when regarded as a matrix with complex entries. However, the eigenvalues might not be real.

- A Hermitian matrix is always diagonalizable, and the eigenvalues are always real.

Base change between orthonormal bases. Let $\beta = (u_1, \ldots, u_n)$ be an orthonormal basis and $\gamma = (v_1, \ldots, v_n)$ be an arbitrary basis for V. Suppose

$$v_i = a_{1i}u_1 + a_{2i}u_2 + \cdots + a_{ni}u_n, \qquad i = 1, 2, \ldots, n.$$

Then $P = [\mathbf{1}_V]_\gamma^\beta = (a_{ij})_{n \times n}$. By Proposition 7.2.13, we have that

$$\langle v_j, \, v_i \rangle = a_{1j}\overline{a_{1i}} + a_{2j}\overline{a_{2i}} + \cdots + a_{nj}\overline{a_{ni}}$$

for all i and j. Hence $P^*P = \left(\langle v_j, \, v_i \rangle \right)_{n \times n}$. This shows that $P^*P = I_n$ if and only if $\langle v_j, \, v_i \rangle = \delta_{ij}$ for all i and j if and only if γ is an orthonormal basis. We summarize this result in Proposition 7.3.23.

Definition 7.3.22. A square matrix satisfying $P^*P = I_n$ is called a **unitary** matrix.

The inverse of a unitary matrix is its conjugate transpose.

Proposition 7.3.23. *Let β be an orthonormal basis and let γ be a basis for V. Then $P = [\mathbf{1}_V]_\gamma^\beta$ is a unitary matrix if and only if γ is an orthonormal basis.*

Remember that the standard basis of $\mathbb{R}^n$ or of $\mathbb{C}^n$ is orthonormal. Using this result and the spectral theorem we have the following immediate result.

Corollary 7.3.24. *Let A be a normal matrix in $M_n(\mathbb{C})$. Then there exists a unitary matrix $P \in M_n(\mathbb{C})$ such that P^*AP is a diagonal matrix.*

Corollary 7.3.25. *Let A be a Hermitian matrix in $M_n(F)$ where $F = \mathbb{R}$ or $\mathbb{C}$. Then there exists a unitary matrix $P \in M_n(F)$ such that P^*AP is a real diagonal matrix.*

Example 7.3.26. Let

$$A = \begin{pmatrix} 1 & -i \\ i & 1 \end{pmatrix}.$$

Find a unitary matrix P such that P^*AP is a real diagonal matrix.

Solution. This problem asks us to find an orthonormal basis to diagonalize A. This is possible because A is self-adjoint.

The characteristic polynomial of A is

$$\det \begin{pmatrix} 1-t & -i \\ i & 1-t \end{pmatrix} = t(t-2).$$

We find a basis for each of the eigenspaces:

$$E_0 = \mathrm{Sp}\left(\begin{pmatrix} i \\ 1 \end{pmatrix}\right), \quad E_2 = \mathrm{Sp}\left(\begin{pmatrix} 1 \\ i \end{pmatrix}\right).$$

The basis

$$\left\{ \begin{pmatrix} i \\ 1 \end{pmatrix}, \begin{pmatrix} 1 \\ i \end{pmatrix} \right\}$$

is orthogonal. We normalize it to obtain an ordered orthonormal basis

$$\left(\frac{1}{\sqrt{2}} \begin{pmatrix} i \\ 1 \end{pmatrix}, \frac{1}{\sqrt{2}} \begin{pmatrix} 1 \\ i \end{pmatrix} \right)$$

for $\mathbb{C}^2$. Thus

$$P = [\mathbf{1}_{\mathbb{C}^2}]_\beta^\alpha = \frac{1}{\sqrt{2}} \begin{pmatrix} i & 1 \\ 1 & i \end{pmatrix}$$

is unitary. We also have

$$P^{-1}AP = P^*AP = \begin{pmatrix} 0 & \\ & 2 \end{pmatrix}$$

is diagonal.   ◇

Exercises 7.3

1. Compute the adjoint of the following linear endomorphisms.

   (a) $T: \mathbb{R}^3 \to \mathbb{R}^3$, $T(x, yz) = (x + y - z,\ 2x + 2y + 2z,\ x + 2y - 3z)$.

   (b) $T: \mathbb{C}^4 \to \mathbb{C}^4$, $T(x, y, z, w) = (x + iy,\ y + iz,\ z + iw,\ w + ix)$.

   (c) $T: \mathscr{P}_2 \to \mathscr{P}_2$, $T(p(x)) = xp'(x) - (xp(x))'$ assuming the standard basis is an orthonormal basis.

(d) $T\colon M_3(\mathbb{R}) \to M_3(\mathbb{R})$, $T(A) = A + A^t$ assuming the standard basis is an orthonormal basis.

2. Prove Corollary 7.3.9.

3. Let T be a linear endomorphism on V. Show that $T + T^*$ is self-adjoint.

4. Let T be a self-adjoint operator on V and let W be a subspace of V. If W is invariant under T, show that $W^\perp$ is also invariant under T.

5. Find an unitary matrix $P \in M_3(\mathbb{R})$ such that $P^t \begin{pmatrix} 0 & 1 & 1 \\ 1 & 0 & -1 \\ 1 & -1 & 0 \end{pmatrix} P$ is diagonal.

6. Let $A \in M_n(\mathbb{R})$. Suppose the columns of A are assumed to be orthogonal in $\mathbb{R}^n$. Are the rows also orthogonal in $\mathbb{R}^n$?

7. Show that a square matrix of size n is unitary if and only if its columns form an orthonormal basis for the standard inner product space $\mathbb{R}^n$ if and only if its rows form an orthonormal basis for the standard inner product space $\mathbb{R}^n$. What if $\mathbb{R}$ is replaced by $\mathbb{C}$?

8. We say a linear transformation $T\colon V \to W$ between two inner product spaces is an **isometry** if $\langle v, v \rangle = \langle T(v), T(v) \rangle$ for all $v \in V$. In other words, an isometry preserves length. Show that an isometry is always injective.

9. Let $A \in M_n(\mathbb{R})$ be a symmetric matrix such that $A^m = \mathbf{0}$ for some positive integer m. Show that A is the zero matrix. Is the assertion still true if A is not symmetric? What if A is symmetric but A has entries in $\mathbb{C}$?

10. Let A and $B \in M_n(\mathbb{R})$ be symmetric matrices. Show that $A = B$ if $v^t A v = v^t B v$ for all column n-vectors $v \in \mathbb{R}^n$. Is the assertion still true if A and B are not assumed to be symmetric?

11. Show that a unitary matrix is diagonalizable by working through the following steps.

(a) Let U be a unitary matrix of size n. Let the columns of U form the ordered basis $\mathscr{B}$ for $\mathbb{C}^n$. Let u be a normalized eigenvector of A of eigenvalue λ. Expand $\{U\}$ to an orthonormal basis $\mathscr{B}'$ for $\mathbb{C}^n$. Let P be the base change matrix from $\mathscr{B}'$ to $\mathscr{B}$. Show that both P and $P^{-1}UP$ are unitary.

(b) Show that $P^{-1}UP$ has the form

$$\begin{pmatrix} \lambda & \\ & A \end{pmatrix},$$

where A is an $(n-1) \times (n-1)$ matrix. Show that A is a unitary matrix.

(c) Show by induction that U is diagonalizable over $\mathbb{C}$.

12. A unitary matrix with real entries is also called an **orthogonal** matrix. Show that any square matrix A can be decomposed as QR where Q is an orthogonal matrix Q and a upper triangular matrix R. This is the so-called **QR decomposition**. (Hint: Use Gram-Schmidt process.)

13. Consider the real skew-symmetric matrix A over $\mathbb{C}$.

(a) Show that A is diagonalizable over $\mathbb{C}$ and all the eigenvalues belong to $\mathbb{R}i = \{ai : a \in \mathbb{R}\}$.

(b) Show that A is never diagonalizable over $\mathbb{R}$ if $A \neq 0$.

The following problems are not on normal or self-adjoint endomorphisms. Thus the respective results are not as strong as what we have discussed in this section.

15. The scalar 0 is an eigenvalue of the matrix $A = \begin{pmatrix} 0 & 0 \\ 1 & 0 \end{pmatrix}$. By Proposition 7.3.10, 0 is also an eigenvalue of A^{t}. Find the eigenspaces with eigenvalue 0 of A and of A^{t} respectively to show that they need not be the same.

16. Prove the following theorem by Schure:

> *Let T be an linear endomorphism on a finite dimensional inner product space V. Suppose that the characteristic polynomial of T splits. Then there exists an orthonormal basis β for V such that the matrix $[T]_\beta$ is upper triangular.*

(Hint: Since the characteristic polynomial of T splits, T has an eigenvector. Hence T has a unit eigenvector z. Let $W = \mathrm{Sp}\,(z)$. Then $W^\perp$ is T-invariant. Proceed by induction on $\dim V$. Cf. Pb (1), §5.3.)

17. Let A be a matrix whose characteristic polynomial splits. The previous problem says that we can find a a unitary matrix U and an upper triangular matrix B such that $A = UBU^*$ is upper triangular. This is called a **Schure decomposition**. Let

$$A = \begin{pmatrix} 5 & -2 & 2 \\ 4 & 1 & 4 \\ 4 & -4 & 9 \end{pmatrix}.$$

   (a) Find the characteristic polynomial and the minimal polynomial of A

   (b) Find a Schure decomposition for A.

7.4   Applications of the spectral theorem

Here we provide a few examples to demonstrate how the spectral theorem is used.

Rigid transformations. Let T be an isometry (see Exercise 8, §7.3) on the standard inner product space $\mathbb{R}^n$ (the Euclidean space). Not only does an isometry on $\mathbb{R}^n$ preserve the length, it preserves the angle. Note that

$$\langle v,\ w\rangle = \frac{1}{2}(\langle v+w,\ v+w\rangle - \langle v,\ v\rangle - \langle w,\ w\rangle)$$
$$= \frac{1}{2}(\langle T(v+w),\ T(v+w)\rangle - \langle T(v),\ T(v)\rangle - \langle T(w),\ T(w)\rangle)$$
$$= \langle T(v),\ T(w)\rangle$$

for all $v, w \in \mathbb{R}^n$. Hence the angle between v and w, and the angle between $T(v)$ and $T(w)$ can both be given by

$$\cos\theta = \frac{\langle v, \, w \rangle}{\|v\| \, \|w\|} = \frac{\langle T(v), \, T(w) \rangle}{\|T(v)\| \, \|T(w)\|}.$$

In general, a linear endomorphism on $\mathbb{R}^n$ preserving length and angle is called a **rigid transformation**. Clearly, rigid transformations preserve shapes of objects. Moreover, an isometry is a rigid transformation stabilizing the origin. In fact, the converse is also true as shown in the following proposition.

Proposition 7.4.1. *Suppose $T \colon \mathbb{R}^n \to \mathbb{R}^n$ is a rigid transformation stabling the origin. Then T is an isometry.*

Proof. Since T is a rigid transformation, we have

$$\langle v, \, w \rangle = \langle T(v) - T(O), \, T(w) - T(O) \rangle$$

for all $v, w \in \mathbb{R}^n$. By assumption, $T(O) = O$. Hence

(7.3) $$\langle v, \, w \rangle = \langle T(v), \, T(w) \rangle$$

for all $v, w \in \mathbb{R}^n$. We now find

$$
\begin{aligned}
&\langle T(v+w) - T(v) - T(w), \, T(v+w) - T(v) - T(w) \rangle \\
={} & \langle T(v+w), \, T(v+w) \rangle - \langle T(v+w), \, T(v) \rangle - \langle T(v+w), \, T(w) \rangle \\
& - \langle T(v), \, T(v+w) \rangle + \langle T(v), \, T(v) \rangle + \langle T(v), \, T(w) \rangle \\
& - \langle T(w), \, T(v+w) \rangle + \langle T(w), \, T(v) \rangle + \langle T(v), \, T(v) \rangle \\
={} & \langle v+w, \, v+w \rangle - \langle v+w, \, v \rangle - \langle v+w, \, w \rangle - \langle v, \, v+w \rangle \\
& + \langle v, \, v \rangle + \langle v, \, w \rangle - \langle w, \, v+w \rangle + \langle w, \, v \rangle + \langle v, \, v \rangle \\
={} & \langle v+w - v - w, \, v+w - v - w \rangle \\
={} & 0.
\end{aligned}
$$

Hence $T(v+w) - T(v) - T(w) = O$, or $T(v+w) = T(v) + T(w)$. Similarly, we may show that $T(av) = aT(v)$ for all $a \in \mathbb{R}$ and $v \in \mathbb{R}^n$. Thus T is an linear endomorphism. From (7.3), we have

$$\langle v, \, v \rangle = \langle T(v), \, T(v) \rangle$$

for all $v \in \mathbb{R}^n$. We conclude T is an isometry. $\qquad\square$

Rigid transformation can thus be considered as the composite of an isometry preceded and/or followed by a **translation** (see Figure 7.1).

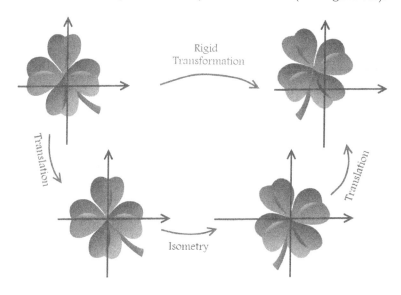

Figure 7.1: A rigid transformation on the real plane

Since translations are easier to handle, we will concentrate on isometries.

Proposition 7.4.2. *Let P be a square real matrix of size n. The linear endomorphism L_P is an isometry on the standard inner product space $\mathbb{R}^n$ if and only if P is unitary.*

Proof. Since L_p measures angle and length, it sends the standard basis (an orthonormal basis) to an orthonormal basis. Thus its matrix is unitary by Proposition 7.3.23.

Conversely, suppose P is unitary of size n. Let $\alpha = \{e_1, \ldots, e_n\}$. Then by Proposition 7.3.23 or by Exercise 7 in §7.3,

$$B = \{T(e_1), T(e_2), \ldots, T(e_n)\}$$

is also an orthonormal basis for $\mathbb{R}^n$. Let $v = \sum_{i=1}^n a_i e_i$ where $a_i \in \mathbb{R}$. Then

$$\langle T(v),\ T(v) \rangle = \left\langle \sum_{i=1}^n a_i T(e_i),\ \sum_{i=1}^n a_i T(e_i) \right\rangle$$

$$= \sum_{i=1}^{n} a_i^2 = \langle v,\ v \rangle$$

for all $v \in \mathbb{R}^n$. We have that T is an isometry. $\qquad \square$

Thus, a square real unitary matrix now has double meaning: It can be viewed as an isometry, or it can be viewed as a base change between two orthonormal bases. Note that $\det P = \pm 1$

Isometries on $\mathbb{R}^2$. Suppose given an isometry T on $\mathbb{R}^2$ and let e_1, e_2 be mapped to f_1 and f_2 respectively. By definition, f_1 and f_2 would remain an orthonormal basis of $\mathbb{R}^2$. Let θ be the angle from e_1 to f_1 (counterclockwise). We have the following two situations:

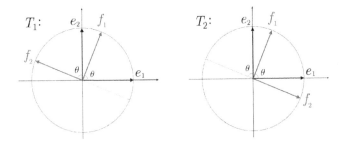

Figure 7.2: Two types of isometries

In Figure 7.2, the isometry T_1 represents a *counterclockwise* **rotation** by θ. Let $\alpha = (e_1, e_2)$. Clearly,

$$(7.4) \qquad [T_1]_\alpha = \begin{pmatrix} \cos\theta & -\sin\theta \\ \sin\theta & \cos\theta \end{pmatrix},$$

and $\det[T_1]_\alpha = 1$. While T_2 may be obtained by applying a **reflection**

$$R: \qquad f_1 \longmapsto g_1 = e_1, \qquad f_2 \longmapsto g_2 = -e_2,$$

followed by a countercloskwise rotation by θ. Hence
(7.5)

$$[T_2]_\alpha = [T_1]_\alpha [R]_\alpha = \begin{pmatrix} \cos\theta & -\sin\theta \\ \sin\theta & \cos\theta \end{pmatrix} \begin{pmatrix} 1 & 0 \\ 0 & -1 \end{pmatrix} = \begin{pmatrix} \cos\theta & \sin\theta \\ \sin\theta & -\cos\theta \end{pmatrix}.$$

Note that $\det[R]_\alpha = \det[T_2]_\alpha = -1$. Hence an isometry of determinant 1 is a rotation while an isometry of determinant -1 is the composite of a reflection and a rotation.

There is a second view point. The unitary matrix $P_1 = [T_1]_\alpha$ can also represent a base change. Let $\beta = (f_1, f_2)$. Then $P_1 = [1_{\mathbb{R}^2}]_\beta^\alpha$ is the base change from β to α. When we apply

$$\begin{pmatrix} x \\ y \end{pmatrix} = P_1 \begin{pmatrix} u \\ v \end{pmatrix},$$

After applying P_1, the coordinates (u, v) expressed in terms of f_1 and f_2 is transformed to the coordinate in terms of e_1 and e_2. We may think of it as a clockwise rotation of the axes by θ. Similarly,

$$P_2 = P_1 \begin{pmatrix} 1 & 0 \\ 0 & -1 \end{pmatrix}$$

may be viewed as reflecting the second axis along the first axis followed by a clockwise rotation of the axes by θ.

Both are valid viewpoints. One may choose whatever one feels comfortable about.

Quadratic forms in two variables. When studying quadratic curves (also called **conic sections**) such as

$$x^2 - 3xy + 3y^2 = 7$$

or

$$3x^2 + 6xy - 4y^2 - 3x + 5y = 1,$$

there is difficulty caused by the cross-product term $-3xy$ and $6xy$. We introduce a method of rotating the coordinate system so as to eliminate the cross-product term. This is really the second viewpoint in the previous section. However, we will forego the whole geometric aspect and treat the problem in a purely algebraic fashion.

A **quadratic form** is a **homogeneous** quadratic function of the form

$$\sum_{i,j=1}^{n} a_{ij} x_i x_j.$$

A quadratic form in two variables may be expressed as

$$\varphi(x,y) = ax^2 + 2bxy + cy^2 = \begin{pmatrix} x & y \end{pmatrix} \begin{pmatrix} a & b \\ b & c \end{pmatrix} \begin{pmatrix} x \\ y \end{pmatrix}$$

where a, b and $c \in \mathbb{R}$.

By the spectral theorem, we may find an ordered orthonormal basis $\beta = (f_1, f_2)$ such that

$$P^{\mathrm{t}} \begin{pmatrix} a & b \\ b & c \end{pmatrix} P = \begin{pmatrix} \lambda_1 & 0 \\ 0 & \lambda_2 \end{pmatrix}$$

where $P = [1_{\mathbb{R}^2}]_\beta^\alpha$ and α is the standard basis of $\mathbb{R}^2$. Remember that $P^{\mathrm{t}} = [1_{\mathbb{R}^2}]_\alpha^\beta$. It is often preferable (but not necessary) to choose β such that $\det P = 1$ (you may replace f_2 by $-f_2$ if necessary). This way we are only rotating the axes, which is easier to visualize.

Let the coordinate of $\mathbf{v} = (x,\ y) \in \mathbb{R}^2$ with respect to β be $(u,\ v)$. Then

$$\begin{pmatrix} x \\ y \end{pmatrix} = P \begin{pmatrix} u \\ v \end{pmatrix} \qquad \text{and} \qquad \begin{pmatrix} x & y \end{pmatrix} = \begin{pmatrix} u & v \end{pmatrix} P^{\mathrm{t}}.$$

It follows that

$$\varphi(x,y) = ax^2 + 2bxy + cy^2 = \begin{pmatrix} x & y \end{pmatrix} \begin{pmatrix} a & b \\ b & c \end{pmatrix} \begin{pmatrix} x \\ y \end{pmatrix}$$

$$= \begin{pmatrix} u & v \end{pmatrix} P^{\mathrm{t}} \begin{pmatrix} a & b \\ b & c \end{pmatrix} P \begin{pmatrix} u \\ v \end{pmatrix} = \begin{pmatrix} u & v \end{pmatrix} \begin{pmatrix} \lambda_1 & 0 \\ 0 & \lambda_2 \end{pmatrix} \begin{pmatrix} u & v \end{pmatrix}$$

$$= \lambda_1 u^2 + \lambda_2 v^2.$$

Hence the β-coordinate $(u,\ v)$ of $\mathbf{v}$ in the graph of the quadratic equation

$$ax^2 + 2bxy + cy^2 + dx + ey = f$$

satisfies the quadratic equation in u, v

$$\lambda_1 u^2 + \lambda_2 v^2 + d'u + e'v = f',$$

which is easy to graph. The orthonormal basis β will be called an ordered set of **principal axes** of the quadratic form $\varphi(x,y)$. Graphically, f_1 and f_2 are the principal axes of the given conic section.

Example 7.4.3. Sketch the graph of $x^2 + 4xy + y^2 = 1$.

Solution. The original equation can be rewritten as

$$(x, y)A \begin{pmatrix} x \\ y \end{pmatrix} = 7, \quad \text{where} \quad \begin{pmatrix} 1 & 2 \\ 2 & 1 \end{pmatrix}.$$

We use the result in Example 7.3.20:

$$\beta = \left(\begin{pmatrix} 1/\sqrt{2} \\ 1/\sqrt{2} \end{pmatrix}, \begin{pmatrix} 1/\sqrt{2} \\ -1/\sqrt{2} \end{pmatrix} \right)$$

is an orthonormal basis such that

$$[L_A]_\beta = \begin{pmatrix} 3 & \\ & -1 \end{pmatrix}.$$

Now if we let

$$P = \begin{pmatrix} 1/\sqrt{2} & 1/\sqrt{2} \\ 1/\sqrt{2} & -1/\sqrt{2} \end{pmatrix},$$

we have $\det P = -1$. Hence we will use instead

$$\beta = \left(\begin{pmatrix} 1/\sqrt{2} \\ 1/\sqrt{2} \end{pmatrix}, \begin{pmatrix} -1/\sqrt{2} \\ 1/\sqrt{2} \end{pmatrix} \right).$$

Let

$$P = \begin{pmatrix} 1/\sqrt{2} & -1/\sqrt{2} \\ 1/\sqrt{2} & 1/\sqrt{2} \end{pmatrix} = \begin{pmatrix} \cos \frac{\pi}{4} & -\sin \frac{\pi}{4} \\ \sin \frac{\pi}{4} & \cos \frac{\pi}{4} \end{pmatrix}$$

be the unitary matrix such that

$$\begin{pmatrix} 3 & \\ & -1 \end{pmatrix} = P^t A P.$$

The principal axes of the given quadratic form are

$$f_1 = \begin{pmatrix} 1/\sqrt{2} \\ 1/\sqrt{2} \end{pmatrix} \quad \text{and} \quad f_2 = \begin{pmatrix} -1/\sqrt{2} \\ 1/\sqrt{2} \end{pmatrix},$$

and this set of principal axes are obtained by rotating the x-axis and the y-axis by $\pi/4$ counterclockwise. With respect to principal axes, the coordinate $(u, v) = (x, y)P$ satisfies

$$1 = (x, y)A \begin{pmatrix} x \\ y \end{pmatrix} = (u, v)P^t A P \begin{pmatrix} u \\ v \end{pmatrix} = (u, v) \begin{pmatrix} 3 & -1 \end{pmatrix} \begin{pmatrix} u \\ v \end{pmatrix} = 3u^2 - v^2.$$

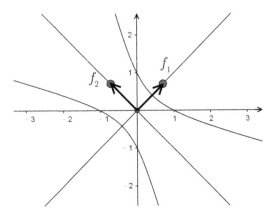

Figure 7.3: The graph of $x^2 + 4xy + y^2 = 1$

This conic section is usually expressed as

$$\frac{u^2}{(1/\sqrt{3})^2} - \frac{v^2}{1^2} = 1.$$

Its graph is a hyperbola whose transverse axis aligns with the u-axis. Its graph is shown in Figure 7.3.   ◇

Example 7.4.4. Sketch the graph of $16x^2 - 24xy + 9y^2 - 30x - 40y = 0$.

Solution. First, we need eliminate the $-24xy$ from the quadratic term. The quadratic term may be expressed as

$$(x, y) A \begin{pmatrix} x \\ y \end{pmatrix} \quad \text{where } A = \begin{pmatrix} 16 & -12 \\ -12 & 9 \end{pmatrix}.$$

The characteristic polynomial of A is

$$\det \begin{pmatrix} 16 - t & -12 \\ -12 & 9 - t \end{pmatrix} = t(t - 25).$$

The eigenvalues of A are 0 and 25. We find a basis for each of the eigenspaces:

$$E_{25} = \text{Null} \left(\begin{pmatrix} -9 & -12 \\ -12 & -16 \end{pmatrix} \right) = \text{Sp} \left(\begin{pmatrix} 4 \\ -3 \end{pmatrix} \right),$$

$$E_0 = \mathrm{Null}\left(\begin{pmatrix} 16 & -12 \\ -12 & 9 \end{pmatrix}\right) = \mathrm{Sp}\left(\begin{pmatrix} 3 \\ 4 \end{pmatrix}\right).$$

We have found an orthogonal basis

$$\left\{ \begin{pmatrix} 4 \\ -3 \end{pmatrix}, \begin{pmatrix} 3 \\ 4 \end{pmatrix} \right\}$$

for $\mathbb{R}^2$. We normalize it to obtain an orthonormal basis

$$B = \left\{ f_1 = \begin{pmatrix} 4/5 \\ -3/5 \end{pmatrix}, f_2 = \begin{pmatrix} 3/5 \\ 4/5 \end{pmatrix} \right\}$$

for $\mathbb{R}^2$. Choose

$$P = \begin{pmatrix} 4/5 & 3/5 \\ -3/5 & 4/5 \end{pmatrix}.$$

Note that $\det P = 1$. We have

$$\begin{pmatrix} 25 & \\ & 0 \end{pmatrix} = P^{\mathrm{t}} A P.$$

So if we make the change of variable

$$\begin{pmatrix} x \\ y \end{pmatrix} = P \begin{pmatrix} u \\ v \end{pmatrix} = \begin{pmatrix} \dfrac{4}{5}u + \dfrac{3}{5}v \\ -\dfrac{3}{5}u + \dfrac{4}{5}v \end{pmatrix},$$

the original quadratic term is transformed into

$$16x^2 - 24xy + 9y^2 = (x, y) A \begin{pmatrix} x \\ y \end{pmatrix} = (u, v) P^{\mathrm{t}} A P \begin{pmatrix} u \\ v \end{pmatrix}$$

$$= (u, v) \begin{pmatrix} 25 & \\ & 0 \end{pmatrix} \begin{pmatrix} u \\ v \end{pmatrix} = 25u^2.$$

The original conic section becomes

$$16x^2 - 24xy + 9y^2 - 32x - 48y$$

$$= 25u^2 - 3\left(\frac{4}{5}u + \frac{3}{5}v\right) - 40\left(-\frac{3}{5}u + \frac{4}{5}v\right)$$

$$= 25u^2 - 50v = 0$$

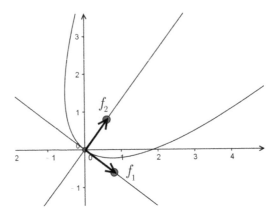

Figure 7.4: The graph of $16x^2 - 24xy + 9y^2 - 30x - 40y = 0$

$$\implies \quad u^2 = 2v.$$

With respect to the principal axes, the graph of $u^2 = 2v$ is easy. We can see that this is a parabola. The graph is shown in Figure 7.4.   ◇

How to construct an inner product.   Let $\beta = (v_1, \ldots, v_n)$ be an arbitrary basis for the F-vector space where $F = \mathbb{R}$ or $\mathbb{C}$. To construct an inner product in V, it suffices to assign

$$\pi_{ij} = \langle v_i, \, v_j \rangle, \qquad \text{for } i, j = 1, \ldots, n.$$

However, randomly assigning these values will not make a valid inner product. By condition (iii) in the definition of inner products (Definitions 7.1.1 and 7.1.6), it is required that

$$\pi_{ji} = \overline{\pi}_{ij} \qquad \text{and} \qquad \pi_{ii} > 0, \qquad \text{for all } i, j.$$

Thus, if we let $\Pi = \left(\pi_{ij} \right)_{n \times n}$, we should have that Π is self-adjoint with positive real diagonals. Now in keeping with conditions (i)–(iii), we can define the inner product to be

$$(7.6) \qquad\qquad \langle v, \, w \rangle = \sum_{i,j=1}^{n} \alpha_i \overline{\beta_j} \pi_{ij}.$$

where $v = \alpha_1 v_1 + \cdots + \alpha_n v_n$ and $w = \beta_1 v_1 + \cdots + \beta_n v_n$ with $\alpha_i, \beta_i \in F$.

To satisfy condition (iv) in the definition of inner products, further conditions must be imposed on Π. We need to check that

$$\langle v,\ v \rangle = \sum_{i,j=1}^{n} \alpha_i \overline{\alpha}_j \pi_{ij} = \begin{pmatrix} \alpha_1 & \cdots & \alpha_n \end{pmatrix} \Pi \begin{pmatrix} \overline{\alpha}_1 \\ \vdots \\ \overline{\alpha}_n \end{pmatrix}$$

is a positive real number for all $v = \alpha_1 v_1 + \cdots + \alpha_n v_n \neq O$.

Definition 7.4.5. A Hermitian $n \times n$ matrix Π over $F = \mathbb{C}$ (or $\mathbb{R}$) is called **positive-definite** if

$$v \Pi v^* \in \mathbb{R} \qquad \text{and} \qquad v \Pi v^* > 0$$

for all nonzero vector $v = (\alpha_1, \ldots, \alpha_n) \in F^n$.

Note that the diagonals of a positive-definite matrix are positive and real since $\pi_{ii} = e_i \Pi e_i{}^*$.

From our discussion above, we have the following result.

Proposition 7.4.6. *We can construct an inner product as in (7.6) if and only if the the matrix $\Pi = \left(\pi_{ij} \right)_{n \times n}$ is a positive-definite matrix.*

Proposition 7.4.7. *A Hermitian matrix is positive-definite if and only if all its eigenvalues are positive.*

Proof. Let Π be a Hermitian matrix. By the spectral theorem, we may find a unitary matrix P such that

$$\Pi = P \begin{pmatrix} \lambda_1 & & \\ & \ddots & \\ & & \lambda_n \end{pmatrix} P^*.$$

Let $(\beta_1, \ldots, \beta_n) = (\alpha_1, \ldots, \alpha_n)P$. It follows that

$$\begin{pmatrix} \overline{\beta}_1 \\ \vdots \\ \overline{\beta}_n \end{pmatrix} = \overline{(\beta_1, \ldots, \beta_n)^{\mathrm{t}}} = \overline{((\alpha_1, \ldots, \alpha_n)P)^{\mathrm{t}}}$$

$$= \overline{P^{\mathrm{t}}(\alpha_1, \ldots, \alpha_n)^{\mathrm{t}}} = P^* \begin{pmatrix} \overline{\alpha}_1 \\ \vdots \\ \overline{\alpha}_n \end{pmatrix}.$$

Thus,

$$(\alpha_1, \ldots, \alpha_n)\Pi \begin{pmatrix} \overline{\alpha}_1 \\ \vdots \\ \overline{\alpha}_n \end{pmatrix}$$

$$= (\alpha_1, \ldots, \alpha_n)P \begin{pmatrix} \lambda_1 & & \\ & \ddots & \\ & & \lambda_n \end{pmatrix} P^* \begin{pmatrix} \overline{\alpha}_1 \\ \vdots \\ \overline{\alpha}_n \end{pmatrix}$$

$$= (\beta_1, \ldots, \beta_n) \begin{pmatrix} \lambda_1 & & \\ & \ddots & \\ & & \lambda_n \end{pmatrix} \begin{pmatrix} \overline{\beta}_1 \\ \vdots \\ \overline{\beta}_n \end{pmatrix}$$

$$= \lambda_1 \left\| \beta_1 \right\|^2 + \cdots + \lambda_n \left\| \beta_n \right\|^2.$$

Since $(\alpha_1, \ldots, \alpha_n) = O$ if and only if $(\beta_1, \ldots, \beta_n) = O$, we see that Π is positive definite if and only if $\lambda_i > 0$ for all i. $\qquad \square$

Exercises 7.4

1. Show that all the eigenvalues of a unitary matrix are of unit length.

2. Let A be an orthogonal matrix of size n. Determine whether $A - 2I_n$ is always invertible.

3. Sketch the graph of the quadratic equation $2x^2 + 3xy - 2y^2 = 10$.

4. Find a set of principal axes for the quadratic form
$$\varphi(x, y, z) = 3x^2 + 2y^2 + 3z^2 - 2xy - 2yz.$$

5. We say a conic section is **degenerate** if its graph is empty, a finite set or a union of two lines. Suppose given a non-degenerate conic section
$$Ax^2 + Bxy + Cy^2 + Dx + Ey + F = 0.$$

    (a) Show that the **discriminant** $\Delta = B^2 - 4AC$ is an invariant under rigid transformation.

    (b) Show that the conic section is

$$\begin{cases} \text{an ellipse or a circle,} & \text{if } \Delta < 0; \\ \text{a parabola,} & \text{if } \Delta = 0; \\ \text{a hyperbola,} & \text{if } \Delta > 0. \end{cases}$$

6. Show that a real quadratic form $\sum\limits_{i,j=1}^{n} a_{ij}x_i x_j$ in n variables may be expressed as

$$\left(x_1, \ldots, x_n \right) A \begin{pmatrix} x_1 \\ \vdots \\ x_n \end{pmatrix}$$

where A is a symmetric matrix.

7. Show that the unit circle in any inner product space is a circle or an ellipse in the Euclidean space.

8. Give an example of a positive-definite matrix in which some entries are negative.

9. Give an example of a positive-definite matrix in which all entries are positive.

10. Show that every positive-definite matrix is invertible. Show that the inverse matrix of a positive-definite matrix is also positive-definite.

11. Let A be a square matrix of size n. The **principal minors** of A refer to the determinants of the upper-left $i \times i$ corners of A for $i = 1, \ldots, n$. Show that a Hermitian matrix is positive-definite if and only if all of its principal minors are positive. (Hint: Use induction on n.)

7.5   The method of least squares

    The French mathematician Legendre (Adrien-Marie Legendre, 1752–1833) presented the method of least squares in the appendix of his 1806 book *Nouvelles Me'thodes pour la Détermination des Orbites des Comètes.* Later in 1809, Gauss (Johann Carl Friedrich Gauss, 1777–1855) claimed in his book *Theoria Motus Corporum Coelestium in sectionibus conicis solem ambientium* that he had been using this method since 1795 and thus was able to locate the missing asteroid Ceres in 1801. Since in science, priority is always determined by the publication date, the honor of inventing the method of least squares naturally went to Legendre.

The regression lines.   The method of least squares is often used in data fitting. Suppose given n points

$$(x_1, y_1), \ (x_2, y_2), \ \ldots, \ (x_n, y_n) \in \mathbb{R}^2$$

which do not necessarily lie on a straight line. We want to find a linear equation $y = ax + b$ such that it fits the given points as well as possible.

Let

$$d_i = \left| y_i - (ax_i + b) \right|$$

be the *vertical deviation* of the point (x_i, y_i) from the line L given by $y = ax + b$. If the point happens to lie on L, then naturally the vertical deviation is 0. In general, we want the total deviation to be as small as possible. On the other hand, we want the computation involved to be as simple as possible. As a good compromise, the method of least squares aims to find a an b such that the sum of squares of vertical deviations

$$E(a, b) = \sum_{i=1}^{n} \left[y_i - (ax_i + b) \right]^2$$

is minimized.

Since $E(a, b)$ is a quadratic function in a and b, to minimize $E(a, b)$ we need to solve

$$\frac{\partial E}{\partial a} = \sum_{i=1}^{n} -2 \left[y_i - (ax_i + b) \right] x_i = 0;$$

$$\frac{\partial E}{\partial b} = \sum_{i=1}^{n} -2 \left[y_i - (ax_i + b) \right] = 0.$$

This is equivalent to solving

$$a \sum_{i=1}^{n} x_i^2 + b \sum_{i=1}^{n} x_i = \sum_{i=1}^{n} x_i y_i;$$

(7.7)
$$a \sum_{i=1}^{n} x_i + nb = \sum_{i=1}^{n} y_i.$$

Let
$$\overline{x} = \frac{x_1 + x_2 + \cdots + x_n}{n} \quad \text{and} \quad \overline{y} = \frac{y_1 + y_2 + \cdots + y_n}{n}$$

be the *averages* (or the *arithmetic means*) of the x- and y-coordinates of the n given points respectively. Using Crammer's rule, we have that

$$a = \frac{n \sum_{i=1}^{n} x_i y_i - \sum_{i=1}^{n} x_i \sum_{i=1}^{n} y_i}{n \sum_{i=1}^{n} x_i^2 - \left(\sum_{i=1}^{n} x_i\right)^2}$$
$$= \frac{\left[\sum_{i=1}^{n} x_i y_i\right] - n\overline{x}\,\overline{y}}{\left[\sum_{i=1}^{n} x_i^2\right] - n\overline{x}^2}$$

Notice that

$$\sum_{i=1}^{n}(x_i - \overline{x})(y_i - \overline{y}) = \sum_{i=1}^{n} x_i y_i - \overline{x} \sum_{i=1}^{n} y_i - \overline{y} \sum_{i=1}^{n} x_i + n\overline{x}\,\overline{y}$$
$$= \sum_{i=1}^{n} x_i y_i - \overline{x}(n\overline{y}) - \overline{y}(n\overline{x}) + n\overline{x}\,\overline{y} = \left[\sum_{i=1}^{n} x_i y_i\right] - n\overline{x}\,\overline{y}$$

We leave it as an exercise to show that

$$\sum_{i=1}^{n}(x_i - \overline{x})^2 = \left[\sum_{i=1}^{n} x_i^2\right] - n\overline{x}^2.$$

We conclude that

(7.8)
$$\begin{cases} a = \dfrac{\sum_{i=1}^{n}(x_i - \overline{x})(y_i - \overline{y})}{\sum_{i=1}^{n}(x_i - \overline{x})^2}, \\ b = \overline{y} - a\overline{x}, \end{cases}$$

using (7.7).

The line $y = ax + b$ is called the **least squares line** or the **regression line** determined by the n points $(x_1, y_1), (x_2, y_2), \ldots, (x_n, y_n)$.

Example 7.5.1. Find the least squares line of the points $(1, -1)$, $(2, -3)$, $(3, -7)$, $(6, 9)$.

Solution. The averages of the x- and y-coordinates are

$$\overline{x} = \frac{1 + 2 + 3 + 6}{4} = 3,$$

$$\bar{y} = \frac{(-1) + (-3) + (-7) + (-9)}{4} = -5,$$

respectively. So

$$\sum_{i=1}^{4}(x_i - \bar{x})^2 = (1-3)^2 + (2-3)^2 + (3-3)^2 + (6-3)^2 = 14;$$

$$\sum_{i=1}^{4}(x_i - \bar{x})(y_i - \bar{y}) = (-2)4 + (-1)2 + 0(-2) + 3(-4) = -20.$$

Thus

$$a = \frac{-20}{14} = -\frac{10}{7}, \qquad b = -5 - \left(-\frac{10}{7}\right)3 = -\frac{5}{7}.$$

The least squares solution is $y = -\dfrac{10}{7}x - \dfrac{5}{7}$.   ◇

Example 7.5.2. The production numbers of a company for the last six-month period were as follows:

	Mar	Apr	May	June	July	Aug
Units of production p	120	140	155	135	110	102
Total cost C (in dollars)	6300	6500	6670	6450	6100	5750

Determine a least squares line (called the cost-volume formula) for the given data.

Solution. The average of the units of production is $\bar{p} = 127$, and the average of cost $\overline{C} = 6295$. Let $C = ap + b$ be the least squares solution corresponding to the given data. Applying (7.8), we obtain that

$$a = \frac{31310}{1980} \approx 16, \quad \text{and} \quad b = 6295 - \frac{3131 \cdot 127}{198} \approx 4287.$$

We may choose the regression line to be $C = 16p + 4287$.   ◇

Overdetermined systems of linear equations.   Let $m > n$. Suppose given the system of linear equations with real coefficients:

$$a_{11}x_1 + a_{12}x_2 + \cdots + a_{1n}x_n = b_1$$
$$a_{21}x_1 + a_{22}x_2 + \cdots + a_{2n}x_n = b_2$$

$$\vdots \qquad\qquad \vdots$$

$$a_{m1}x_1 + a_{m2}x_2 + \cdots + a_{mn}x_n = b_m$$

with m equations in n variables. Such a system generally does not have a solution. When this indeed happens we say such a system is *overdetermined.* Now rewrite the system in matrix form

(7.9) $A\mathbf{x} = \mathbf{b}$

where

$$A = \left(a_{ij}\right)_{m \times n}, \qquad \mathbf{x} = \begin{pmatrix} x_1 \\ \vdots \\ x_n \end{pmatrix}, \qquad \mathbf{b} = \begin{pmatrix} b_1 \\ \vdots \\ b_m \end{pmatrix}.$$

Now multiply by A^t on both sides of (7.9). We get

(7.10) $A^t A\mathbf{x} = A^t\mathbf{b}.$

This system has a unique solution if $A^t A$ is invertible:

$$\mathbf{x}_0 = (A^t A)^{-1} A^t \mathbf{b}.$$

This is a not a "true" solution to the system (7.9). However, it is in some sense the "nearest" solution we can get for the system. More precisely,

$$\|A\mathbf{x}_0 - \mathbf{b}\| \le \|A\mathbf{x} - \mathbf{b}\| \qquad \text{for all } \mathbf{x} \in \mathbb{R}^n.$$

In order for $\|A\mathbf{x}_0 - \mathbf{b}\|$ to be minimum, it must be orthogonal to the range of A, which is the column space of A. This implies that

$$A^t(A\mathbf{x}_0 - \mathbf{b}) = 0.$$

In other words, we have (7.10).

Since the square of the norm of a vector is the sum of the squares of the coordinates, $\mathbf{x}_0$ is also called the *least squares solution* of the system (7.9).

Example 7.5.3. Find the least squares solution to the following determined system:

$$x + y = 4$$
$$-x + y = 2$$
$$2x + 3y = 6$$

Solution. The coefficient matrix and the constant vector are

$$A = \begin{pmatrix} 1 & 1 \\ -1 & 1 \\ 2 & 3 \end{pmatrix} \quad \text{and} \quad \mathbf{b} = \begin{pmatrix} 4 \\ 2 \\ 6 \end{pmatrix}.$$

One can easily compute that

$$A^t A = \begin{pmatrix} 1 & -1 & 2 \\ 1 & 1 & 3 \end{pmatrix} \begin{pmatrix} 1 & 1 \\ -1 & 1 \\ 2 & 3 \end{pmatrix} = \begin{pmatrix} 6 & 6 \\ 6 & 11 \end{pmatrix}$$

$$A^t \mathbf{b} = \begin{pmatrix} 1 & -1 & 2 \\ 1 & 1 & 3 \end{pmatrix} \begin{pmatrix} 4 \\ 2 \\ 6 \end{pmatrix} = \begin{pmatrix} 14 \\ 24 \end{pmatrix}.$$

Hence the least squares solution must satisfy

$$6x + 6y = 14$$
$$6x + 11y = 24$$

and we conclude that the least squares solution $(1/3, 2)$.   ◇

We now compare the least squares solution with the previously discussed least squares line. To find the least squares line for the n points (x_1, y_1), (x_2, y_2), ..., (x_n, y_n) is clearly to find the least squares solution to the overdetermined system

$$ax_1 + b = y_1$$
$$ax_2 + b = y_2$$
$$\vdots \qquad \vdots$$
$$ax_n + b = y_n.$$

The coefficients matrix of this system is

$$A = \begin{pmatrix} x_1 & 1 \\ x_2 & 1 \\ \vdots & \vdots \\ x_n & 1 \end{pmatrix}.$$

The least squares solution (a, b) must satisfy the system

$$A^t A \begin{pmatrix} a \\ b \end{pmatrix} = \begin{pmatrix} \sum_{i=1}^{n} x_i^2 & n\bar{x} \\ n\bar{x} & n \end{pmatrix} \begin{pmatrix} a \\ b \end{pmatrix} = A^t \begin{pmatrix} y_1 \\ \vdots \\ y_n \end{pmatrix} = \begin{pmatrix} \sum_{i=1}^{n} x_i y_i \\ n\bar{y} \end{pmatrix}.$$

This is exactly the same system as in (7.7).

Definition 7.5.4. For a matrix A, the matrix $(A^t A)^{-1} A^t$ is called the **pseudoinverse** of A, and is denoted by $\mathrm{pinv}(A)$.

Example 7.5.5. In Physics, Hook's law asserted that when a force is applied to a spring, the length of the spring will be a linear function of the force. Below is a table of experimental data:

Force F (in Kilo)	3	6	9	12
Length L (in cm)	14	20	38	48

Find the least squares solution $L = a + bF$.

Solution. Substituting all the data into the equation $L = a + bF$, we get the system

$$a + 3b = 14$$
$$a + 6b = 20$$
$$a + 9b = 38$$
$$a + 12b = 48.$$

The coefficient matrix here is

$$A = \begin{pmatrix} 1 & 3 \\ 1 & 6 \\ 1 & 9 \\ 1 & 12 \end{pmatrix}.$$

The equation for the least squares solution is

$$A^t A \begin{pmatrix} a \\ b \end{pmatrix} = A^t \begin{pmatrix} 14 \\ 20 \\ 38 \\ 48 \end{pmatrix}$$

or equivalently

$$4a + 30b = 120$$
$$30a + 270b = 1080.$$

The least squares solution is $a = 0$ and $b = 4$, and the least squares line is $L = 4F$. ◇

Fitting curves of higher degrees. The method of least square is not only used to determine the regression line for n given points. It can also be used to determine polynomial functions of higher degrees for those n points.

Example 7.5.6. Find the least squares parabola $y = a + bx + cx^2$ for the following points

$$(1, 3), \quad (2, 7), \quad (3, 13), \quad (4, 21)$$

Solution. Substitute these points into the equation $y = a + bx + cx^2$, we obtain the system of linear equations

$$\begin{aligned} a + b + c &= 3 \\ a + 2b + 4c &= 7 \\ a + 3b + 9c &= 13 \\ a + 4b + 16c &= 21. \end{aligned}$$

Thus we need to solve the overdetermined system $A\mathbf{x} = \mathbf{b}$ where

$$A = \begin{pmatrix} 1 & 1 & 1 \\ 1 & 2 & 4 \\ 1 & 3 & 9 \\ 1 & 4 & 16 \end{pmatrix} \quad \text{and} \quad \mathbf{b} = \begin{pmatrix} 3 \\ 7 \\ 13 \\ 21 \end{pmatrix}.$$

since

$$\text{pinv}(A) = (A^t A)^{-1} A^t = \frac{1}{20} \begin{pmatrix} 45 & -15 & -25 & 15 \\ -31 & 23 & 27 & -19 \\ 5 & -5 & -5 & 5 \end{pmatrix},$$

the least squares solution is

$$
\begin{pmatrix}
45 & -15 & -25 & 15 \\
-31 & 23 & 27 & -19 \\
5 & -5 & -5 & 5
\end{pmatrix}
\begin{pmatrix}
3 \\ 7 \\ 13 \\ 11
\end{pmatrix}
=
\begin{pmatrix}
1 \\ 1 \\ 1
\end{pmatrix}.
$$

The fitting parabola for the four given points is thus $y = 1 + x + x^2$.   ◇

Example 7.5.7. Newton's Laws of Motion implies that an object thrown upwards vertically at a velocity v will be at a height $s(t) = vt - \dfrac{1}{2}gt^2$ after t seconds, where g is the acceleration due to gravity. Here is a table of observed data:

t	0.5	1.0	1.5	2.0
$s(t)$	23.7	45.1	64.0	80.4

Estimate v and g.

Solution. Let $s(t) = a_0 + a_1 t + a_2 t^2$ be the least squares curve to the given data. The table of data gives us an overdetermined system of equations $A\mathbf{x} = \mathbf{y}$ where

$$
A = \begin{pmatrix}
1 & 0.5 & 0.25 \\
1 & 1.0 & 1.00 \\
1 & 1.5 & 2.25 \\
1 & 2.0 & 4.00
\end{pmatrix},
\quad
\mathbf{x} = \begin{pmatrix} a_0 \\ a_1 \\ a_2 \end{pmatrix}
\quad \text{and} \quad
\mathbf{y} = \begin{pmatrix} 23.7 \\ 45.1 \\ 64.0 \\ 80.4 \end{pmatrix}.
$$

Solving the equation

$$
A^t A \mathbf{x} = A^t \mathbf{y},
$$

we may obtain the solution $s(t) = -0.2 + 50.3t - 5t^2$. Hence $v \approx 50.3$ and $g \approx 10$.   ◇

Exercises 7.5

1. Find the regression line determined by the points $(1, 1)$, $(4, 2)$, $(2, 3)$ and $(4, 4)$.

2. Find the regression line determined by the points $(-3, -0.5)$, $(-1, -0.5)$, $(1, 0.9)$, $(2, 1.1)$ and $(3, 3.3)$.

3. Find the regression line determined by the points $(0, 90)$, $(5, 197)$, $(10, 335)$ and $(15, 394)$.

4. An insurance company has calculated the following percentages of people dying at various ages from 20 to 50.

Age in years	20	25	30	35	40	45	50
death percentage	7	7	10	12	16	20	26

   Predict the percentage of deaths at 27 years of age based on the least squares line.

5. Find the pseudoinverse of the matrix

$$A = \begin{pmatrix} 1 & 3 \\ 2 & 4 \\ -1 & 1 \\ 4 & 2 \end{pmatrix}.$$

6. Find the least squares curve $y = a + bx + cx^2$ for the following points $(1, 9)$, $(2, 4)$, $(3, 3)$ and $(4, 5)$.

7.6 The L^2-approximation of a continuous function

Let $C([0, 1])$ be the vector space of all continuous real-valued functions defined on the closed interval $[0, 1]$. There are different ways to define norms on $C([0, 1])$. In this section we discuss the so-called L^2-norm.

The space $C([0, 1])$ as an inner product space. Remember that in Example 7.1.4 we have seen that $C([0, 1])$ is an inner product space with the inner product given as

$$\langle f, g \rangle = \int_0^1 f(x) f(x) \, dx.$$

The norm defined with respect to this inner product is called the L^2-*norm* of $C([0, 1])$. With this inner product the set

(7.11) $S_n = \{\cos 2k\pi x, \ \sin 2k\pi x : k = 0, 1, 2, \ldots, n\}$

is orthogonal:

(i) $\langle \cos 2m\pi x, \ \sin 2n\pi x \rangle = \displaystyle\int_0^1 \cos 2m\pi x \ \sin 2n\pi x \ dx$

$= \dfrac{1}{2} \displaystyle\int_0^1 \sin 2(m + n)\pi x + \sin 2(m - n)\pi x \ dx = 0$

(ii) $\langle \cos 2m\pi x, \ \cos 2n\pi x \rangle = \displaystyle\int_0^1 \cos 2m\pi x \ \cos 2n\pi x \ dx$

$= \dfrac{1}{2} \displaystyle\int_0^1 \cos 2(m + n)\pi x + \cos 2(m - n)\pi x \ dx = 0$

$= \begin{cases} 0, & \text{if } m \neq n; \\ 1/2, & \text{if } m = n. \end{cases}$

(iii) $\langle \sin 2m\pi x, \ \sin 2n\pi x \rangle = \displaystyle\int_0^1 \sin 2m\pi x \ \sin 2n\pi x \ dx$

$= \dfrac{1}{2} \displaystyle\int_0^1 \cos 2(m - n)\pi x - \cos 2(m + n)\pi x \ dx = 0$

$= \begin{cases} 0, & \text{if } m \neq n; \\ 1/2, & \text{if } m = n. \end{cases}$

Now an interesting problem arises. For a given function $f(x)$ in $C([0, 1])$, how to find a vector in $\text{Sp}(S_n)$ which is as close as possible to $f(x)$ with respect to the L^2-norm. More precisely, we want to find real constants a_0, $a_1, \ldots, a_n, b_1, \ldots, b_n$ so that the function

$$\varphi(x) = f(x) - \sum_{k=0}^n a_k \cos 2k\pi x - \sum_{k=1}^n b_k \sin 2k\pi x$$

has a minimal L^2-norm. This can be achieved if $\varphi(x)$ is orthogonal to $\text{Sp}(S_n)$. In other words, we must have

$$\langle \varphi(x), \ \cos 2k\pi x \rangle = 0 \quad \text{and} \quad \langle \varphi(x), \ \sin 2k\pi x \rangle = 0$$

for $k = 0, 1, \ldots, n$. From the orthogonality of S_n, it can be easily verified that the coefficients $a_0, a_1, \ldots, a_n, b_1, \ldots, b_n$ must be the *Fourier coefficients* of $f(x)$, *i.e.*,

$$a_0 = \int_0^1 f(x) \, dx;$$

$$a_k = 2 \int_0^1 f(x) \cos 2k\pi x \, dx, \qquad k = 1, 2, \ldots, n;$$

$$b_k = 2 \int_0^1 f(x) \sin 2k\pi x \, dx, \qquad k = 1, 2, \ldots, n.$$

The Bernoulli polynomials. Some special polynomials can be expressed as trigonometric series through their Fourier coefficients. Here we provide an example.

The **Bernoulli polynomials** $B_n(x)$, $n = 1, 2, 3, \ldots$, are defined by

$$(7.12) \qquad \frac{te^{tx}}{e^t - 1} = \sum_{n=0}^{\infty} \frac{B_n(x)t^n}{n!}, \qquad \text{for } |t| < 2\pi.$$

A direct calculation shows that the first few Bernoulli polynomials are

$$B_0(x) = 1;$$
$$B_1(x) = x - \frac{1}{2};$$
$$B_2(x) = x^2 - x - \frac{1}{6};$$
$$B_3(x) = x^3 - \frac{3}{2}x^2, \text{ etc.}$$

Observe that if we differentiate both sides of (7.12) with respect to x, we obtain that

$$\frac{t^2 e^{xt}}{e^t - 1} = \sum_{n=0}^{\infty} \frac{B_n(x)t^{n+1}}{n!} = \sum_{n=1}^{\infty} \frac{B_{n-1}(x)t^n}{(n-1)!} = \sum_{n=0}^{\infty} \frac{B_n'(x)t^n}{n!}.$$

By comparing the coefficients of t^n, we can see that

$$B_0'(x) = 0 \quad \text{and} \quad B_n'(x) = nB_{n-1}(x) \quad \text{for } n \geq 1.$$

To determine $B_n(x)$, we also need to find the **Bernoulli numbers**

$$B_n = B_n(0)$$

for each n. To do so we substitute x by 0 in the identity (7.12) and obtain

$$t = \left(\sum_{i=0}^{\infty} \frac{B_i t^i}{i!} \right) \left(\sum_{j=1}^{\infty} \frac{t^j}{j!} \right).$$

Compare the coefficients of t on both sides, we see that

(7.13) $$B_0 = 1,$$

while the coefficients of t^{n+1} for $n \geq 1$ give us the relations

$$\frac{B_n}{n!} \cdot \frac{1}{1!} + \frac{B_{n-1}}{(n-1)!} \cdot \frac{1}{2!} + \cdots + \frac{B_0}{0!} \cdot \frac{1}{(n+1)!} = 0, \qquad \text{for } n \geq 1.$$

Multiply both sides by $(n+1)!$, we get that

(7.14) $$\binom{n+1}{n} B_n + \binom{n+1}{n-1} B_{n-1} + \cdots + \binom{n+1}{0} B_0 = 0, \qquad \text{for } n \geq 1.$$

Hence the Bernoulli numbers may be evaluated recursively by the relations (7.13) and (7.14).

By calculating the Fourier coefficients of Bernoulli polynomials, one may obtain the following formulas:

- $$\sum_{k=1}^{\infty} \frac{\cos(2k\pi x)}{k^{2m}} = \frac{(-1)^{m+1}(2\pi)^{2m} B_{2m}(x)}{2(2m)!};$$

- $$\sum_{k=1}^{\infty} \frac{\sin(2k\pi x)}{k^{2m+1}} = \frac{(-1)^{m+1}(2\pi)^{2m+1} B_{2m+1}(x)}{2(2m+1)!}$$

for $0 \leq x \leq 1$.

In the following example we verify the formula for $B_2(x)$.

Example 7.6.1. Find the Fourier coefficients of $B_2(x) = x^2 - x + \frac{1}{2}$ and then express the sum

$$\sum_{k=1}^{\infty} \frac{\cos(2k\pi x)}{k^2}$$

in terms of $B_2(x)$ when $0 \leq x \leq 1$.

Solution. First we have

$$\int_0^1 B_2(x)\, dx = \frac{1}{3}\int_0^1 B_3'(x)\, dx = \frac{1}{3}\left(B_3(1) - B_3(0)\right) = 0.$$

Next, we calculate other Fourier coefficients through integration by parts.
Indeed we have

$$2\int_0^1 B_2(x)\cos(2k\pi x)\, dx$$

$$= \frac{1}{k\pi}B_2(x)\sin(2k\pi x)\Big|_0^1 - \frac{2}{k\pi}\int_0^1 B_1(x)\sin(2k\pi x)\, dx$$

$$= \frac{1}{(k\pi)^2}B_1(x)\cos(2k\pi x)\Big|_0^1 - \frac{1}{(k\pi)^2}\int_0^1 \cos(2k\pi x)\, dx$$

$$= \frac{1}{(k\pi)^2}\left(B_1(1) - B_1(0)\right) = \frac{1}{(k\pi)^2}.$$

On the other hand, we have

$$2\int_0^1 B_2(x)\sin(2k\pi x)\, dx$$

$$= \frac{1}{k\pi}B_2(x)\cos(2k\pi x)\Big|_0^1 + \frac{2}{k\pi}\int_0^1 B_1(x)\cos(2k\pi x)\, dx$$

$$= \frac{1}{k\pi}B_1(x)\sin(2k\pi x)\Big|_0^1 - \frac{1}{(k\pi)^2}\int_0^1 \sin(2k\pi x)\, dx$$

$$= 0.$$

It follows that for $0 \le x \le 1$,

$$B_2(x) = \sum_{k=1}^\infty \frac{\cos(2k\pi x)}{(k\pi)^2}, \quad \text{or} \quad \sum_{k=1}^\infty \frac{\cos(2k\pi x)}{k^2} = B_2(x)\pi^2.$$

◇

Exercises 7.6

1. Through the recursive formula

$$\begin{cases} B_0 = 1, \\ \dbinom{n+1}{n}B_n + \dbinom{n+1}{n-1}B_{n-1} + \cdots + \dbinom{n+1}{0}B_0 = 0, \qquad \text{for } n \ge 1, \end{cases}$$

evaluate the first few Bernoulli numbers.

2. Prove that the function

$$F(t) = \frac{t}{e^t - 1} + \frac{t}{2}$$

is an even function, i.e., $F(-t) = F(t)$. Hence $B_{2k+1} = 0$ for all positive integers k.

3. Show that $\cot u$ has the power series expansion

$$\cot u = \sum_{k=0}^{\infty} (-1)^k \frac{B_{2k}}{(2k)!} 2^{2k} u^{2k-1}$$

and $\tan u$ has the power series expansion

$$\tan u = \sum_{k=1}^{\infty} (-1)^{k+1} \frac{B_{2k}}{(2k)!} (2^{2k} - 1) 2^{2k} u^{2k-1}.$$

Review Exercises for Chapter 7

1. Let $a, b \in \mathbb{C}$ and let

$$A = \begin{pmatrix} a & b & & & & & b \\ b & a & b & & & & \\ & b & a & b & & & \\ & & \ddots & \ddots & \ddots & & \\ & & & b & a & b & \\ & & & & b & a & b \\ b & & & & & b & a \end{pmatrix}_{n \times n}$$

for $n \geq 3$. Find all the eigenvalues of A.

2. Let $A = B + C$ where B is a positive-definite real symmetric matrix and C is a real skew-symmetric matrix. Show that $\det B \leq \det A$.

3. We say a Hermitian matrix A is **negative-definite** if

$$v^* A v < 0$$

for any nonzero column n-vector v. Show that the following three conditions are equivalent for any Hermitian matrix A.

(i) The matrix A is negative-definite.

(ii) The matrix A is diagonalizable with only negative real eigenvalues.

(iii) For the matrix A, all of the principal minors of even order are positive and all of the principal minors of odd order are negative.

Index

Printed in the United States
By Bookmasters